KU-097-062

This book is the first study involving original analyses of potential global effects of climate change. The project, mainly funded by the U.S. Environmental Protection Agency, involved over 150 scientists from across the globe, and thus is the largest study of its kind. Common climate change scenarios were used to examine the impacts on agriculture, water resources, coastal resources, forests and human health. The studies focused on the impacts of climate change in the developing countries although some global analyses were also conducted. In addition Egypt was used as a case study and is the first integrated analysis of a single country. This book will be of great use to climate change researchers and policy makers. Interested individuals will also find this book helpful in understanding potential impacts of climate change.

As Climate Changes:
International Impacts and Implications

As Climate Changes:
International Impacts and Implications

Kenneth M. Strzepek and Joel B. Smith, Editors

Published for the U.S. Environmental Protection Agency
Office of Policy, Planning and Evaluation
Climate Change Division by
Cambridge University Press

Project Coordinators:

Laurence S. Kalkstein
Stephen Leatherman
Robert J. Nicholls
Martin L. Parry
William E. Riebsame
Cynthia Rosenzweig
Herman H. Shugart
Thomas M. Smith
Kenneth M. Strzepek

Project Manager:

Joel B. Smith

Published by the Press Syndicate of the University of Cambridge
The Pitt Building, Trumpington Street, Cambridge CB2 1RP
40 West 20th Street, New York, NY 10011-4211, USA
10 Stamford Road, Oakleigh, Melbourne 3166, Australia

First published 1995

Printed in Great Britain at the University Press, Cambridge

A catalogue record for this book is available from the British Library

Library of Congress cataloguing in publication data

As Climate Changes: International Impacts and Implications / Kenneth M. Strzepek and Joel B. Smith, editors.
p. cm.
Includes index
ISBN 0 521 46224 X (hc). – ISBN 0 521 46796 9 (pb)
1. Climatic changes. 2. Climatic changes – Developing countries.
3. Climatic changes – Egypt. I. Strzepek, Kenneth M. II. Smith, Joel B. III. United States. Environmental Protection Agency. Climate Change Division.
QC981.8.C5A86 1995
551.6–dc20 95–6090 CIP

ISBN 0 521 46224 x hardback
ISBN 0 521 46796 9 paperback

SE

Contents

List of authors

Chapter 1: Introduction

Joel B. Smith
Hagler Bailly Consulting Inc.

Laurence S. Kalkstein
Department of Geography
University of Delaware, USA

Chapter 2: World Food Supply

Cynthia Rosenzweig
Columbia University and Goddard Institute for Space Studies

Martin L. Parry
University College London

Günther Fischer
International Institute for Applied Systems Analysis

Chapter 3: Complex River Basins

W. E. Riebsame
K. M. Strzepek
J. L. Wescoat, Jr.
R. Perritt
G. L. Gaile
J. Jacobs
R. Leichenko
C. Magadza
H. Phien
B. J. Urbiztondo
P. Restrepo
W. R. Rose
M. Saleh
L. H. Ti
C. Tucci
D. Yates
Institute of Behavioral Science
University of Colorado, USA

Chapter 4: Global Sea-Level Rise

Robert J. Nicholls
School of Geography and Environmental Management
Middlesex University, UK

Laboratory for Coastal Research
University of Maryland, USA

Stephen P. Leatherman
Laboratory for Coastal Research
University of Maryland, USA

Chapter 5: Human Health

Laurence S. Kalkstein
Department of Geography
University of Delaware, USA

Guanri Tan
Department of Atmospheric Sciences
Zhongshan University
Guangzhou, People's Republic of China

Chapter 6: Global Forests

T. M. Smith
P. N. Halpin
H. H. Shugart
Department of Environmental Sciences
University of Virginia, USA

C. M. Secrett
International Institute for Environment and Development

Chapter 7: An Assessment of Integrated Climate Change Impacts on Egypt

Kenneth M. Strzepek
University of Colorado at Boulder, USA
International Institute for Applied Systems Analysis, Laxenburg, Austria

S. Chibọ Onyeji
University of Colorado at Boulder, USA
International Institute for Applied Systems Analysis, Laxenburg, Austria

Magdy Saleh
Cairo University, Egypt

David N. Yates
University of Colorado at Boulder, USA
International Institute for Applied Systems Analysis, Laxenburg, Austria

Chapter 8: Adaption Policy

Joel B. Smith
Hagler Bailly Consulting, Inc.

Jeffrey J. Carmichael
Department of Economics
University of Colorado at Boulder, USA

James G. Titus
Office of Policy, Planning and Evaluation
U.S. Environmental Protection Agency

Collaborating Researchers and their Areas of Study

Chapter 2: World Food Supply

Argentina
Osvaldo Sala
Dept. Ecologia, University of Buenos Aires

Australia
Wayne S. Meyer
Dept. of Crop and Soil Sciences, Michigan State University

Doug Godwin
Alton Park

Bangladesh
Z. Karim
Bangladesh Agricultural Research Council

Mahbub Ahmed
Bangladesh Agricultural Research Council

Brazil
Otavio J. F. de Siqueira
Centro de Pesquisa Agropeuaria de Terras, Baxias de Clima Temperado

Canada
Michael Brklacich
Land Resource Research Centre, Agriculture Canada, Ottawa

China
Zhiqing Jin
Jiangsu Academy of Agricultural Sciences

Egypt
H. M. Eid
Soil and Water Research Institute, ARC, Giza, Ministry of Agriculture

France
Richard Delécolle
INRA, Centre de Recherches Agronomiques

India
D. Gangadhar Rao
Central Research Institute for Dryland Agriculture

Japan
Hiroshi Seino
National Institute of Agro-Environmental Sciences

Mexico
Diana Liverman
Dept. of Geography, Pennsylvania State University, USA

Leticia Menchaca
National University of Mexico

Pakistan
Ata Qureshi
Climate Institute, Washington, D.C., USA

Philippines
Cristano R. Escaño
Council for Agric., For. and Nat. Resources Research and Development

Leandro Buendia
Council for Agric., For. and Nat. Resources Research and Development

Russia
Gennady Menzhulin
State Hydrological Institute

Larisa A. Koval
State Hydrological Institute

Spain
Ana Iglesias
CIT-INIA

Taiwan
Tien-Yin Chou
Feng Chai University

Thailand
M. L. C. Tongyai
Soil Science Division, Dept. of Agriculture, Ministry of Agriculture & Co-operatives

Uruguay
Walter E. Baethgen
International Fertilizer Development Center

USA
Brian Baer
Michigan State University

Bruce Curry
Agriculture Engineering Dept., University of Florida, USA

Gerrit Hoogenboom
Dept. of Agricultural Engineering, University of Georgia, USA

Jim Jones
Agriculture Engineering Dept., University of Florida, USA

Joe T. Ritchie
Dept. of Crop and Soil Sciences, Michigan State University, USA

Upendra Singh
International Fertilizer Development Center, Muscle Shoals, Alab., USA

Gordon Tsuji
IBSNAT Project, University of Hawaii, USA

Zimbabwe
Thomas E. Downing
University of Oxford

Paul Muchena
Plant Protection Research Institute, Harare

Global Food Trade
Günther Fischer
International Institute for Applied Systems Analysis

Klaus Frohberg
University of Bonn

Chapter 3: Complex River Basins

Indus
Robin Leichenko
Dept. of Geography, University of Colorado, USA

Khalid Mohtadullah
Former General Manager Planning, Water and Power Development Authority, Lahore, Pakistan

Syed Wali Waheed
Environmental Engineer, Planning Division, Water and Power Development Authority, Lahore, Pakistan

James Wescoat
Dept. of Geography and Inst. of Behavioral Science, University of Colorado, USA

Uruguay
José Asunción Barbosa
Comisión Mixta Paraguaya-Argentino, Asunción

Richard Perritt
Dept. of Geography, Auburn University, Alabama, USA

Victor Pochat
Dept. Técnico, Comisión Mixta Argentina Paraguaya

Carlos Eduardo Morelli Tucci
Institute of Hydraulic Research, Federal University of Rio Grande do Sul, Brazil

Nile
Magdy Saleh
Dept. of Irrigation and Hydraulic Engineering, Cairo University

Juan Valdes
Texas A&M University, USA

David Yates
International Institute for Applied Systems Analysis, Austria

Mekong
Prachoom Chomchai
Mekong Secretariat, Bangkok

Jeff Jacobs
Dept. of Geography, University of Colorado, USA

Huynh Ngoc Phien
Dept. of Computer Sciences, Asian Institute of Technology, Bangkok

Le Huu Ti
Mekong Secretariat, Bangkok

Zambezi
Gary Gaile
Dept. of Geography and Inst. of Behavioral Science, University of Colorado, USA

Chris Magadza
University of Zimbabwe Lake Kariba Research Station, Kariba, Zimb.

David Mazvida
Zambezi River Authority, Kariba

Pedro Restrepo
Center for Advanced Decision Support for Water and Environmental Systems, University of Colorado, USA

B. J. Urbiztondo
Center for Advanced Decision Support for Water and Environmental Systems, University of Colorado, USA

Robert Rose
Dept. of Geography, University of Colorado, USA

Chapter 4: Global Sea-level Rise

Argentina
Karen Clemens Dennis
Laboratory for Coastal Research, Dept. of Geography, Univ. of Maryland

Felix H. Mouzo
Buenos Aires

Carlos R. Orona
Buenos Aires

Enrique J. Schnack
Laboratorio de Oceanografia Costera, La Plata

Bangladesh
Syed Iqbal Ali
Bangladesh Centre for Advanced Studies, Dhaka

Saleemul Huq
Bangladesh Centre for Advanced Studies, Dhaka

A. Atiq Rahman
Bangladesh Centre for Advanced Studies, Dhaka

Brazil
Dieter Muehe
Dept. of Geography, Universidade Federal do Rio de Janeiro

Claudio Freitas Neves
Coppe-Programa de Engenharia Oceania, Universidade Federal do Rio de Janeiro

China
S. Chen
Institute of Estuarine and Coastal Research, East China Normal University, Shanghai

Mukang Han
Dept. of Geography, Peking University, Beijing

Jianjun Hou
Dept. of Geography, Peking University, Beijing

C. Liu
Dept. of Geography, Peking University, Beijing

J. Shen
VIMS, Gloucester Point, Virginia, USA

Boacan Wang
Institute of Estuarine and Coastal Research, East China Normal University, Shanghai

Lun Wu
Dept. of Geography, Peking University, Beijing

K. Zhang
Institute of Estuarine and Coastal Research, East China Normal University, Shanghai

Z. Zhang
Dept. of Geography, Shandong Normal University, Jinan

G. Zhao
Tianjin Seismological Bureau, Tianjin

Egypt
S. Desouki
Institute of Graduate Studies and Research, University of Alexandria

Kh. Dewidar
Institute of Graduate Studies and Research, University of Alexandria

M. El-Raey
Institute of Graduate Studies and Research, University of Alexandria

O. Frihy
Institute of Coastal Research, Alexandria

S. Nasr
Institute of Graduate Studies and Research, University of Alexandria

Hong Kong
Wyss Yim
Earth Science Unit, University of Hong Kong

India
Virendra Asthana
School of Environmental Sciences, Jawaharalal Nehru Univ., New Delhi

Malaysia
Say-Chong Lee
Drainage and Irrigation Dept., Coastal Engineering Technical Center, Kuala Lumpur

Zamali B. Midun
Drainage and Irrigation Dept., Coastal Engineering Technical Center, Kuala Lumpur

Nigeria
Larry F. Awosika
Nigerian Institute for Oceanography and Marine Research, Victoria Island, Lagos

Gregory French
Laboratory for Coastal Research, Dept. of Geography, Univ. of Maryland

A. C. Ibe
Nigerian Institute for Oceanography and Marine Research, Victoria Island, Lagos

C. E. Ibe
Nigerian Institute for Oceanography and Marine Research, Victoria Island, Lagos

Senegal
Karen Clemens Dennis
Laboratory for Coastal Research, Dept. of Geography, Univ. of Maryland

Isabelle Niang-Diop
Dept. de Geologie, Université Cheikh Anta Diop, Dakar-Fann

Uruguay
Claudio R. Volonté
Canelones

Venezuela
José Arismendi
Instituto de Ingenieria, Caracas

Claudio R. Volonté
Canelones

Chapter 5: Human Health

China
Guanri Tan
Dept. of Atmospheric Sciences, Zhongshan University, Guangzhou

Egypt
Rifky Faris
Depts. of Community, Environmental, and Occupational Medicine, Ain Shams University, Cairo

Global
Karl A. Western
National Institutes of Health, National Institute of Allergy and Infectious Diseases, Maryland, USA

West Africa
Soumbey Alley
Biostatistics and Information Systems Unit, World Health Org., Onchocerciasis Control Programme in West Africa, Burkina Faso

Chapter 6: Global Forests

Costa Rica
Sean O'Brien
Env. Science Dept., University of Virginia, USA

Global
Miguel Acevedo
University of Andes, Venezuela

Michael Antonovski
International Institute for Applied Systems Analysis, Austria

Gordon Bonan
National Center for Atmospheric Research, Colorado, USA

Wolfgang Cramer
University of Trondheim, Norway

William Emanuel
Env. Sciences Div., Oak Ridge National Laboratory, USA

Sergei Golovanov
International Institute for Applied Systems Analysis, Austria

Patrick Halpin
Env. Sciences Dept., University of Virginia, USA

Rik Leemans
Biosphere Project, IIASA, Austria

Phillipe Martin
MacQuarie Univ., School of Earth Sciences, Sydney, Australia

Nedialko Nikolov
Higher Institute of Forestry and Forest Tech., Sofia, Bulgaria

Colin Prentice
Institute of Ecological Botany, University of Uppsala, Sweden

Luc Sirois
Rimouski, Quebec, Canada

Alan Solomon
School of Forestry, Michigan Tech University, USA

Brian Walker
CSIRO, Division of Wildlife & Ecology, Australia

David Weinstein
Ecosystems Research Center, Cornell Univ., New York, USA

F. Ian Woodward
Dept. of Botany, Cambridge University, UK

Chapter 7: Integrated Impacts on Egypt

Bayoumi Attia
Ministry of Public Works and Water Resource, Egypt

Helmy Eid
Ministry of Agriculture, Egypt

Hans Lofgren
American University of Cairo

Jeffrey Niemann
Massachusetts Institute of Technology, USA

Gerald O'Mara
International Institute for Applied Systems Analysis, Austria

Chibo Onyeji
International Institute for Applied Systems Analysis, Austria

Sofian Saidin
University of Colorado at Boulder

Magdy Saleh
Cairo University

David Yates
International Institute for Applied Systems Analysis, Austria

Chapter 8: Adaptation Policy

Jeffrey J. Carmichael
Dept. of Economics, University of Colorado, USA

James G. Titus
Office of Policy, Planning and Evaluation, U.S. Environmental Protection Agency

Climate Data

Roy Jenne
National Center for Atmospheric Research, Colorado, USA

Preface

Since the Industrial Revolution, human activities have led to a significant increase in the atmospheric concentrations of greenhouse gases. There is strong scientific evidence that the continued addition of greenhouse gases into the atmosphere will alter global climate, increasing temperatures and changing rainfall and other weather patterns. Global mean temperature is estimated to increase by about 0.3°C per decade during the next century, resulting in a likely increase of about 1°C above today's levels by 2025 and 3°C before the end of the next century. By the middle of the next century these predicted changes will expose ecosystems to temperatures higher than any seen for the last 150,000 years. The rate of increase in global temperatures will be greater than those which have occurred naturally over the last 10,000 years.

In 1989 the U.S. Environmental Protection Agency (EPA) published a report to Congress entitled *The Potential Effects of Global Climate Change on the United States*. This report identified the vulnerability of physical and biological systems to climate change, including agriculture, forests, water resources, sea level rise, biodiversity and wildlife, energy demand, air pollution, and human health. The results of the report, as well as more recent scientific findings, suggest that global climate change will result in a world very different than that which exists today.

As a natural extension to the 1989 report, EPA initiated a series of studies to assess the potential global implications of climate change, with a particular focus on developing countries. This book is the culmination of that effort. The studies concentrated on the potential international impacts of climate change on coastal resources, agriculture, forests, rivers, and human health. The studies found that developing countries are significantly vulnerable to climate change. However it is not possible to say that climate change will result in negative impacts for all physical and biological systems in all developing countries. The particular vulnerabilities of a country depend upon its own set of physical and economic circumstances. But all countries will incur some negative impacts, and will certainly spend scarce and valuable resources to avoid the detrimental effects of climate change.

An important feature of this effort was the direct involvement of developing country scientists. Scientists from over 30 countries, including 18 developing countries, actively collaborated in the studies reported here, ensuring that the best available data was used for each country. Another important outcome was the transfer of analytic methods and skills to developing country scientists. The successful cooperation between researchers in the developed and developing countries in this effort helped lay the foundation for the current U.S. Country Studies Program

It is anticipated that the insights reported in this book will be of interest to the entire international community, particularly the Intergovernmental Panel on Climate Change (IPCC), the International Negotiating Committee (INC), and the Conference of the Parties to the Framework Convention on Climate Change. Hopefully, it will spur continued collaboration among scientists from different countries, and further efforts to assess the potential global implications of climate change.

Joel D. Scheraga
Chief, Adaptation Branch
Climate Change Division
U.S. Environmental Protection Agency
Washington, D.C.

Acknowledgments

This volume is the result of work by hundreds of intenational researchers with input from local, regional, and national decision-makers. The authors of the chapters were those directly involved in performing the quantitative analysis and writing the results. The primary international collaborators are listed by the chapter in which they participated. The list of people who helped to make this a better work through a variety of contributions is too large to include here, but to all of them we are grateful.

We want to thank Dennis Tirpak, Director, Climate Change Division, Office of Policy, Planning and Evaluation, USEPA, for his initiation and support of this work and Dr. Joel Scheraga, Chief, Adaptation Branch, Climate Change Division, USEPA for his technical contributions and his support of the production of this document. We also thank Wayne Cheney and Joe Wensman of the United States Bureau of Reclamation for their collaboration on the Nile and Egypt studies and togethre with Dexter Hinckley, Climate Change Division, USEPA for their logistical support on the completion of this document.

We want to thank Bev LeSuer, University of Colorado, for her editing, and Shelley Preston, Bruce Company, for the many graphics she prepared for this volume. Finally, this book would never have gone to press without the outstanding efforts of Perrin Bowling. International Institute for Applied Systems Analysis. Her administration and technical skills were wonderfully blended in editing, graphic design, figure and table preparation, and text processing. Her organizational skills, eye for detail, and firmness to meet deadlines proved to be the key ingredient in preparing the final manuscript for the publisher. To Perrin, our sincerest and warmest thanks.

Kenneth M. Strzepek
Joel B. Smith
Boulder, CO, USA

Executive Summary

JOEL B. SMITH, KENNETH M. STRZEPEK, LAURENCE S. KALKSTEIN, ROBERT J. NICHOLLS, THOMAS M. SMITH, WILLIAM E. RIEBSAME, and CYNTHIA ROSENZWEIG

Six studies of potential climate change impacts on coastal resources, agriculture, rivers, health, and vegetation in developing countries were commissioned by the U.S. Environmental Protection Agency to improve understanding of the potential impacts of climate change and involved extensive collaboration with developing country scientists. The studies are not comprehensive, but were geographically broad and cover a number of important, climate-sensitive systems. The agriculture and forests studies also examined global changes. Since regional climate change is uncertain, the studies used three to four common scenarios derived from general circulation models along with incremental changes in temperature, precipitation, and sea-level rise. The studies account for some technological improvements and adaptive responses.

On the whole, the studies found significant vulnerability of developing countries to climate change. Crop production could decrease in many countries, low-lying land, particularly in deltas could be inundated by sea-level rise, some river basins could have increased risks of water scarcity, and heat stress could increase in some cities. Some of this vulnerability is the result of relatively fewer economic resources available for developing countries for adapting to such impacts as sea-level rise or diminished water supplies. In addition, climate change could cause developing countries to fall farther behind developed countries. For example, agricultural output in many developing countries may be reduced by climate change compared to what it would be without it, while many developed nations could have increased output. Whether global agricultural production will increase or decrease depends on the magnitude of climate change, the CO_2 fertilization effect, and the affordability of adaptive measures.

The sensitivity of ecosystems to climate change does not depend on whether they exist in developed or developing countries. Widespread ecological disruptions could occur because rates of climate change could far exceed the ability of species to adapt. Many isolated species, such as those in reserves, could be threatened with extinction. Boreal ecosystems may be at a relatively greater risk of disruption than lower latitude forests. Some forests, such as tropical rain forests, could expand, if land were available. Sea-level rise could inundate many coastal wetlands.

Yet, not all impacts of climate change in developing countries would be negative. Some areas may have enhanced water supplies, increased crop yields, or higher species productivity. Furthermore, adaptation may offset many of the negative impacts of climate change. Farm level practices can be adjusted, coastal defenses built, and water supplies allocated more efficiently. Taking action to anticipate climate change could further reduce vulnerability. Whether these measures would offset the negative impacts of climate change will depend on the rate and severity of climate change and availability of financial resources. But, these adaptive investments will consume resources that could have gone to other uses and may still not restore societal welfare to conditions that would prevail if climate change does not happen.

INTRODUCTION

If, as many scientists predict, average global temperatures increase 1.5 to 4.5°C and sea-level rises up to a meter over the next century, the welfare of many societies and ecosystems around the world could be adversely affected. Coastal development, agriculture, water supplies, and other societal systems are at risk. Ecosystems would be exposed to temperatures higher than any seen for over 125,000 years.

Since they have a relatively poor resource base, high dependence on climate sensitive systems, and rapid population growth, developing countries could be more vulnerable to climate change than developed countries. Yet, very few studies of potential climate change impacts have focused on developing countries. Five major assessments were conducted of the potential effects of climate change on agriculture, forests, river basins, coastal resources, and human health in developing countries. The analyses were done by or in collaboration with scientists in developing countries. The projects, which were funded by the U.S. Environmental Protection Agency, were developed to: (1) provide information to policymakers about sensitivities to climate change; and (2) to collaborate with scientists in developing countries on assessing climate change impacts. In addition, a study of Egypt was conducted to integrate the impacts of climate change on coastal resources, water supplies, and agriculture.

Structure of the studies

Developing country scientists collaborated with principal investigators in the United States and Britain to examine potential climate change impacts in specific countries, using consistent methodologies, state-of-the-art models, and common climate change scenarios. More than 150 scientists, 85 of whom are from developing countries, were involved in this project. The projects were initiated in 1989–90.[1] Each study forms the basis for a chapter in this volume.

All of the projects focused on case studies in developing countries. Two projects, agriculture and forests also examined, respectively, changes in global food production and trade and changes in global forests. With the exception of Egypt, results from different studies were generally not integrated (e.g. combining changes in water supply and crop yields).

Climate change scenarios

Given the uncertainties about regional climate change, regional impacts cannot accurately be predicted. However, sensitivities of systems to climate change can be identified by using scenarios of global and regional climate change. A 'climate change scenario' is defined as a physically plausible set of changes in meteorological variables, consistent with generally accepted projections of global temperature change (i.e., 1.5 to 4.5°C for an effective doubling of CO_2). *The scenarios are not predictions, nor do they bound all of the potential changes in regional climates.*

Following consultation with climate modelers, climatologists, and impacts researchers, a combination of scenarios from general circulation models (GCMs) and uniform, arbitrary changes in climate were chosen to create regional doubled CO_2 climate change scenarios.

[1] The research for this project was supported and managed by the U.S. Environmental Protection Agency. The U.S. Agency for International Development and the U.S. Bureau of Reclamation provided support and co-funding for, respectively, the agriculture project and the Nile River study as part of the rivers project.

Table 1. *GCM estimates of increases in global temperature and precipitation*

GCM	Temperature Increase (°C)	Precipitation Increase (%)
GISS	4.2	11.0
GFDL	4.0	8.7
UKMO	5.2	15.0
OSU	2.8	7.8
LOWEND*	2.4	5.0

Note:
* Transient Run A of the GISS GCM for the decade of the 2030s.

The GCMs used here are from the United Kingdom Meteorological Office (UKMO), the Geophysical Fluid Dynamics Laboratory (GFDL), the Goddard Institute for Space Studies (GISS), and Oregon State University (OSU). The average global temperature and precipitation changes for each GCM are displayed in Table 1. A scenario with a lesser amount of warming than these four scenarios (GISS-A) was defined as the average change in the 2030s of the GISS-A transient run and was used as a scenario for doubled CO_2 conditions. Note that the UKMO projects more warming than the IPCC's most likely warming range of 1.5 to 4.5°C for CO_2 doubling, while the other models project warming between 2.4 and 4.2°C. Average monthly changes for each grid box were combined with observed daily climate data, from 1951–80, to create scenarios. The arbitrary scenarios were created by adjusting the observed climate data set by +2 or +4°C and no change, +20 percent, and −20 percent in precipitation. The arbitrary scenarios were only used in regional analysis, not in the global analysis. Sea-level rise scenarios are +0.5 and +1.0 meters. All of the scenarios only examined sensitivity to changes in average conditions, not changes in variability. These impacts were assumed to happen in the latter half of the next century.

Kalkstein (1991) compared the estimates of current climate from the four GCMs with real-time climate observations. In most areas of the world, the models reasonably simulate normal temperature and precipitation. The models exhibited the largest errors in coastal regions, mountains, and hot deserts. The one region with particularly large errors was south Asia, where the models did not duplicate the summer monsoon. For that area, annual averages (rather than monthly averages) from the models were used for precipitation.

The ultimate effect of climate change will be the result not only of the physical effects of a changed climate, but also of adaptations by society and nature. Such responses to climate change as migration, planting of different crops, and more efficient use of water, can offset some of the negative impacts of climate change or take advantage of opportunities created by climate change. The studies in this volume examined the sensitivities of systems to a variety of adaptations, although no study can account for all possible adjustments.

STUDY FINDINGS

Agriculture

Global population is estimated to grow from the current 5 billion to 10 billion by the middle of the next century. This increase in population will present challenges to global food supplies. Climate change could exacerbate or ease these challenges. Shifts in regional production patterns could make it more difficult for some nations to feed themselves and significantly alter the relative competitiveness of food exporters.

This project involved three components:

(1) *Crop modeling*
Agricultural scientists in 18 countries estimated potential changes in national crop yields using the crop models developed by the U.S. Agency for International Development's International Benchmark Sites Network for Agrotechnology Transfer (IBSNAT). The models account for the beneficial physiological effects of increased atmospheric CO_2 concentrations (555 ppm by 2060) on crop growth and water use (using experimental results which may overstate the positive effects of CO_2). The studies did not account for potential reductions in water supplies for irrigation. Crop model results were aggregated by weighting regional yield changes to estimate changes in national yields. The changes in national yields in other crops and regions were then estimated based on the crop model results, other studies, and GCMs.

(2) *World food trade*
The national yield changes were used as inputs into the International Institute for Applied Systems Analysis' world food trade model, called the Basic Linked System (BLS). This general-equilibrium model simultaneously estimates annual changes in global and regional crop production, prices, trade patterns, and the number of people at risk of hunger.[2] Arable land was assumed to increase and technology was assumed to improve. Mid-range United Nations estimates for population and economic growth through 2060 were used to establish a base case without climate change.

[2] The number of people at risk of hunger is based on Food and Agricultural Organization (FAO) estimates and methodology and is an indicator of the ability of individuals to grow or procure food. It is not a projection of the number of people actually suffering from hunger.

(3) *Vulnerable regions*
Finally, case studies were conducted of specific countries that may be vulnerable to increases in hunger because of the combination of climate change and other factors such as population growth.

CROP PRODUCTION

Whether global production would increase or decrease depends on the magnitude of climate change, the effect of CO_2 fertilization, and the degree to which adaptive measures are adopted. All of the climate change scenarios, however, resulted in decreased yields in developing countries and generally increased yields in the developed countries. Figure 1 shows the changes in national grain crop yields simulated for the GISS and UKMO climate change scenarios with the physiological effects of increased CO_2. The 5.2°C average global warming in the UKMO scenario appears to significantly reduce crop yields in most areas of the world compared to the 4.2°C average warming in the GISS scenario. Results for any country are national averages and should not be interpreted as applying to specific crops or regions within countries.

After economic adjustments were accounted for (through the BLS), disparities in crop production between developed and developing countries were estimated to increase. Production in developing countries was estimated to decline from base case (2060 with no climate change) levels, while total production in the developed countries was estimated to increase except in the extreme UKMO scenario (see Table 2). Estimates of changes in global production of cereal crops range from an increase of 2.5 percent to a decrease of 8 percent.

CEREAL PRICES AND THE NUMBER OF HUNGRY PEOPLE

A reduction in agricultural output would most likely increase food prices over the baseline scenario, while a decrease has the opposite effect. In the GCM-based scenarios, world cereal prices were estimated to range from a decrease of 17 percent in base prices to an increase of 145 percent by 2060 (see Figure 2).

A reduction in agricultural output and subsequent increase in prices would raise malnourishment. With no climate change and population and technology increases, the BLS model estimated that 641 million people would be at risk of hunger in 2060. Under the GISS, GFDL, and UKMO climate change scenarios, an additional 63 to 369 million people in developing countries would be at risk of hunger, while the number of potentially hungry people would decrease by 84 million under the GISS-A scenario (see Figure 3).

Table 2. *Changes in cereal production in 2060 (with direct effects of CO_2)*

Region	Base Case (MMT)	GISS (%)	GFDL (%)	UKMO (%)	GISS-A (%)
Global	3286	−1.2	−2.8	−7.6	+2.5
Developed countries	1449	+11.3	+5.2	−3.5	+10.6
Developing countries	1836	−11.0	−9.2	−10.8	−3.9
Africa	296	−19.3	−20.9	−11.1	n/a
Central & South America	274	−23.7	−15.7	−5.8	n/a
South & Southeast Asia	1133	−6.0	−3.9	−11.8	n/a
Southwest Asia	133	−9.8	−14.3	−11.2	n/a

Notes: Base Case is estimated production in 2060 assuming no climate change and no increase in CO_2 levels.
n/a = not available

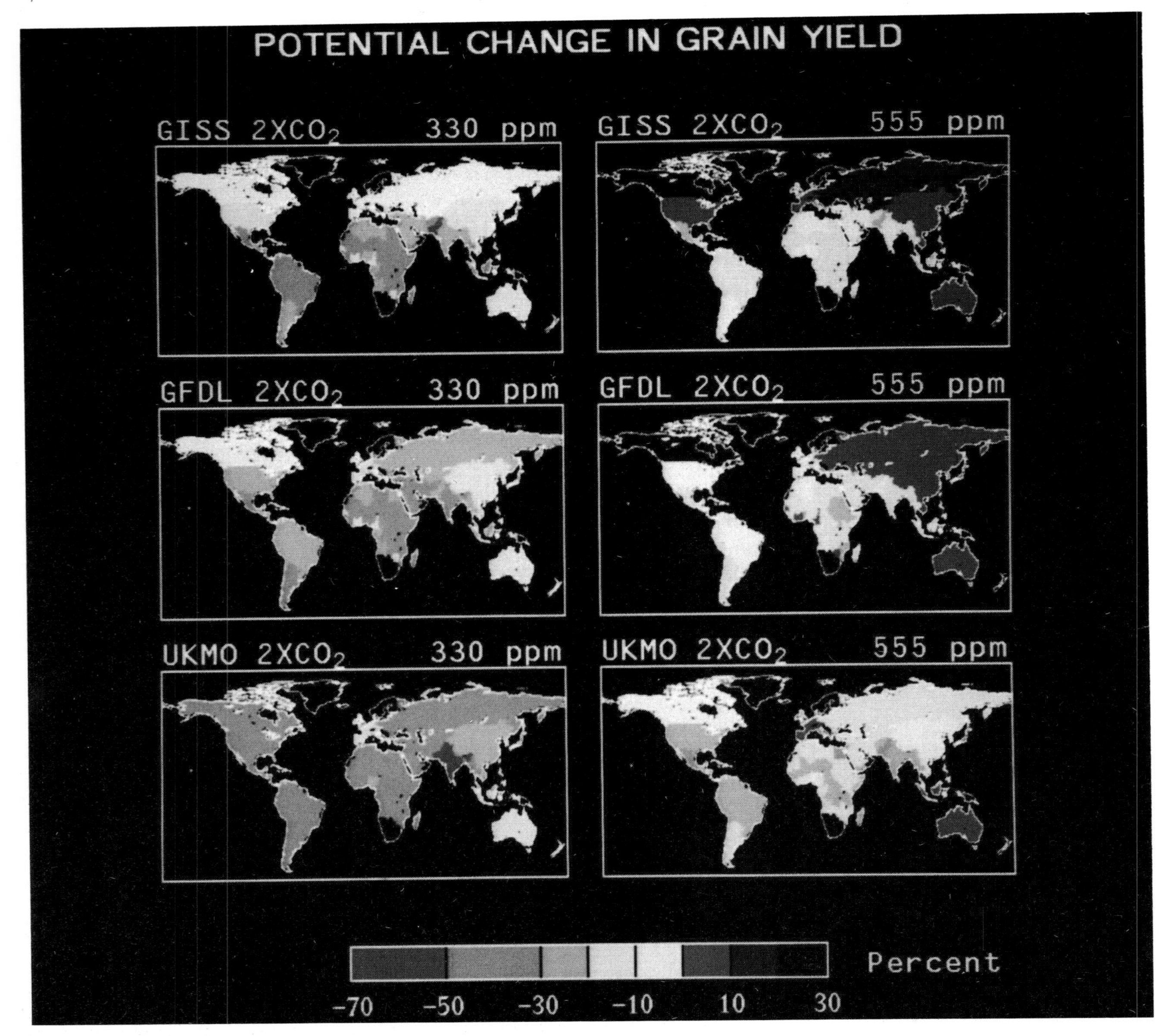

Figure 1 Estimated change in average national grain yield (wheat, rice, coarse grains, and protein feed) for the GISS, GFDL, and UKMO climate change scenarios. The left-hand column shows effects of climate change alone (330 ppm CO_2); the right-hand column shows the combined effects of climate change and direct CO_2 effects (555 ppm CO_2). Results shown are averages for countries and groups of countries in the BLS world food trade model; regional variations within countries are not reflected (Rosenzweig and Parry, 1994).

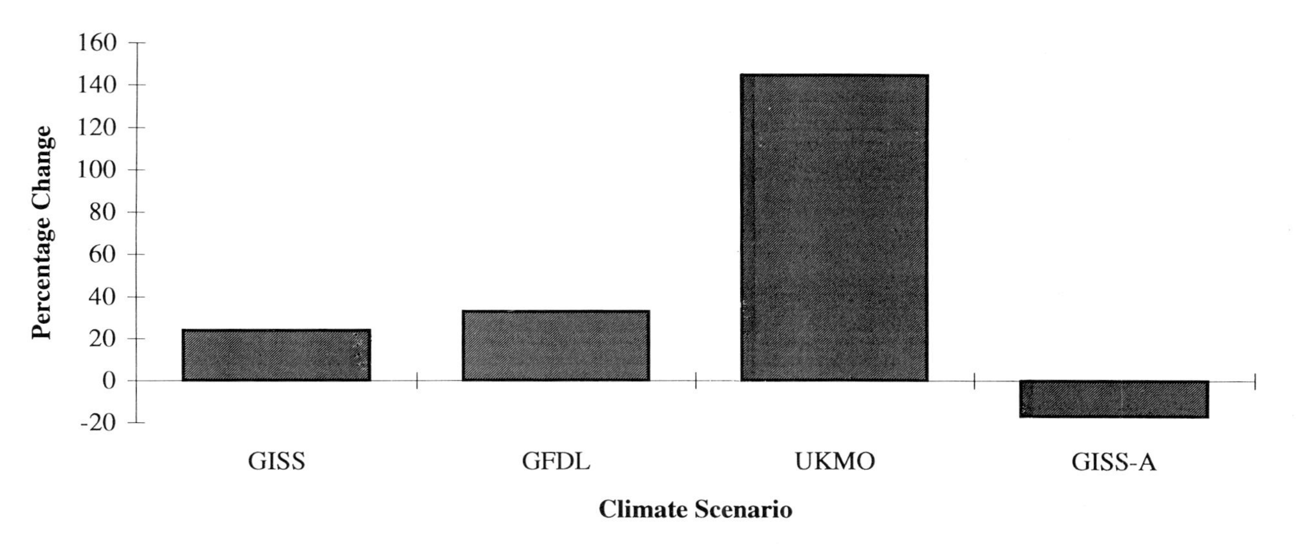

Figure 2 Change in cereal price index in 2060 calculated by the Basic Linked System under the four climate change scenarios. Base case is 121; prices are relative to 1970, which is 100.

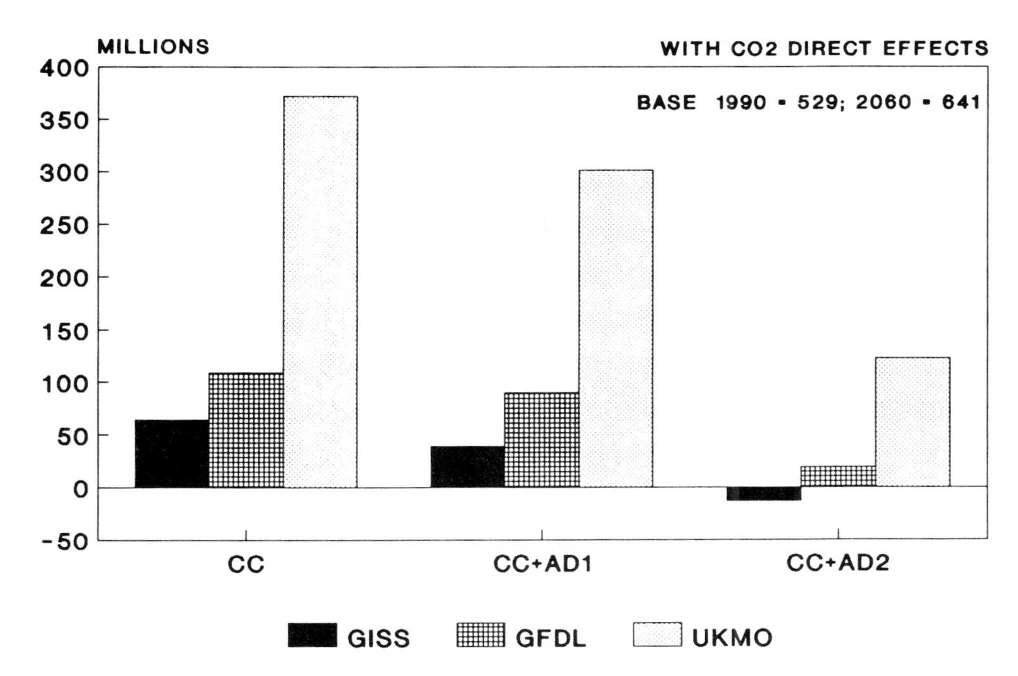

Figure 3 Change in number of people at risk of hunger in 2060 calculated by the Basic Linked System under climate change scenarios for no adaptation and Adaptation Levels 1 and 2 (AD1 and AD2). Reference scenario for 2060 assumes no climate change (529 million people at risk of hunger in 1990; 641 million people are at risk of hunger in 2060).

SENSITIVITY OF RESULTS TO ASSUMPTIONS ABOUT POPULATION AND ECONOMIC GROWTH AND TRADE LIBERALIZATION

The impact of climate change compared to several policy choices was also examined under the GISS, GFDL, and UKMO scenarios. The relative impacts on the number of people at risk of hunger of lower population growth rates, full trade liberalization, and lower economic growth rates, assuming climate change, are shown in Figure 4.[3] Reducing population growth results in fewer people at risk of hunger

[3] The base case assumes a global population of 10.3 billion by 2060, while the low population growth scenario assumes 8.6 billion people. The base case assumes that trade barriers are reduced by half by 2020 and do not change after that, while the full trade liberalization scenario eliminates trade barriers by 2020. The base case assumes 1.85 percent GDP growth per year from 1990–2060, while the lower economic growth rate scenario assumes 1.72 percent GDP growth. The low economic growth rate scenario results in 10 percent lower GDP in 2060 than the base case.

than fully liberalizing trade by 2020. On the other hand, a decrease in the average annual economic growth rate results in more people at risk of hunger than in the base case scenario.

FARM LEVEL ADAPTATION MEASURES

The effects of farm-level adaptation measures, such as switching planting dates or crop varieties, and installing irrigation were examined. Whether such responses would be affordable or whether water supplies would be available was not analyzed. As shown in Table 3, adaptation appears to fully mitigate the global reduction in cereal output in two of the three scenarios. Yet, yield declines in developing countries are only partially mitigated by the adaptation measures. As is shown in Table 4, when adaptation is included, the number of people at risk of hunger in the GISS scenario decreases compared to the base case.

The sensitivity of the results to slowing the effective climate change impacts until 2100, rather than 2060 was also tested. This generally reduced the negative impacts by about one-third.

SENSITIVITY ANALYSES

Sensitivity analyses were conducted assuming a uniform 2 and 4°C increase in temperature in all regions, with no change in precipitation and beneficial effects of CO_2. In all likelihood, regional temperature change will vary, with higher latitude areas having, on the whole, the largest temperature increases. Precipitation will increase in many regions, although not necessarily during the growing season.

Table 3. *Changes in cereal production in 2060 with farm-level adaptation (MMT)*

GCM Scenario	GCM Alone	GCM + Adaptation
	(% Change from 2060 Base Case)	
GISS		
Global	−1.2	+1.1
Developing	−11.0	−6.6
GFDL		
Global	−2.8	−0.1
Developing	−9.2	−5.6
UKMO		
Global	−7.6	−2.4
Developing	−10.8	−5.7

Note: Base global production is 3286 MMT in 2060 and is 1836 MMT for developing countries. This assumes both low and high cost adaptation measures are implemented.

Table 4. *Changes in number of people at risk of hunger in 2060 with farm-level adaptation (MMT)*

GCM Scenario	GCM Alone	GCM + Adaptation
	(millions of people)	
GISS – Global	+10	−12
GFDL – Global	+108	+18
UKMO – Global	+369	+119

Note: A negative number is a reduction in the number of people at risk of hunger compared to the base case.

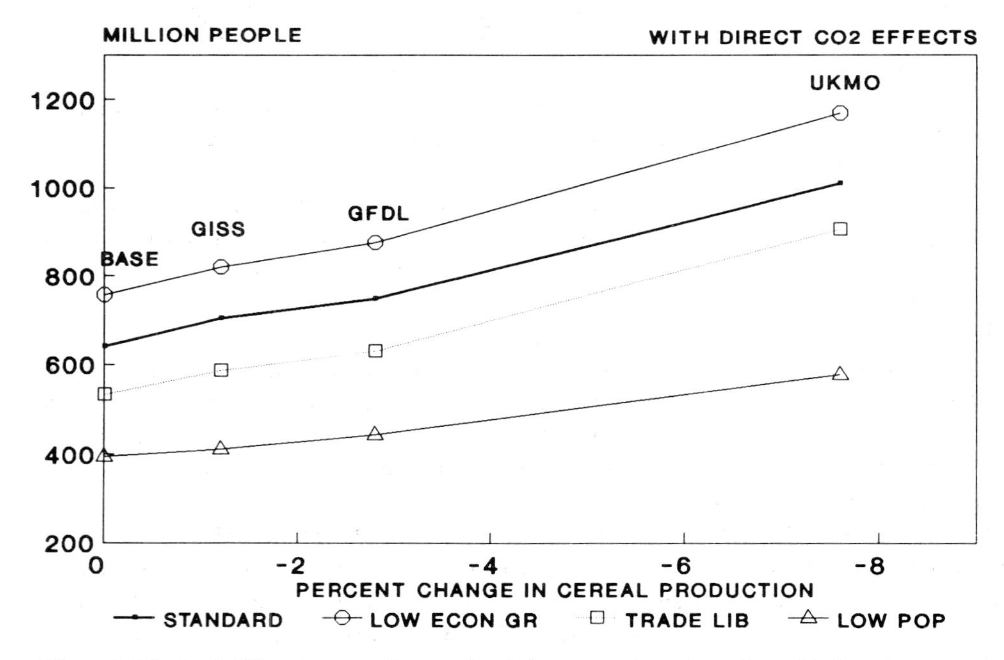

Figure 4 Effects of different assumptions and policies on number of people at risk from hunger calculated by the Basic Linked System under the climate change scenarios.

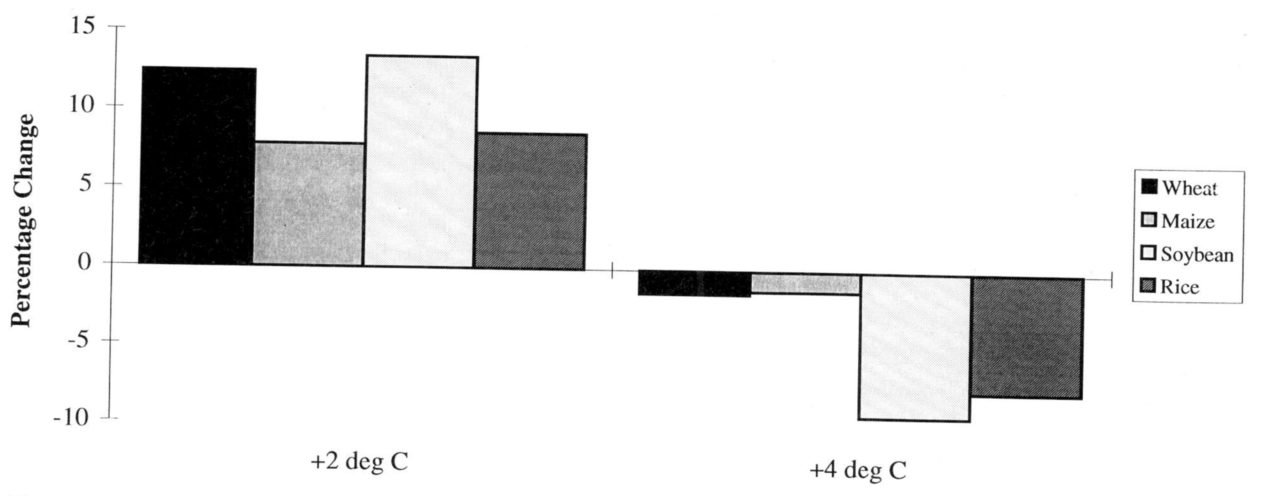

Figure 5 Aggregated IBSNAT crop model yield changes for +2°C and +4°C temperature increase. Country results are weighted by contribution of national production to world production. Direct effects of CO_2 on crop growth and water use are taken into account.

As shown in Figure 5, aggregate global yields are estimated to increase at 2°C, but decline at 4°C. This result is consistent with the difference in direction between the GISS-A scenario and the other GCM based scenarios. Thus, the effects of climate change on crop yields could be positive at first, and then turn negative.[4]

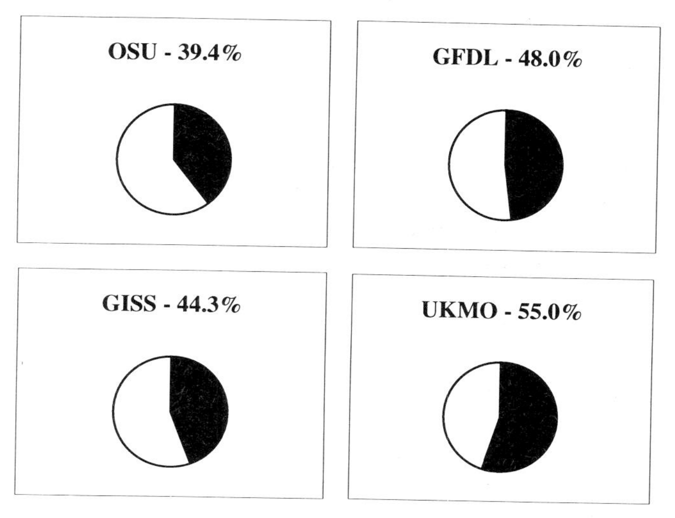

Figure 6 Percentage of the Earth's terrestrial ecosystem zones that will change under the four climate-change scenarios.

Forests

Of the sectors studied in this report, the world's forests may be the most vulnerable. The reason is that the other sectors are heavily managed and societal response measures can be quickly implemented. In contrast, most of the world's forests are not intensively managed, and their natural ability to adapt to climate change through migration or physiological change is limited.

The forests study examined the potential impacts of climate change at the global and regional levels. It is the first study to examine both levels of potential impacts using consistent scenarios. The project analyzed global distribution of major terrestrial ecosystems, changes in the distribution of boreal and tropical systems, and the socio-economic implications of changes in tropical and sub-tropical systems.

Bioclimate models were used to estimate potential shifts in eco-climatic zones associated with the distribution of major ecosystem types (such as boreal forests or tropical rainforests). These estimates only account for shifts in climate zones. They do not examine whether forests could survive or how long it would take to migrate to new locations. Dieback of forests in areas where the climate becomes too warm or dry could be on the order of decades, whereas migrating hundreds of kilometers to new, more suitable zones could take centuries to millennia. Thus, the inhabited range of many forests could be significantly reduced.

[4] If CO_2 levels are not as high or the fertilization effect less positive than assumed in this analysis, the net effect of even the GISS-A scenario could be a decline in crop yields.

VEGETATION COVER

The four GCM scenarios show a major impact on the potential distribution of terrestrial vegetation. Figure 6 shows that 40 to 55 percent of the Earth's terrestrial land surface could have a change in eco-climatic zone and major type of vegetation.

The corresponding potential for changes in areal coverage of major ecosystems as defined by these eco-climatic zones is shown in Table 5. The calculations in this table are based on the assumption that forests have migrated to and established themselves in new areas. As noted above, dieback will be much more rapid than migration. All four scenarios show

Table 5. *Equilibrium changes in surface area of major ecosystem types* (1000 km²)*

Ecosystem	Present Cover	Changes			
		OSU	GISS	GFDL	UKMO
Tundra	938.9	−301.7	−314.3	−514.9	−573.2
Desert	3699.0	−618.4	−962.1	−630.0	−980.1
Grassland	1923.1	+380.2	+694.0	+969.6	+810.2
Dry forest	1815.7	+4.6	+487.0	+608.4	+1296.3
Mesic forest	5172.1	+560.6	+119.9	−402.5	−519.0

Note:
* Assumes climate has stabilized long enough for forests to completely adapt and all areas that could become forest types do transform to those types.

decreases in desert and tundra and increases in grassland and dry forests. Two scenarios show increases in wet (mesic) forests, while the other two estimate decreases.

With almost half of the Earth's surface changing eco-climate zones, there is a significant potential for a major shift in productivity of the world's forests. Assuming climate does not further change after carbon dioxide levels are doubled, and the forests have centuries to adapt, total forest cover could increase. However, if dieback in areas that become drier happens on a timescale of decades, while migration to new areas takes centuries, forest cover could be reduced over the next century.

BOREAL FORESTS

Climate change could cause a major dieback of boreal (northern) forests in the next century. Under the four GCM scenarios, current boreal forest cover would be reduced by 15 to 45 percent. Figure 7 shows potential shifts in location of boreal forests in North America. Boreal forests that are estimated to shift to grasslands or shrublands may die back relatively quickly. Transitions to warmer forests types or migration up to 600 km into the tundra will probably proceed much more slowly.

TROPICAL FORESTS

Wet tropical forests could decline in some areas because they may become drier. Although other areas could become suitable for expansion of tropical forests, migration or transplantation is unlikely, given human settlement patterns. Figure 8 displays reductions in wet tropical forests in Africa due to moisture stress. If these forests are unable to migrate or be transplanted to new areas, under the four GCM scenarios, their areal cover could be reduced by 5 to 26 percent.

BIODIVERSITY IN NATURE RESERVE SITES

With major changes in global vegetation and habitat, many threatened or endangered species in isolated habitats could face reduction in their populations or even extinction. Although some species may thrive under warmer climate conditions, others may find their habitat so altered they can no longer survive. Changes in major vegetation types associated with the 243 Man and the Biosphere (MAB) nature reserve sites were examined (see Figure 9). The reserve system appears to be very sensitive to climate change as one-half to four-fifths of the reserves could face a major change in eco-climate zone.

Sea-level

While the magnitude of sea-level rise is uncertain, climate change is likely to expand ocean waters and melt glaciers. Rising sea-level can inundate and erode land, exacerbate coastal flooding, and increasing the salinity of estuaries and aquifers.

The sea-level rise study was comprised of national studies (and more detailed case studies) by in-country scientists and five national studies using aerial videotape-assisted vulnerability analysis (or AVVA), in collaboration with in-country scientists. AVVA was developed because detailed information on coastal characteristics, most particularly elevation, was not available for many developing countries. The coastlines of Senegal, Nigeria, Venezuela, Uruguay, and Argentina were videotaped at low altitudes from small airplanes and the video record was compared with selected measurements from the ground to characterize the coastline.

Land loss was then estimated together with the cost of various protection options, including no protection (or retreat), and protecting medium to highly developed areas. Tourist beaches were assumed to have sand pumped on them (beach nourishment), while seawalls would be constructed to protect other developed areas. Local engineering cost estimates were used to calculate adaptation costs. No change in populations levels and patterns of coastal development was assumed.

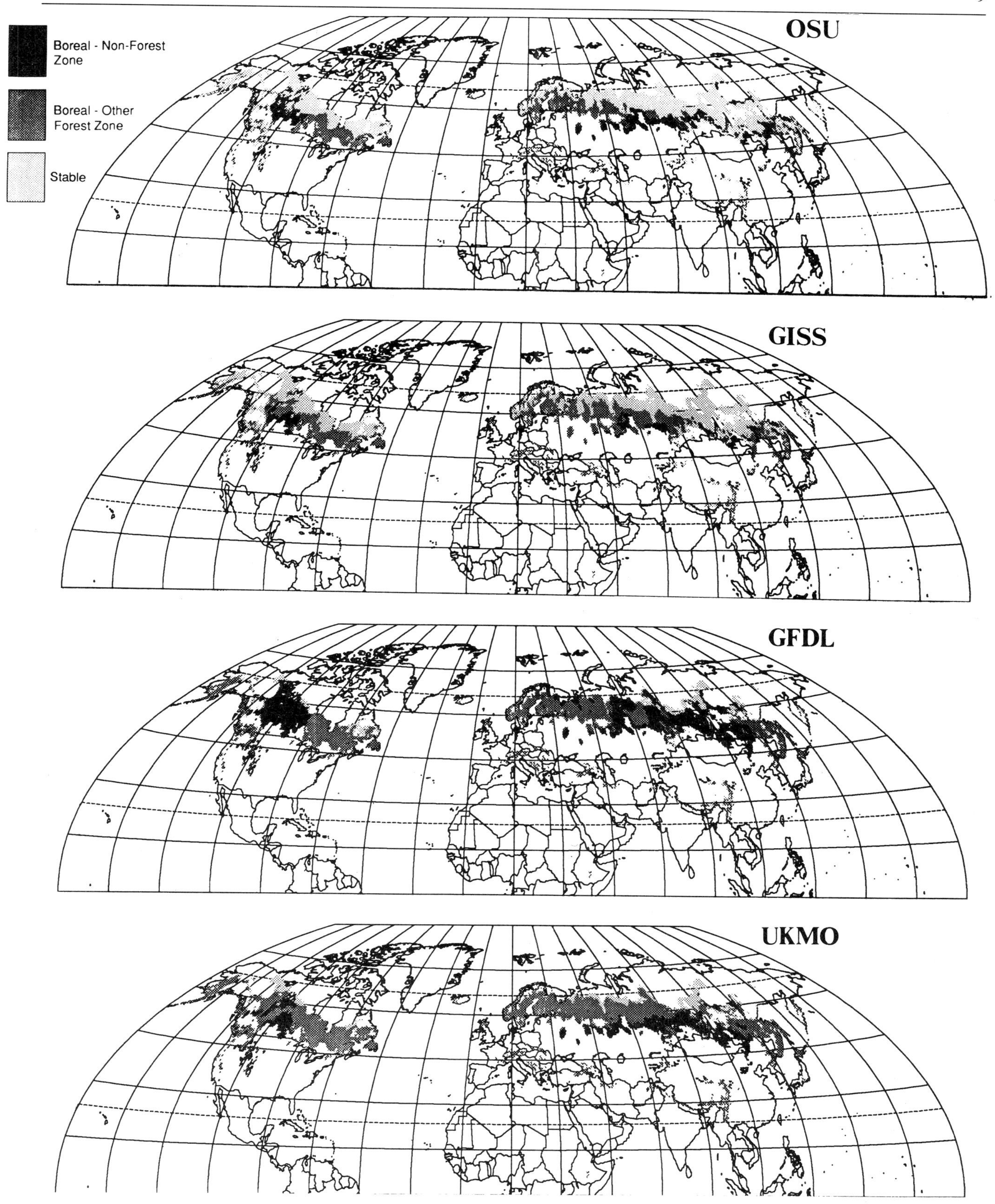

Figure 7 Changes in the current distribution of eco-climate zones (Holdridge 1967) within the boreal forest region under four climate change scenarios. Categories of change include (1) shift from boreal forest to nonforest cover (i.e., grassland or shrubs), (2) shift to climate associated with warmer forest types (i.e., cool temperature forest), and (3) areas that remain in the boreal forest eco-climatic zone. Map also includes areas of montane conifer forest.

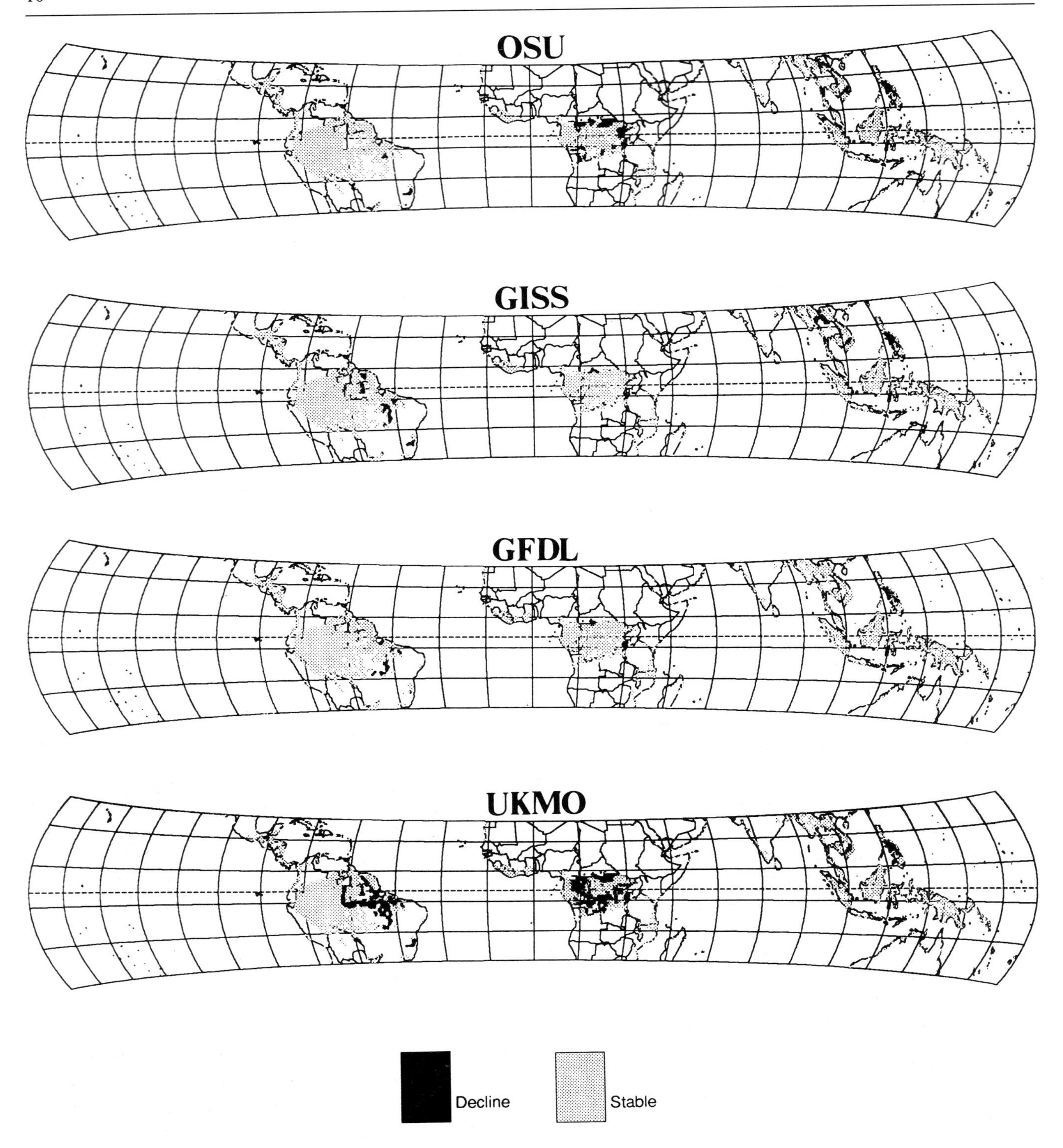

Figure 8 Changes in the current distribution of tropical mesic forest under four climate change scenarios. Areas of decline defined by shift to drier eco-climatic zone (Holdridge 1967).

Table 6. *Land and population at risk from the one-meter sea-level-rise scenario*

Country	Land at Risk in sq. km	Land at Risk %	Population at Risk in millions	Population at Risk %
Argentina	>3,430 to 3,492	>0.1	n/a	n/a
Bangladesh	25,000	17.5	13	11
China	125,000*	1.3	72.0*	6.5
Egypt	4,200 to 5,250	12 to 15**	6.0	10.7
Malaysia	7,000	2.1	n/a	n/a
Nigeria	18,398 to 18,803	2.0	3.2	3.6
Senegal	6,042 to 6,073	3.1	0.1 to 0.2	1.4 to 2.3
Uruguay	94	<0.1	0.01	0.4
Venezuela	5,686 to 5,730	0.6	0.06	0.3
TOTAL	194,852 to 196,498		94.4 to 94.5	

Notes:
n/a – not available
* In China, land and population at risk includes people at risk to increased flooding due to sea-level rise (Han *et al.*, 1994).
** This is the percent of arable land.

LAND AND POPULATION AT RISK

Table 6 presents total land loss and population at risk from a one-meter rise assuming no coastal protection measures are taken. Results are expressed in ranges to reflect uncertainties. Over 190,000 km^2 of land is at risk of inundation or flooding, most of it in deltas. The current population of these areas is over 94 million people. Countries with large agricultural populations in deltas are particularly vulnerable to sea-level rise.

Major results using a one-meter scenario are as follows:

- *Bangladesh.* Sixteen percent of land used for rice production could be inundated and 13 million people could be displaced. The Sunderbans, the second largest mangrove swamp in the world and one of the last refuges of the Bengal tiger, could be lost.
- *China.* Over 72 million people, with major cities such as Shanghai, and tens of thousands of square kilometers of agricultural land are at increased risk of flooding or complete inundation, mainly in the four major coastal plains of China.
- *Egypt.* Sea-level rise could inundate 12–15 percent of the existing agricultural land by 2100 (4 percent by 2060) and could displace over six million people, including half the population of Alexandria.

In many cases, protection is technically feasible, but large populations would be living below high tide and would face catastrophic consequences in the event of failure of coastal defenses.

COASTAL WETLANDS

Most of the inundated land is coastal wetlands. This pattern of inundation of wetlands is likely to be repeated elsewhere with major implications for other coastal resources such as fisheries and biodiversity of coastal ecosystems. Known options to counter these losses are limited, and assuming accelerated sea-level rise, a major reduction in coastal wetlands at a global scale appears almost inevitable.

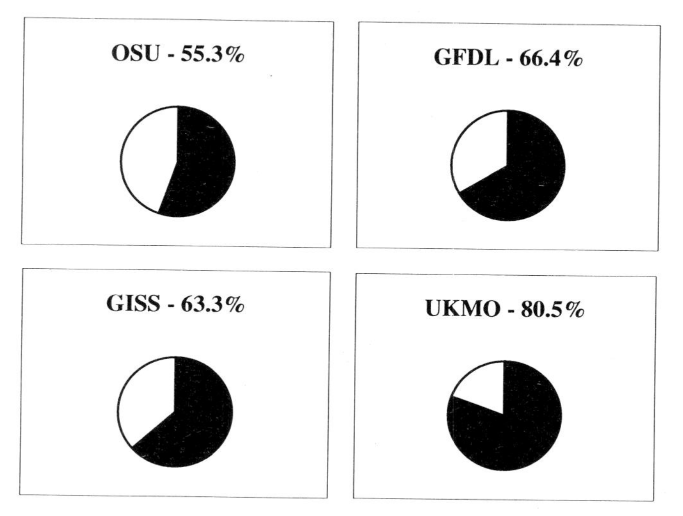

Figure 9 Percentage changes in major vegetation types associated with the 243 MAB nature reserve sites.

NOURISHMENT OF SANDY BEACHES

Table 7 displays the cost of protecting developed areas against rising sea-levels. A large portion of these costs is from nourishment which is more expensive per kilometer than building seawalls. Uruguay, which has relatively small land loss, has very high protection costs since tourism is very important and beach nourishment would be likely in many

Table 7. *Costs of Important Areas Protection (IAP) and Total Protection (TP) (in millions of U.S. dollars)*

Country	Rise (m)	IAP	TP
Argentina	0.5	337–883	829–2,150
	1.0	580–1,298	1,829–3,328
Nigeria	0.5	223–319	609–888
	1.0	558–688	1,424–1,766
Senegal	0.5	146– 575	407–1,422
	1.0	255–845	973–2,156
Uruguay	0.5	2,068–2,140	2,154–2,728
	1.0	2,903–2,993	3,124–3,790
Venezuela	0.5	454–960	719–1,613
	1.0	999–1,517	1,717–2,634

areas. In contrast, tourism is not a major part of Nigeria's existing economy, so it is unlikely that they would use much beach nourishment. In all cases, the broad range of costs is mostly due to uncertainties regarding how much sand will need to be pumped onto beaches. Significant growth in coastal tourism is apparent in all five countries and future requirements for beach nourishment are likely to increase.

RELATIVE COSTS OF PROTECTION

Figure 10 shows the relative vulnerability of five countries to the one-meter scenario if medium to highly developed areas are protected. Protection costs are shown relative to gross investment in 1990 and assuming this expenditure occurs uniformly over a 50 year period. Since gross investment includes depreciation, the percentage of available capital diverted to coastal defenses may be understated. Conversely, real resources available for investment could increase in the future. Senegal and most particularly, Uruguay, face much larger relative protection costs than Nigeria, Venezuela and Argentina. However, Senegal and Nigeria will experience substantial land loss despite the protection while Uruguay will have minimal losses (in relation to national land area).

Rivers

The Intergovernmental Panel on Climate Change concluded that relatively small changes in climate can cause or exacerbate water resource problems, especially in arid and semi-arid river basins. Though many river basin studies have been conducted, most have focused on developed regions. This study assessed potential impacts and adaptations in international basins shared by developing countries: the Indus, Mekong, Uruguay, Zambezi, and Nile basins. Each case estimated runoff and water resource management impacts with simulation models typically used by water resource planners. In some cases, models were already available; in others, new models were developed in cooperation with researchers in the basins. Capacity for adapting to estimated changes was assessed via workshops involving water managers in the basins.

Even though all five basins are in lower latitudes and have relatively high current temperatures, their sensitivity to climate change scenarios varies. Figure 11 shows changes in runoff for GCM scenarios. Predictions about the future direction of runoff in any of these basins are not possible as different scenarios yield very different results. Although higher temperatures tend to decrease runoff (except in the

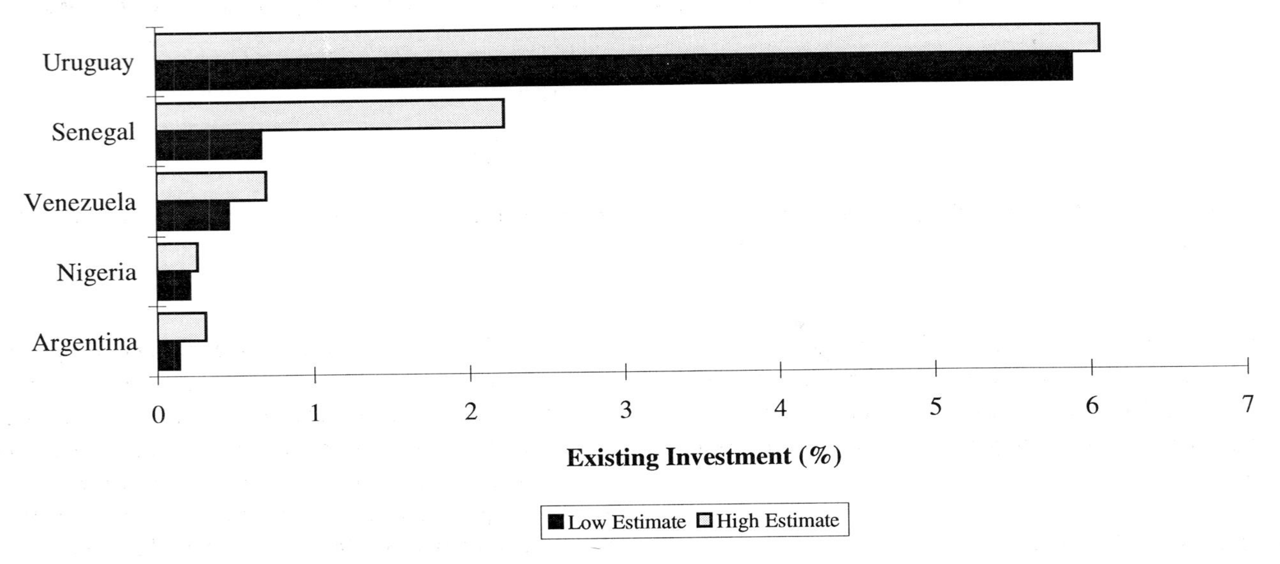

Figure 10 A vulnerability profile for Important Areas Protection and a 1 m sea-level-rise scenario. The protection costs are assumed to occur uniformly over 50 years (2051 to 2100) and are expressed as a percentage of gross investment in 1990 (World Bank, 1992).

Table 8. *Summary of climate and hydrological changes for case study basins*

GCM	Uruguay	Mekong	Indus	Zambezi	Nile
GISS					
Temperature (°C)	+4.5	n/a	+4.7	+4.0	+3.4
Precipitation (%)	97	97	110	109	131
Runoff (%)	88	100	111	89	130
GFDL					
Temperature	+3.6	n/a	+4.5	+3.7	+3.1
Precipitation	119	97	120	102	105
Runoff	122	100	116	83	23
UKMO					
Temperature	+5.6	n/a	n/a	+5.3	+4.7
Precipitation	104	107	n/a	119	122
Runoff	106	115	n/a	118	88

Note: Modified scenarios were used in the Indus, as described in the Indus section in the text. Note also that selection and averaging of grid cells and runoff stations, and annual averaging of some marked seasonal changes, for this summary table obscure the complex effect of temperature changes, seasonality shifts, and subbasin contributions in the final discharge changes. For example, runoff increases proportionally greater than precipitation in the Uruguay under GFDL because much of the rainfall increase occurs in a single month.

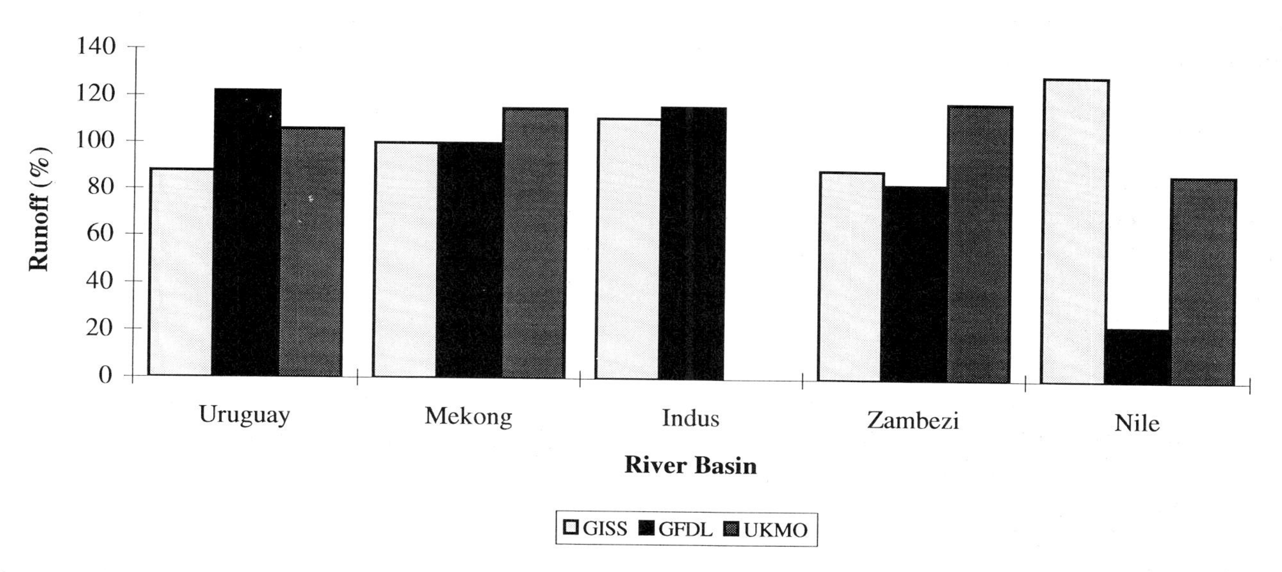

Figure 11 The impact of climate change scenarios on mean annual runoff in the case-study river basins expressed as a percentage of current runoff.

Indus), an increase in precipitation can offset the higher temperatures in each basin. In several basins, at least one scenario results in increased potential for both flooding and water scarcity; that is, larger high flows and reduced low flows.

THE NILE RIVER BASIN

The Nile appears to be the most sensitive of the basins studied, with the GFDL scenario reducing runoff by 77 percent, and a 4°C warming with no change in precipitation reducing runoff by 90 percent. This extreme sensitivity is probably because the source of the Nile is close to the Equator, where temperatures are already relatively high, and there is significant evaporation from the lower basin, the Sudd swamps, and Lake Nasser at the Aswan Dam. River basin managers in Egypt felt they could readily adapt to a 10 percent reduction in flow through operating changes, but a reduction greater than 25 percent would cause major problems. Changes in flow of the magnitude estimated here would most likely necessitate a re-negotiation of the treaty allocating Nile flows between Egypt and Sudan. The effects of climate change on Egypt are discussed in more depth in the last section of this Executive Summary.

THE UPPER ZAMBEZI RIVER BASIN

Like the Nile, the Upper Zambezi River is quite sensitive to higher temperatures. Flow could decline, even if precipitation increases. Under many of the climate change scenarios, the level of Lake Kariba was estimated to fall and hydropower production would be reduced. When the currently planned Batoka dam was simulated in the model, the levels of Lake Kariba were estimated to return to normal. However, some of the hydropower provided by Batoka would be needed to offset lost hydropower from Kariba, instead of meeting rising electricity demand. The lost power production would have to be met by other sources.

THE URUGUAY RIVER BASIN

The GCM scenarios came to different conclusions about whether average annual flows would increase or decrease in the Uruguay River, but each scenario showed decreased river flow during the dry season. This would result in lower hydropower output, reduced water supplies for irrigation (while demand for irrigation water could increase), and greater water quality problems. Flood potential was estimated to decrease under two of the scenarios, but would reach catastrophic levels under the GFDL scenario. In fact, the GFDL scenario included both increases in flood potential in the wet season and lower flow during the dry season.

THE INDUS RIVER BASIN

Runoff in the Upper Indus Basin was estimated to increase because of melting from the Himalayan glaciers. This would provide more water for irrigation in most areas, but could also increase flooding and exacerbate current waterlogging and salinity problems. Water scarcity in parts of the Punjab and Upper Basin could increase. In some scenarios, climate change could practically offset the potential economic benefits of water development plans (see Figure 12). Climate change and water development could jointly reduce freshwater flows to the Indus delta. If water allocation rules are modified to reduce water scarcity problems, freshwater flows to the delta would be reduced 12 to 78 percent below current levels. Without mitigating measures, the regional ecology and economy could be harmed.

THE MEKONG RIVER BASIN

Analysis of climate change impacts in the Mekong River examined sensitivity only to changes in precipitation from GCM scenarios. Flows were generally estimated to increase or remain the same, although seasonality shifted. Hydropower production and low-flow augmentation from the planned cascade of thirteen dams were estimated to be reduced because of greater seasonal differences in flow. Basin managers said that modification of operating rules and modest structural changes would restore planned performance levels. Floods in the Mekong Delta, which would be intensified by sea-level rise, could start earlier and be larger, adversely affecting fisheries and agriculture.

Human health

Summer heat waves are often associated with excess deaths, particularly in urban areas. The range of many infectious diseases is limited by climate, and a warmer world could shift the location of many of these diseases.

The health study examined potential changes in mortality, particularly for summer temperatures, and shifts in the ranges and prevalence of infectious diseases. Case studies on mortality were performed in China, Egypt, and Canada.[5] Case studies are also being conducted on onchocerciasis (river blindness) and malaria in Africa, but results are not available for this report.

The heat stress/mortality study evaluated daily mortality data, and attempted to determine whether temperature thresholds exist beyond which mortality rises rapidly. If a threshold was found, the relationship between mortality and temperature was used to estimate changes in mortality. The direct use of the historic relationship assumed there is no adaptation – no physiological, behavioral, or infrastructure adjustments to mitigate the change in climate. To account for short-term adaptation, changes in mortality for warm or cool years were examined. This only considered physiological and short term behavioral adjustment, not infrastructure changes. The analysis assumed no change in temperature variability.

HEAT STRESS MORTALITY

Studies of U.S. and Chinese cities have found strong temperature thresholds in mid-latitude cities with high temperature variability; and weak thresholds in southern, humid, cities with low temperature variability. The Chinese cities of Shanghai and Guangzhou appear to have thresholds of 34°C. The increase in mortality at temperatures above the threshold is much greater in Shanghai than in Guangzhou. Shanghai is located north of and has much higher summer temperature variability, than Guangzhou. A similar relationship exists in northern and southern cities in the U.S. Thus, if temperature variability decreases, there could be a

[5] The last project was commissioned by Environment Canada.

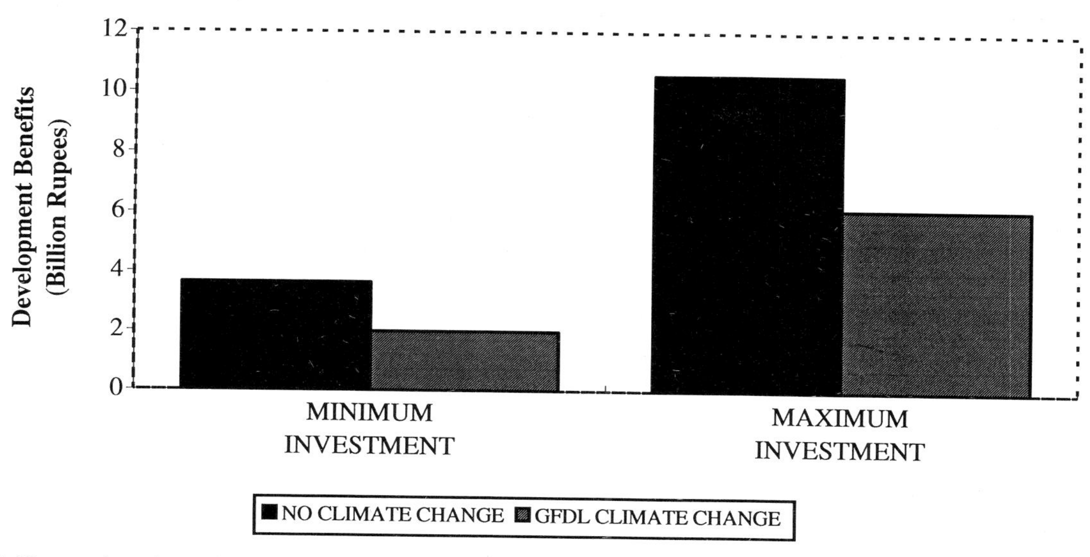

Figure 12 Changes in estimated benefits relative to a no investment scenario from planned Indus Basin water developments caused by the GFDL scenario under two alternative levels of investment through the year 2000. A 3.25 percent annual economic growth rate was assumed.

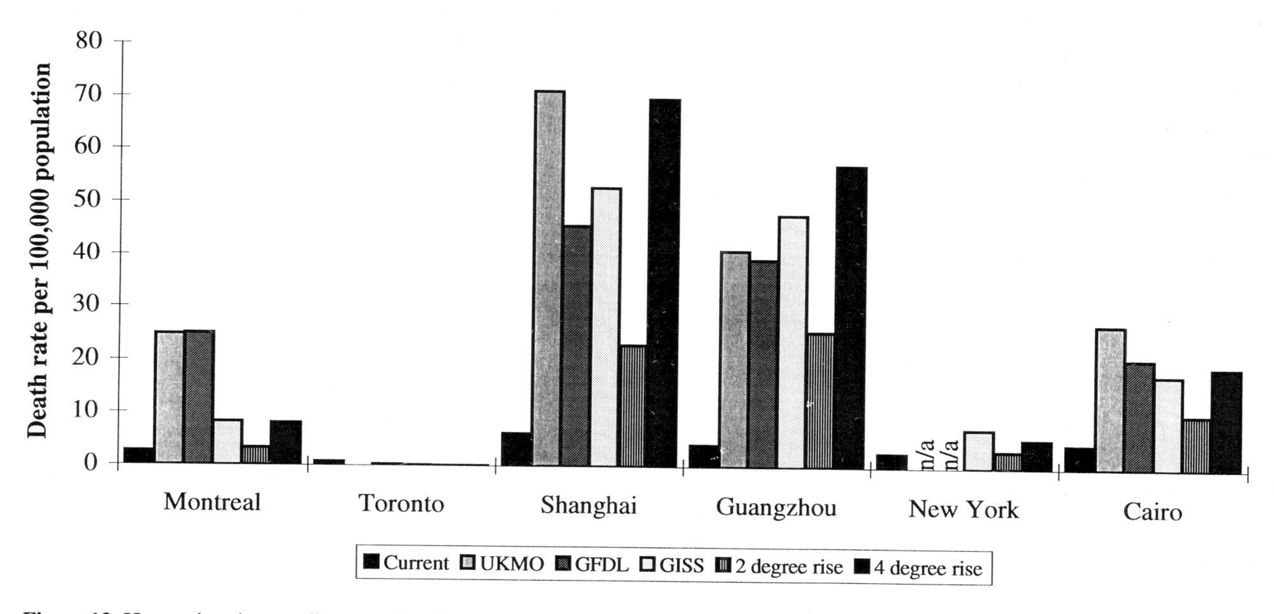

Figure 13 Heat-related mortality rates for six large cities are compared. Acclimatization is not assumed to occur in Shanghai, Guangzhou, or Cairo and thus is not reflected in the mortality rates for those cities. The rates for Montreal, Toronto, and New York, however, include acclimatization.

smaller or no increase in summer heat stress mortality. If it stays the same, mortality could increase.

Heat stress mortality figures for selected developing and developed country cities are displayed in Figure 13. Since there was no difference in mortality for warm or cool years in China and Egypt, we assumed that no acclimatization occurs. For purposes of comparison, only the 'no acclimatization' results are shown for New York and Toronto. The reasons for the lack of a short-term acclimatization response in the three developing country cities are not known, but, may be related to limited access to air conditioning and poorer health care than in developed countries.

Rates of heat stress mortality are already higher in the developing country cities than in the North American cities. On top of this, there is a much greater absolute increase in the heat stress mortality rate due to climate change in developing countries than in developed countries. In New York City, a 4°C warming increases the heat stress mortality rate from 2.7 to 6.5 per 100,000 people, while the same temperature increase raises the mortality rate in Cairo from 4.5 to 19.3

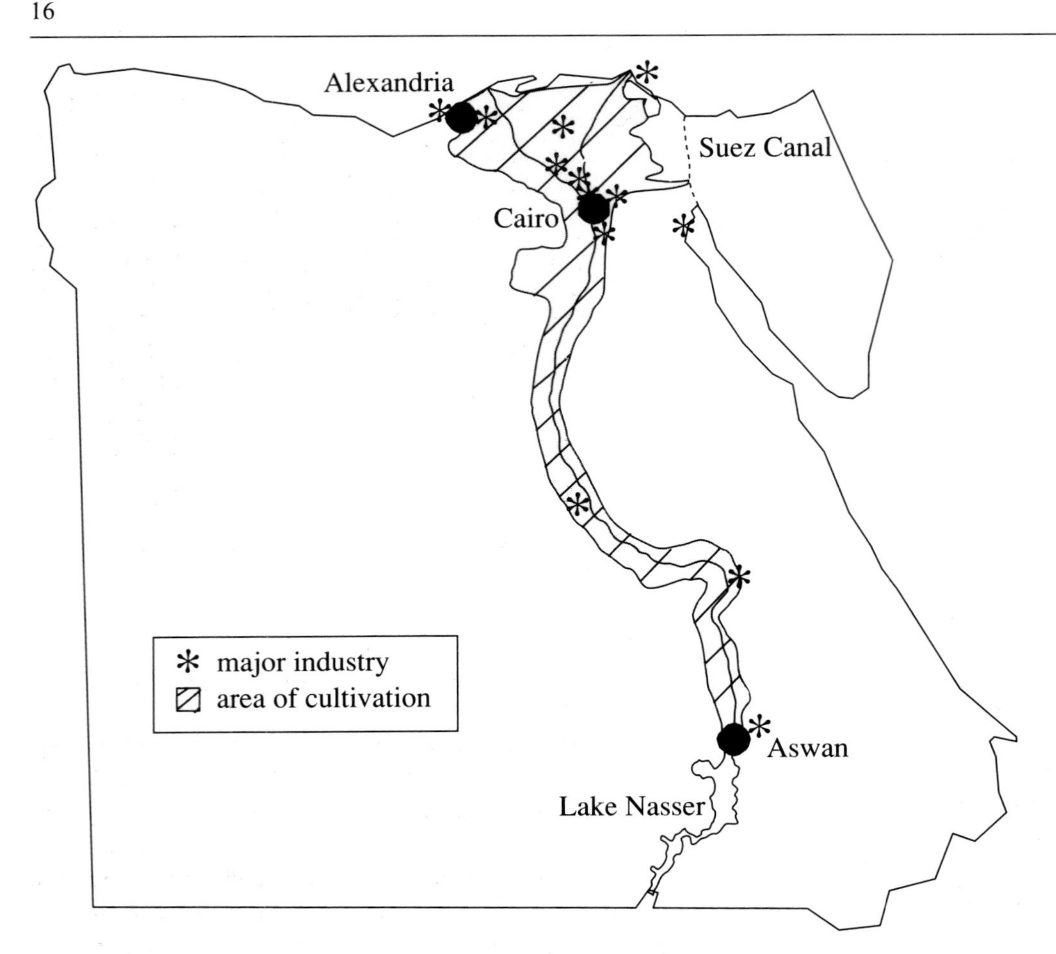

Figure 14 Map of Egypt showing agriculture, population, and economic activity confined to Nile Valley and Delta.

and the rate in Shanghai from 6.2 to 69.8. The elderly will likely comprise a disproportionately large segment of this increase.

Winter results for China and Canada indicate that extreme cold has a much lesser impact on human mortality than extreme heat. Any reduction in winter mortality will probably be considerably smaller than the increase in heat stress mortality.

Integrated impacts – a case study of Egypt

Individual physical and socio-economic sectors will not be affected by climate change in isolation. Crop production will be sensitive not only to changes in crop yields, but also to changes in water supplies and to the inundation and salinization of arable land by rising seas. Water supplies in coastal areas will be affected not only by changes in streamflow, but also by sea-level rise. The combined effects of climate change on Egyptian agricultural production were examined to study how one potentially vulnerable country could be affected.

Agriculture is restricted to only 3 percent of the land area of Egypt and is limited to the very fertile lands of the narrow Nile valley from Aswan to Cairo and the very flat Nile Delta north of Cairo (see Figure 14). Egypt's entire agricultural water supply comes from irrigation, solely from the Nile River. In 1990, agriculture (crops and livestock) accounted for 17 percent of Egypt's gross domestic product, not taking into account agro-industries, such as textiles or food processing.

In this study, an agricultural economic model for the country was used to examine the combined effects of changes in crop yields, water supply, and arable land on Egypt's agricultural economy. The analysis using the economic model uses outputs from the global agriculture study. Therefore, only GCM scenarios were used.

The effects of climate change scenarios on Nile River flow, crop yields (which may increase water use) and amount of arable land are displayed in Table 9. Perhaps most significant is the potential large drop in the flow of the Nile: 77 percent in the GFDL scenario. Crop yields, on average are estimated to decrease, even when the positive effects of CO_2 are considered. Demand for water for irrigation is estimated to increase. A 37 cm rise in sea-level by 2060 (a rate that would result in a one-meter sea-level rise by 2100) would inundate or salinize 4 percent of arable land, entirely in the Nile Delta.

Table 9. *Impacts of climate change on Egypt (percent change)*

Sector	GISS	GFDL	UKMO	LOWEND
	Scenario			
Nile flow	+30	−77	−12	+18
Crop yields*				
Maize	−19	−22	−18	−22
Wheat	−36	−31	−73	−4
Rice	−1	−4	−5	n/a
Soybeans	−8	−30	−33	n/a
Sea-level rise**	−4	−4	−4	−4

Notes:
* Includes direct effects of CO_2 and assumes unlimited water supplies.
** Inundates or salinizes 4% of arable land by 2060, based on a one-meter rise by 2100.
n/a – not available

Table 10. *Changes in agricultural sector of the Egyptian economy (percent change from 2060 baseline)*

Indicator	UKMO	GISS	GFDL	LOWEND
	Scenario			
Total Welfare	−23	−6	−52	−10
Agricultural Trade Balance	−94	−19	−198	−44
Consumption of calories per capita	−3	−1	−1	0

Combined effect on the agricultural sector of the Egyptian economy

Table 10 displays the combined effect of changes in crop production, water supply, and sea-level rise on the agricultural portion of the Egyptian economy. The table displays results for total welfare, which is a measure of the economic loss to consumers and producers, the trade balance for agricultural goods alone, and average daily consumption of calories by Egyptians. A negative number in the trade balance means that the agricultural trade deficit increases.

All four scenarios have a detrimental effect on social welfare in Egypt. This is estimated to happen even in the GISS and GISS-A scenarios under which the water supplies from the Nile would increase by 18 percent and 14 percent, respectively. The increased water supplies cannot compensate for reduction in yields, increased demand for water for irrigation (seen as a decrease in the agricultural water productivity), and the loss of land from sea-level rise. When

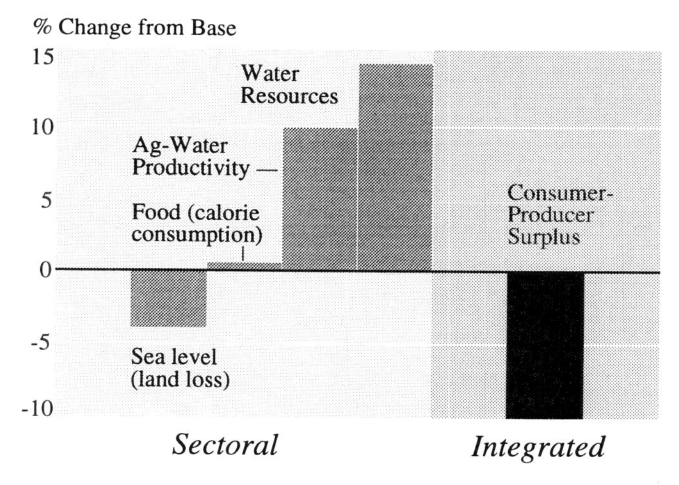

Figure 15 Comparison of sectoral versus integrated impacts for the GISS-A scenario.

water supplies are reduced, as in the UKMO and GFDL scenarios, the effects on the agricultural economy become even more severe.

Consumers are only modestly affected by climate change because their consumption is based on income and price levels. The BLS estimates strong growth in the non-agricultural sector which is not affected by climate change, so Egyptian income levels are relatively unaffected compared to the base scenario. In addition, imports remain high because global food prices stay affordable; but to fund these imports, there must be increased non-agricultural exports, loans, or foreign aid. The most negative impact on consumers happens under the UKMO scenario, due to the largest increase in global food prices.

The reduction in social welfare is mainly from farmers, who are affected by changes in agricultural water productivity, water supply, and domestic and global prices. These impacts result in lower yields, less land under irrigation, and in some cases, reduced prices for exports.

The Egypt case study tested whether building coastal defenses to protect agricultural land would offset losses in total welfare from the climate change scenarios. Building sea walls only increases total welfare by one percent, so on net, total welfare still could decline dramatically. Apparently the reduction in welfare is primarily due to reductions in crop yields, with changes in water supplies significantly affecting the magnitude of economic losses.

The value of an integrated analysis and the complex nature of climate change impacts on developing countries are illustrated in Figure 15 for the GISS-A scenario. In the GISS-A scenario, all sectors except sea-level rise show a positive impact from climate change. However, the model estimated a 10 percent negative impact on social welfare, which is felt entirely by the producers. This is the result of the rest of the world having positive increases in agricultural production, reducing Egypt's comparative advantage, and thereby reducing export volumes and revenues. Thus, even when climate change is bio-physically positive for Egypt, it can result in negative economic impacts.

REFERENCES

Han, M., J. Hou, and L. Wu. 1995. Potential impacts of sea-level rise on China's coastal environment and cities: a national assessment. *Journal of Coastal Research*, Special Issue No. 14:79–95.

Holdridge, L. R. 1967. *Life-Zone Ecology*. San Jose, Costa Rica: Tropical Science Centre.

Kalkstein, L. S. (Ed.), 1991. Global comparisons of selected GCM control runs and observed climate data. US Environmental Protection Agency, Washington DC.

Rosenzweig, C., and M. L. Parry. 1994. Potential impact of climate change on world food supply. *Nature* 367:133–8.

World Bank. 1992. *World Development Report 1992. Development and the Environment*. New York: Oxford University Press.

1 Introduction

JOEL B. SMITH
Hagler Bailly Consulting, Inc.
Formerly Deputy Director, Climate Change Division
U.S. Environmental Protection Agency

LAURENCE S. KALKSTEIN
Department of Geography
University of Delaware

Virtually every nation will be affected by climate change, whether by rising seas, changes in crop yields, variations in water resources, or even changes in death rates. Plants and animals around the world could be greatly influenced by climate change through mass migration or habitat loss, and many species may not survive.

At the June 1992 United Nations Conference on Environment and Development, representatives of over one hundred nations negotiated and signed a framework agreement on climate change (UNGA, 1992). The goal of the conference is to reduce greenhouse gas emissions from developed countries to 1990 levels by the year 2000. Yet many of the nations, particularly developing countries, that participated in these negotiations had little information on how they could be affected by climate change.

The only global assessment of potential climate change impacts to date was produced by the Intergovernmental Panel on Climate Change (IPCC) (Tegart *et al.*, 1990). The assessment was based on a literature review of climate impact studies. The vast majority of climate impact studies have addressed sectoral impacts in particular regions (Scheraga and Sigler, 1991), and most have examined impacts in developed countries. Few studies have examined integrated impacts within a region or country or sectoral impacts across countries.

Among those that have examined climate change impacts across countries are Parry *et al.* (1988), Kane *et al.* (1992), and Emmanuel *et al.* (1985). Parry *et al.* (1988) studied the potential effects of climate change on agriculture in a number of developed countries and concluded that climate change could result in significant shifts in regional crop productivity. Kane *et al.* (1992) used the IPCC's literature review on crop yields across the world to model shifts in global production levels. They found that the overall effect of a moderate climate change scenario on world crop production would be small; reduced output in some areas would be balanced by higher output in other areas. Emmanuel *et al.* (1985) examined global changes in potential vegetation cover, using a single climate change scenario. They found that major biomes would shift poleward and the areal coverage of boreal forests could decline by more than one-third.

Few studies have focused on developing countries. Yet the IPCC concluded in its study that developing countries could bear the greatest risks of climate change because of their poor resource bases and the enormity of problems (such as overpopulation) that already face them (Tegart *et al.*, 1990).

In anticipation of the need for more information on the global impacts of climate change on developing countries, the United States Environmental Protection Agency (EPA) commissioned five major assessments of the potential international effects of climate change. (The study also included an integrated assessment of the impact of climate change on one particular country – Egypt.) The assessments studied the effects of climate change on agriculture, forests, river basins, coastal resources, and human health.[1] These areas were selected because they are sensitive to climate variability and climate change and are of high value to society. Changes in agriculture affect food supplies; changes in forests affect supplies of wood and species diversity; increases or decreases in river flow affect water supplies, hydropower, and flooding; sea-level rise would threaten coastal resources; and human health is sensitive to changes in heat stress and location of infectious diseases. Because climate change impact research

[1] The U.S. Agency for International Development co-funded the agriculture project and the Bureau of Reclamation provided co-funding for the analysis of the Nile River as part of the river basins project.

had already been conducted for each of these areas, new impact models (such as for crop yields) for the most part did not need to be developed, although they were applied to new sites.

The goals of the projects were primarily to provide information to policymakers and natural resource managers about sensitivities of these areas to climate change and to maximize the involvement of in-country scientists in estimating the potential impacts of climate change. It is important for policymakers in developing countries to be informed about the sensitivities of their nations to climate change, not only so that they can better formulate policies concerning emissions of greenhouse gases, but also so that they can examine adaptations to potential impacts. Many of the scientists conducting the analyses were from the developing nations in which the studies were being conducted; in a few cases, researchers from outside the country of interest conducted the analyses. More than 150 scientists in over 30 countries were commissioned for these projects.

The studies on agriculture, forests, river basins, coastal resources, and health used common scenarios, described below, so that comparisons of potential impacts could be made. Climate change will not act in isolation on one sector but will affect all sectors simultaneously. Agriculture, for example, could be influenced not only by changes in temperature and rainfall, but also by changes in water supply for irrigation or by changes in sea level which could result in inundation of low-lying areas.

As mentioned, an integrated assessment of climate change impacts on Egypt was also conducted as part of the EPA-commissioned studies. Egypt was chosen because all of the other studies (on agriculture, water resources, etc.) examined its sensitivity to climate change. With its reliance on the Nile River and with the coastal delta supporting much of the country's agriculture, Egypt is a good site for examining integrated climate impacts.

This volume is an overview of the results of the six studies, with one chapter devoted to each. The final chapter discusses some policy options for adapting to climate change. Since this volume summarizes the work of over one hundred scientists conducted over a three-year period, it was not possible for the chapter authors to reflect fully on all of the findings from the numerous studies. More complete reports of the results for the individual studies have been published or are forthcoming in books and journal articles.

WILL CLIMATE CHANGE?

Since the beginning of the Industrial Revolution, atmospheric carbon dioxide levels have increased 25 percent, methane concentrations have more than doubled, and nitrous oxide concentrations have risen significantly (NAS, 1991).[2] Since these gases hinder the escape of heat energy from the Earth's surface while having little influence on the amount of incoming solar energy, it is widely believed that an increase in their concentrations will eventually raise global temperatures (Houghton *et al.*, 1990; NAS, 1991).

In 1990, the Intergovernmental Panel on Climate Change (IPCC), composed of hundreds of scientists from around the world, reported that the most likely range of warming that would result from a doubling of the carbon dioxide level in the atmosphere over pre-industrial levels would be 1.5 to 4.5°C. Land areas would warm faster than oceans and high northern latitudes would warm more than the global mean in winter. The IPCC concluded that the globe would warm at a rate of 0.2 to 0.5°C per decade, with a most likely rate of 0.3°C per decade. Thus, by 2025, the Earth's temperature would be about 1°C higher than it is today, and by 2100, 3°C higher. Some recent studies have reported a slightly lower estimated rate of change – about 2.5°C by 2100 (Houghton *et al.*, 1992; Wigley and Raper, 1992) – but even that rate of change would be faster than any in the past 10,000 years (Houghton *et al.*, 1990). Recent studies have also indicated that, based on past climate changes, the Earth may warm very suddenly over a short time period (Dansgaard *et al.*, 1993). Such a rapid warming would hinder efforts to adapt to climate change, especially in developing countries where concomitant impacts may be quite significant.

Higher temperatures will most likely raise sea level. This will result from a combination of thermal expansion of ocean water and the melting of glaciers. The degree to which Antarctica will contribute to sea-level rise is uncertain. The IPCC concluded that increased snowfall in Antarctica may offset thermal expansion and melting of mountain and Greenland glaciers, yielding a net sea-level rise of 30–110 cm by the year 2100 (Houghton *et al.*, 1990). Wigley and Raper (1992) revised this estimate downward to 15–90 cm by 2100. As described in a later chapter, even the lower rise in sea level could cause millions of people to be displaced and could ruin thousands of square miles of valuable cropland.

Although there is growing consensus about changes in global climate, considerable uncertainty remains about regional climate changes. Temperatures may not rise in all areas, and the direction of change in precipitation, winds, humidity, and other meteorological variables on a regional scale is unknown. There is also uncertainty about how extreme events, such as hurricanes and droughts, will change. It is quite possible that the number of extremely hot days will increase as will the number of intense precipitation events (Houghton *et al.*, 1990).

[2] Chlorofluorocarbons (CFCs) were introduced in the 1930s, but their contribution to a potential global warming is uncertain since their radiative forcing may be offset by their depletion of ozone (Houghton *et al.*, 1992).

The IPCC reported that global mean surface air temperatures have increased by 0.3 to 0.6°C over the last hundred years. This warming is broadly consistent with climate model forecasts but is also of the same magnitude as natural variability. Whether the warming has been caused by natural variability or by anthropogenic factors is not known (Houghton *et al.*, 1990). Recent studies suggest that increases in anthropogenic aerosols, such as sulfur dioxide, in the lower atmosphere, which increase the reflectivity of sunlight back into space, may temporarily offset a significant portion of the climate forcing from increased greenhouse gas concentrations (Charlson *et al.*, 1992; Penner *et al.*, 1992; Kiehl and Briegleb, 1993).

STRUCTURE OF THE STUDIES

Scope

The focus of these projects was on the potential impacts of climate change on developing countries. However, while some of the projects dealt exclusively with impacts in developing countries, others also included global or higher-latitude impact analyses. For example, the agriculture project estimated potential shifts in global food production and trade, the forest study examined changes in major vegetation patterns across the globe and shifts in tropical and boreal forests, and the health study included a case study on Canadian heat stress. The rivers and sea-level projects dealt exclusively with developing countries.

In each of these projects the investigators sought to identify and evaluate adaptive adjustments in resource management practices or policies that would increase the flexibility of responses to climate change. The management practices they suggested had to be sufficiently robust to withstand a variety of potential climate conditions and provide ancillary environmental or economic benefits. In many of the studies, adaptive responses, such as the additional application of fertilizer, were included in the models.

CLIMATE CHANGE SCENARIOS

Methodologies

Given the uncertainties about regional climate change, it is difficult to predict specific regional impacts. We can, however, identify a range of potential impacts and sensitivities of systems to climate change with the use of climate change scenarios. A climate change scenario is defined as a physically consistent set of changes in meteorological variables that is compatible with generally accepted projections of global temperature change (e.g., 1.5 to 4.5°C for a doubling of carbon dioxide levels). *A scenario is not a prediction, and a set of scenarios may not encompass all of the potential changes in regional climates.*

In August 1989, the EPA sponsored a meeting of climate modelers, climatologists, and climate impact researchers at the National Center for Atmospheric Research (NCAR) to assist in the selection of scenarios of climate change for the international impact studies (ICF, 1989). The choices for scenario selection included the following:

1. GENERAL CIRCULATION MODELS
General circulation models (GCMs) are dynamic computer models of the atmosphere and oceans that have been used to estimate climate change attributed to a doubling of CO_2 and to transient increases in greenhouse gas concentrations. The models yield physically consistent and plausible regional estimates of climate change, but their reliability is limited by the manner in which they simulate oceans, clouds, and soil moisture and by their low resolution (see Schlesinger and Mitchell, 1985, for further discussion). GCMs yield estimates of meteorological parameters for grid boxes of various sizes (from 4° latitude by 5° longitude to 8° × 10°), depending upon which GCM is utilized. Although GCMs generally yield predictions that agree about the direction of global changes in climate variables, they often disagree about regional changes (Smith and Tirpak, 1989). They are generally considered to be unreliable for predictions of regional climate change, although the estimates are plausible and consistent with a doubling of carbon dioxide levels.

2. HISTORIC DATA
A number of climate impact studies have used historic periods with high temperatures as scenarios for future warming. Rosenberg and Crosson (1991) used the 1930s, a hot and dry period in the United States, to estimate potential change in the 2030s in the upper midwestern United States. The advantage of using such data is that they are plausible because the climate events actually occurred. In addition, daily and regional variabilities are represented, as opposed to the monthly averages identified by the GCM grid boxes. The disadvantage of using this approach is that historic data may not be representative of climate changes due to an increase in greenhouse gases. Furthermore, databases of past climate extremes, such as for the 1930s, may not exist for many areas of the world.

3. ARBITRARY SCENARIOS
The sensitivity of resources to climate changes can be tested with scenarios that incorporate arbitrary changes in climate variables, such as +2 or +4°C or +/−20 percent changes in precipitation. The advantages of such scenarios are that a wide range of potential climate changes can be tested and sensitivities to individual parameters (temperature, precipi-

tation, or others) can be compared. The disadvantage of this approach is that the changes incorporated in the scenarios may not be consistent with one another. Temperature and rainfall will not change without commensurate changes in winds, cloud cover, atmospheric pressure, and other parameters. Although it is possible for temperatures on a regional scale to increase while precipitation decreases, such a change cannot happen across the world. Thus, arbitrary scenarios cannot be used to examine global changes, such as were examined in the agriculture and forest studies.

As a result of the August 1989 meeting, the EPA decided to base climate change scenarios on general circulation models (GCMs) and arbitrary changes in climate. The use of historic data was rejected because historic data are not representative of future climate changes and data on past meteorological conditions may not be widely available.

The participants at the NCAR meeting further recommended that the EPA compare GCM estimates of temperature and precipitation for current climate with real-time observations in an attempt to determine GCM reliability for the regions for which impact studies were commissioned. The results of that study (Kalkstein, 1991) are reported below.

To ensure comparability, the scenarios used in all of the projects were required to meet the following guidelines:

- *Climate data for the years* 1951–1980 *were to be used as baseline conditions*
  Thirty years of observations define a broad range of climate conditions and the time period chosen has the virtue of being recent. The 1980s were not included in the baseline because some scientists think the warming that occurred in those ten years may be attributable to anthropogenic sources (ICF, 1989). For many sites, climatological data covering the entire 1951–1980 time period were not available, and shorter records were used. In some cases, researchers used climatological data from outside this period in order to obtain a more complete record.

- *General circulation models were to be used for doubled CO_2 and transient scenarios*[3]
  Monthly average deviations from equilibrium conditions from the doubled CO_2 runs were combined with the baseline data to produce scenarios with 30 years of climate data. The monthly averages were differences between double CO_2 and single CO_2 for temperature and ratios between double CO_2 and single CO_2 for other variables. Temperature differences were added to baseline conditions, while ratio changes were multiplied by baseline data. The data were extracted for the closest GCM grid box to the evaluated site. The forests study interpolated GCM data to 0.5° × 0.5° grid boxes. The GCMs that were used in the study are described in Table 1. At the time these studies were initiated, these were the most recent GCM data sets available for impacts research.

- *Arbitrary scenarios were to be used to test sensitivities*
  We used a 2°C and 4°C warming above the baseline 1951–1980 period to develop plausible arbitrary temperature scenarios. These were combined with three arbitrary precipitation changes that account for a potential range of possibilities: no change in precipitation, 20 percent increase, and 20 percent decrease.

- *The sea level rise scenarios were to use levels ranging from* 0.5 *to* 2.0 *m*
  Most studies focus on a 1.0 m rise in sea level because of the practical difficulty of assessing smaller rises and because it has become an international convention to do so (RSWG, 1991). In addition, a baseline rise of 0.2 m was used, following the analysis of global sea-level rise from 1880 to 1980 by Douglas (1991).

Because of data or resource limitations, many researchers were unable to complete all of these scenarios or occasionally used different baseline data. In some cases, researchers used

[3] The transient runs start in the middle of the twentieth century and end in the middle of the twenty-first century, assuming a constant increase in greenhouse gas concentrations.

Table 1. *General circulation models used to construct climate change scenarios*

GCM	Resolution (lat/lon)	Development year
Oregon State University (OSU)[1]	4° × 5°	1984–1985
Geophysical Fluid Dynamics Laboratory (GFDL)[2]	4.5° × 7.5°	1988
Goddard Institute for Space Studies (GISS)[3]	7.8° × 10°	1982
United Kingdom Meteorological Office (UKMO)[4]	5° × 7.5°	1985–1986

Notes:
[1] Schlesinger and Zhao, 1988.
[2] Mitchell *et al.*, 1990.
[3] GISS model (Hansen *et al.*, 1983); Transient scenarios estimates for the 2030s (Hansen *et al.*, 1988).
[4] Wilson and Mitchell, 1987.

Table 2. *GCM global temperature and precipitation estimates*

GCM	Temperature increase (°C)	Precipitation increase (%)
GISS	4.2	11.0
GFDL	4.0	8.7
UKMO	5.2	15.0
OSU	2.8	7.8
GISS-A	2.4	5.0

Change in Temperature (degrees C) ($2xCO_2$ - $1xCO_2$)

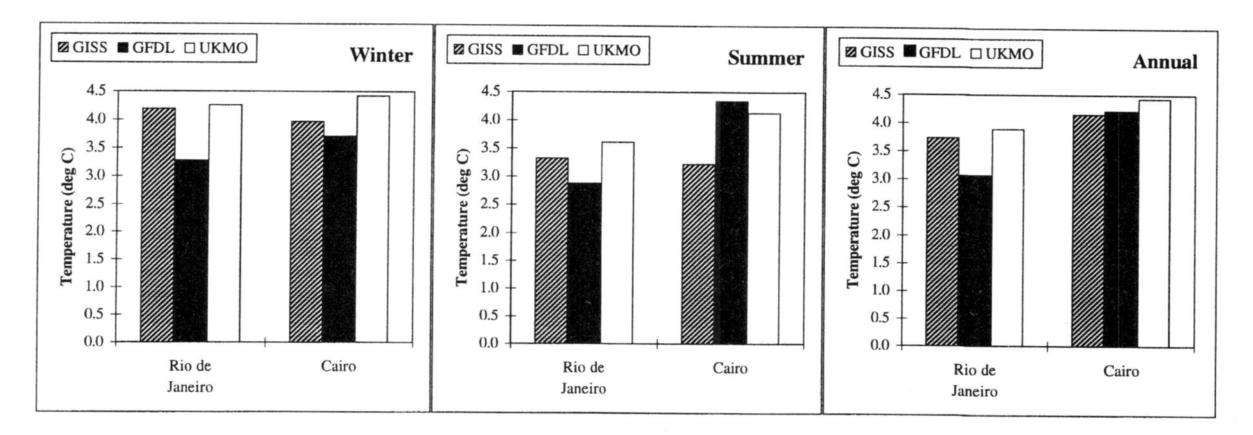

Precipitation Ratio ($2xCO_2/1xCO_2$)

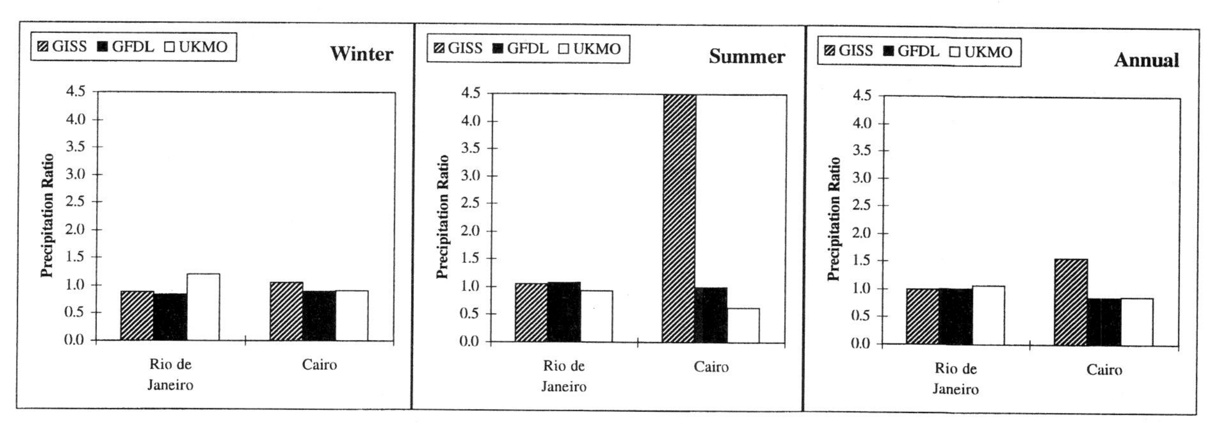

Figure 1 GCM results for grid boxes containing Cairo and Rio de Janeiro for winter, summer, and annual temperature and precipitation changes.

only two or three of the GCM scenarios and may not have had resources to run some of the arbitrary scenarios.

The GCM model projections for global temperature and precipitation changes are summarized in Table 2. In general, the GCM temperature results are from the middle to the upper end of the 1.5 to 4.5°C range identified as most likely to occur by the IPCC. GCMs tend to estimate a 3.5°C warming from CO_2 doubling (Cess *et al.*, 1993). The UKMO warming of 5.2°C is the highest of the evaluated GCMs and should be interpreted as an extreme scenario. All the GCMs indicate a global increase in precipitation of between 7.8 and 15 percent, although it should be noted that many regions show a decrease in precipitation, especially in the warmer seasons. Additionally, these precipitation increases may not translate into greater available soil moisture, as evaporation rates also increase as temperatures rise.

The study assumes that the climate change scenarios will occur by 2060, which would translate to a rate of change of about 4°C per century – a rate that is somewhat faster than the 3°C estimated by IPCC.

GCM results for individual grid boxes and individual months vary considerably (Figure 1). An evaluation of annual, winter, and summer temperature and precipitation changes for GCM grid boxes containing Cairo, Egypt and Rio de Janeiro yield sizable variations between GCMs. Note that these figures are average changes for the entire grid box, which ranges from 4° × 5° to 8° × 10° latitude and longitude. For both cities, all of the GCMs estimate that average temperatures would rise on the order of 3° to 5°C. However, the models often estimate opposite changes in precipitation, particularly for seasons. For example, the UKMO model has winter precipitation increasing 20 percent over Rio de Janeiro, while the GISS and GFDL models estimate decreases of 11 to 16 percent. The models have much greater disagreement about the direction of summer precipitation over Cairo. The ratio of doubled CO_2 to single CO_2 summer precipitation in Cairo is very high in the GISS model because the model estimates that single CO_2 precipitation in the summer is only 0.27 mm/day and precipitation increases by about 1.0 mm/day.

VERIFICATION

At the 1989 NCAR meeting, participants recommended that GCMs be used as a basis for creating regional climate change scenarios and that the EPA commission climatologists who are regional specialists to determine if the outputs of the GCMs were duplicating real-time observations with a reasonable degree of accuracy. It is especially important for the users of the GCMs to know in which areas these models do not simulate current climate realistically so that scenarios and their results are not misinterpreted. In addition, this type of analysis would provide essential information to the climatic modelers who developed the GCMs, as they are constantly striving to improve these models.

This discussion is not an attempt to compare the degree of accuracy of the four GCMs on a global basis, as each has inherent strengths and weaknesses. Nor is this an attempt to determine *why* errors have occurred. Rather, the regional model outputs are simply compared to observed climate, and no attempt is made to examine model structure or assumptions as reasons for discrepancies.

The GCM outputs used in this analysis represent the control ($1 \times CO_2$) model runs, which best approximate mid-twentieth century conditions around the globe. These are compared to a grid-point global average climatic data set developed by the RAND Corporation in Santa Monica, California. The RAND data set represents mid-twentieth century area averages for grid boxes of 4° latitude by 5° longitude and is considered to be among the most comprehensive real-time climatic data sets available (Schutz and Gates, 1974). The RAND data can be readily compared to GCM output, which is also based on a grid pattern, although the sizes of the grids vary among the GCMs.

Comparisons are restricted to mean January, April, July, and October temperatures (°C) and mean winter (December through February) and summer (June through August) season precipitation (mm/day), as these are the statistics compiled in the RAND climate data set. The evaluation includes (1) a coarse geographic grid analysis for the entire globe, which involves comparing the RAND grid with the closest grids for each of the GCMs, and (2) a detailed transect analysis in which RAND latitude and longitude transects are compared with the closest transects for each of the GCMs. All of the climatologists selected several transects for their regions of evaluation and developed detailed figures and tables comparing RAND and GCM output across the transects (Figure 2).

A summary of the findings follows:

1. The models best duplicated reality in mid-latitude and low latitude regions with little relief. They generally exhibited the largest errors in coastal regions (especially where there was a rapid transition from oceans to mountains), hot desert areas, and in areas of complex terrain (i.e., mountains).
2. In North and Central America, temperature differentials between the GCMs and RAND were relatively small, especially for the summer (within 3°C in most cases). The models performed relatively well for the eastern United States and southeastern Canada, especially between longitudes 90°W and 60°W. The greatest temperature disagreements were found for the upper latitudes of North America, particularly north of 60°N.
3. For precipitation, all four models overestimated winter precipitation for western North America, especially from latitudes 30°N to 50°N along a longitudinal transect through 120°W. Summer precipitation was overestimated as well, but to a lesser degree. However, the models duplicated the RAND data set well for eastern North America, where daily disparities were often less than 1 mm.
4. Temperature differentials were relatively small over western Europe but were much larger over most of the interior portions of Asia. For many European grids, summer and winter temperature departures were frequently 2°C or less. However, along longitudinal transects for 90°E and 120°E, positive and negative differentials sometimes exceeded 10°C, especially for winter.
5. Precipitation departures were small for western Europe but were much larger for summer in central Asia. For longitudinal transects at 0°, 30°E, and 60°E, most precipitation departures did not exceed 0.5 mm per day. However, the GCMs overestimated summer precipitation by about 2 mm/day in north central Asia along the 120°E longitudinal transect in the Soviet Union. Further south in mountainous terrain, anomalies of greater than 7 mm/day were noted.
6. The GCMs generally underestimated temperature for interior portions of tropical Asia for all seasons (with the exception of the OSU model, which overestimated temperature). These departures were generally greatest for winter. However, the GCMs replicated real-time temperatures closely for all seasons in southeastern China, the Philippines, and Indonesia.
7. Precipitation differentials were particularly large for the western portions of monsoon Asia in summer, where the GCMs generally underestimated precipitation by a significant margin. These differentials were especially noteworthy for central and western India and eastern Pakistan for all four models. In addition, the GCMs overestimated winter precipitation in these regions; and in some cases they did not duplicate the natural seasonality of summer maximum precipitation in south central and southwestern Asia. For example, within the grid box centered at latitude 22°N and longitude 75°W (far western India), the GFDL model estimated winter precipitation at 2.5 mm/day and summer precipitation at 1.0 mm/day, which misrepresents the natural seasonality of precipitation in the area (RAND values for winter and summer precipitation at this location are 1.2 mm/day and 5.6 mm/day, respectively). However, this is not the case for southeast Asia, where the GCMs

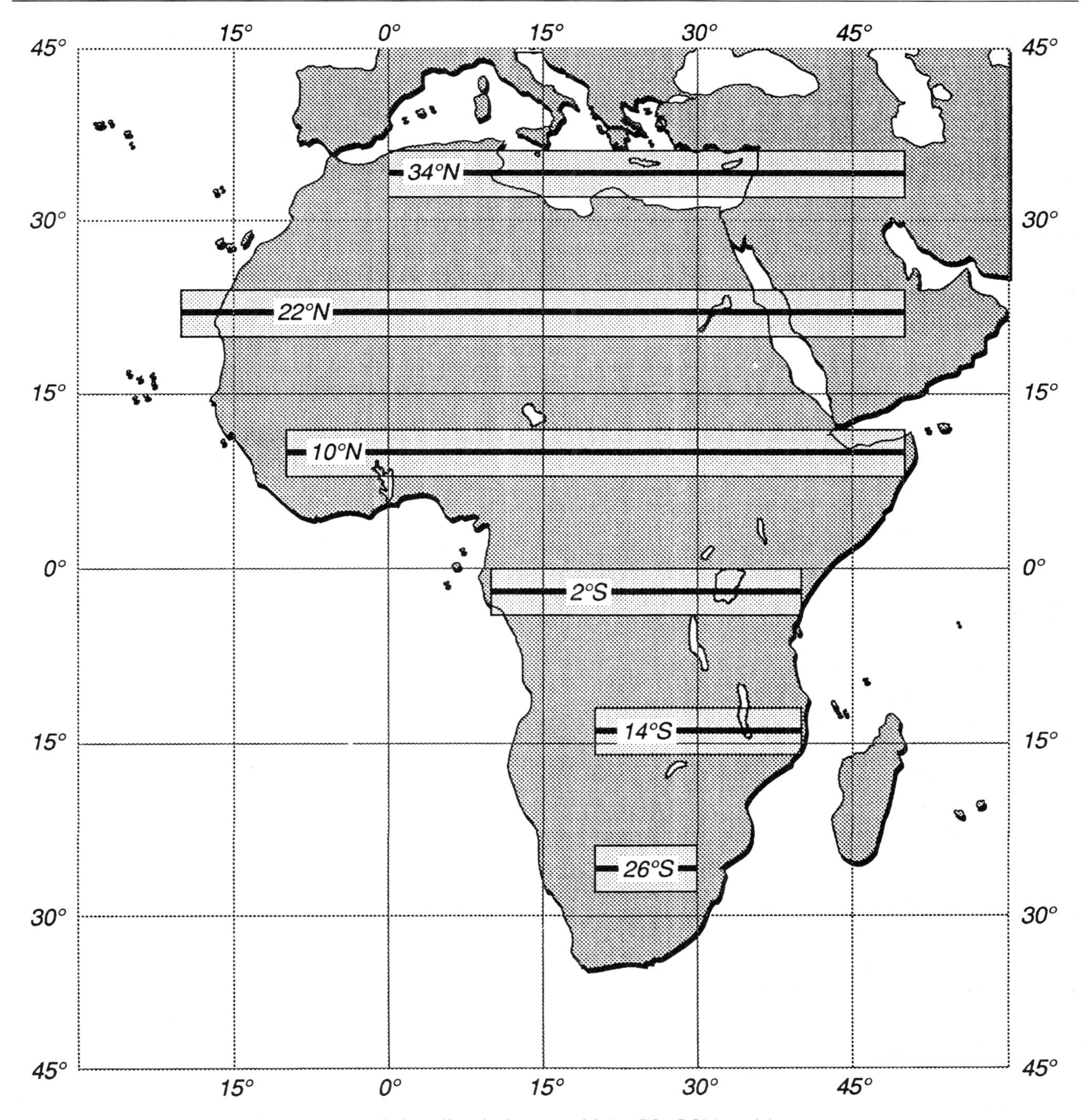

Figure 2 African transects used to compare real-time climatic datasets with 1 × CO_2 GCM model runs.

properly duplicated the seasonality of precipitation and performed quite well.

8. For Africa, temperature differentials rarely exceeded 3°C, especially in summer. However, GISS and GFDL models underestimated winter temperatures for northeastern Africa by up to 9°C. Precipitation departures were relatively small for Africa, especially when compared to Asia. In southern Africa, south of 15°S, precipitation departures were generally less than 1 mm/day. The greatest disparities occurred in central Africa (latitudes 2°S to 14°S; longitudes 20°E to 30°E) during winter, where GISS and GFDL overestimated precipitation.

9. Some sizable temperature differentials were seen for South America, especially in western coastal areas near the Equator, where the GCMs generally overestimated both summer and winter temperatures. Especially large disparities, often exceeding 5°C, were noted along the 60°W longitudinal transect. The GCMs best replicated the RAND data set for the Amazon Basin. Some comparatively large precipitation anomalies were also noted, but the errors were generally random rather than systematic. Some relatively large winter overestimations were found for eastern Brazil with both the GISS and UK models.

CONCLUSIONS

This comparison of GCM and RAND data is not necessarily a true test of GCM performance. The quality and longevity of the real-time climate data vary considerably around the world. This potential quality control problem may influence comparisons between GCM results and the RAND climate data set.

The climatologists point out that since the observed data are aggregated to large gridded regions, which often are not analogous to the RAND grids, the absolute differences between the RAND data set and the models should not necessarily be construed as model error. Rather it is suggested that large departures should be investigated to determine if the disparities can be attributed to theoretical shortcomings in the models or difficulties in the RAND data set (due to large grid sizes) in depicting accurately the Earth's true climate.

Although there are recognized limitations in the climate scenarios, every attempt has been made to develop climate assessments that provide the most accurate determination of impacts possible considering the current state of the science. Skeptics often argue that there is such uncertainty involving the potential impacts of climate change that it is virtually useless to develop assessments of the type presented here. The contributors to this volume believe quite the opposite; we must understand the potential range of situations that might occur if the expected global warming materializes. This knowledge base will assist in the formulation of viable policy options to mitigate climate change impacts. Those who argue that impact analysis should be performed only after climate models are improved or global warming becomes reality neglect to understand that long-term anticipatory responses are required if we are to maintain the present quality of life under the increasingly stressful climate that the majority of scientists expect within the next one hundred years.

REFERENCES

Cess, R. D., *et al.* 1993. Uncertainties in Carbon Dioxide Radiative Forcing in Atmospheric General Circulation Models. *Science* 262:1252–5.

Charlson, R. J., S. E. Schwartz, J. M. Hales, R. D. Cess, J. A. Coakley, Jr., J. E. Hansen, and D. J. Hoffman. 1992. Climate Forcing by Anthropogenic Aerosols. *Science* 255:423–30.

Dansgaard, W., *et al.* 1993. Evidence for General Instability of Past Climate from a 250-kyr Ice-Core Record. *Nature* 364:218–20.

Douglas, B. C. 1991. Global Sea Level Rise. *Journal of Geophysical Research* 94(4):6981–92.

Emmanuel, W. R., H. H. Shugart, and M. P. Stevenson. 1985. Climatic Change and the Broadscale Distribution of Terrestrial Ecosystems Complexes. *Climatic Change* 7:29–43.

Hansen, J., *et al.* 1983. Efficient Three-Dimensional Global Models for Climate Studies: Models I and II. *Monthly Weather Review* 3:609–62.

Hansen, J., *et al.* 1988. Global Climate Changes as Forecast by the GISS 3-D Model. *Journal of Geophysical Research* 93:9341–64.

Houghton, J. T., G. J. Jenkins, and J. J. Ephraums, eds. 1990. *Climate Change: The IPCC Scientific Assessment*. Intergovernmental Panel on Climate Change. Cambridge: Cambridge University Press.

Houghton, J. T., B. A. Callander, and S. K. Varney. 1992. *Climate Change 1992 – The Supplementary Report to the IPCC Scientific Assessment*. WMO/UNEP Intergovernmental Panel on Climate Change. Cambridge: Cambridge University Press.

ICF. 1989. *Scenarios Advisory Meeting: Summary Report*. Report to U.S. Environmental Protection Agency. Fairfax, Virginia: ICF, Inc.

Kalkstein, L. S., ed. 1991. *Global Comparisons of Selected GCM Control Runs and Observed Climate Data*. EPA-21P-2002. Washington, DC: U.S. Environmental Protection Agency.

Kane, S., J. Reilly, and J. Tobey. 1992. An Empirical Study of the Economic Effects of Climate Change on World Agriculture. *Climatic Change* 21:17–36.

Kiehl, J. T., and B. P. Briegleb. 1993. The Relative Roles of Sulfate Aerosols and Greenhouse Gases in Climate Forcing. *Science* 260:311–4.

Mitchell, J. F. B., S. Manabe, T. Tokioka, and V. Meleshko. 1990. Equilibrium Climate Change. In *Climate Change: The IPCC Scientific Assessment*, eds. J. T. Houghton, G. J. Jenkins, and J. J. Ephraums, 131–72. Cambridge: Cambridge University Press.

National Academy of Sciences. 1991. *Policy Implications of Greenhouse Warming – Synthesis Panel*. Washington, DC: National Academy Press.

Parry, M. L., T. R. Carter, and N. T. Konijn. 1988. *The Impact of Climatic Variations on Agriculture*. Boston: Kluwer Academic Publishers.

Penner, J. E., R. E. Dickinson, and C. A. O'Neill. 1992. Effects of Aerosol from Biomass Burning on the Global Radiation Budget. *Science* 256:1432–4.

Response Strategies Working Group (RSWG). 1991. *Assessment of the Vulnerability of Coastal Areas to Sea Level Rise*. Advisory Group on Assessing Vulnerability to Sea Level Rise and Coastal Zone Management. Geneva, Switzerland: Intergovernmental Panel on Climate Change.

Rosenberg, N. J., and P. R. Crosson. 1991. *Processes for Identifying Regional Influences of and Responses to Increasing Atmospheric CO_2 and Climate Change – The MINK Project: An Overview*. Report prepared for the U.S. Department of Energy. DOE/RL/01830T-H5. Washington, DC: Resources for the Future.

Scheraga, J. D., and M. B. Sigler, eds. 1991. *Inventory of Studies on Climate Change Impacts and Adaptation Options*. Washington, DC: U.S. Environmental Protection Agency.

Schlesinger, M. E., and J. F. B. Mitchell. 1985. Model Projections of the Equilibrium Climatic Response to Increased Carbon Dioxide. In *Projecting the Climatic Effects of Increasing Carbon Dioxide*, eds. M. D. McCracken and F. M. Luther, 280–319. DOE/ER-0237. Washington, DC: U.S. Department of Energy.

Schlesinger, M. E., and Z. C. Zhao. 1988. Seasonal Climate Changes Induced by Doubled CO_2 as Simulated by the OSU Atmospheric GCM/Mixed Layer Ocean Model. *Journal of Climate* 2(5):459–95.

Schutz, C., and W. L. Gates. 1974. *Global Climatic Data for Surface 800mb, 400mb: Monthly*. ARPA Report # R-915–ARPA. Santa Monica, Calif.: RAND Corporation.

Smith, J. B., and D. A. Tirpak. 1989. *The Potential Effects of Global Climate Change on the United States*. U.S. EPA Report to Congress. EPA 230-05-89-050. Washington, DC: U.S. Environmental Protection Agency.

Tegart, W. J. McG., G. W. Sheldon, and D. C. Griffiths. 1990. *Climate Change: The IPCC Impacts Assessment*. Intergovernmental Panel on Climate Change. Canberra: Australian Government Publishing Service.

United Nations General Assembly (UNGA). 1992. *Report of the Intergovernmental Negotiating Committee for a Framework Convention on Climate Change*. A/AC.237/18 (Part II)/Add.1, 15 May 1992. New York: United Nations.

United States Environmental Protection Agency (U.S. EPA). 1990. *Progress Reports on International Studies of Climate Change Impacts*. Washington, DC: U.S. EPA, Office of Policy, Planning and Evaluation. (Draft).

Wigley, T. M. L., and S. C. B. Raper. 1992. Implications for Climate and Sea level of Revised IPCC Emissions Scenarios. *Nature* 357: 293–300.

Wilson, C. A., and J. F. B. Mitchell. 1987. A Doubled CO_2 Climate Sensitivity Experiment with a Global Climate Model Including a Simple Ocean. *Journal of Geophysical Research* 92:13315–43.

2 World Food Supply

CYNTHIA ROSENZWEIG
Columbia University and Goddard Institute for Space Studies

MARTIN L. PARRY
University College London

GÜNTHER FISCHER
International Institute for Applied Systems Analysis

SUMMARY

Global assessment

THE ESTIMATION OF POTENTIAL CHANGE IN CROP YIELD

Agricultural scientists in 18 countries estimated potential changes in national grain crop yields using compatible crop models and consistent climate change scenarios. The crops modeled were wheat, rice, maize, and soybean. Wheat, rice, and maize account for approximately 85 percent of world cereal exports; soybean accounts for about 67 percent of world trade in protein cake equivalent. The direct effects of carbon dioxide level on crop yields were taken into account. Site-specific estimates of yield changes were aggregated to national levels for the modeled crops. These national crop yield changes were then extrapolated to provide yield change estimates for other countries and other crops.

THE ESTIMATION OF WORLD FOOD TRADE RESPONSES

The national grain crop yield estimates from the climate change scenarios were used as inputs for a world food trade model, the Basic Linked System (BLS). Outputs from simulations by the BLS provided estimates of changes in cereal production, food prices, and the number of people at risk of hunger. The assessment of the implications of climate change on world food supply took the following into account:

The uncertainty about the level of climate change expected

The effects of three climate change scenarios were tested using climate conditions predicted for doubled levels of atmospheric carbon dioxide by three general circulation models: the Goddard Institute for Space Studies (GISS), the Geophysical Fluid Dynamics Laboratory (GFDL), and the UK Meteorological Office (UKMO) models (see Box 2 for analysis of a scenario with lesser warming).

Different adaptive responses

Two levels of farmer adaptation were considered, based on different assumptions about shifts in crop planting dates, changes in crop variety, level of irrigation, and fertilizer application.

The uncertainty about future trade policy and levels of economic and population growth

Alternative pathways of development including partial and full trade liberalization, low and moderate economic growth, and low and medium population growth were considered in the presence and absence of climate change.

Key findings of global assessment

CLIMATE CHANGE WITH CONTINUATION OF CURRENT TRENDS

When the three projections of climate change were imposed on the world food system up to the year 2060, given a continuation of current trends in economic growth rates, partial trade liberalization, and medium population growth rates, it was estimated that:

- Assuming a minor level of farm-level adaptation (e.g., minor shifts in planting dates and minor changes in crop variety), the net effect of climate change would be to reduce global cereal production by up to 5 percent. This modest global reduction could be largely overcome by more major forms of adaptation, such as the installation of irrigation.
- Climate change would increase the disparities in cereal production between developed and developing countries.

Production in the developed world may well benefit from climate change, whereas production in developing nations may decline. Adaptation at the farm level would do little to reduce the disparities, with the developing world suffering the losses.

- Cereal prices and thus the population at risk of hunger in developing countries could increase despite adaptation. Even a high level of farmer adaptation in the agricultural sector would not entirely prevent such adverse effects.

CLIMATE CHANGE WITH ALTERNATIVE ASSUMPTIONS

When alternative assumptions – lowering of trade barriers, low economic growth, and low population growth – were tested in the absence and the presence of climate change, the following projections were made.

In the absence of climate change:

- Full trade liberalization and low population growth would have a beneficial effect in reducing the population at risk of hunger, as compared to the reference case (partial trade liberalization and medium population growth), whereas low economic growth would increase the number of people at risk of hunger.
- The greatest benefits would accrue from following a low population growth pathway into the future.

In the presence of climate change:

- The beneficial effects of full trade liberalization and low population growth would be equal to or (in the case of population, significantly) greater than the adverse effects of climate change. Therefore, there may be much to be gained from altering the conditions of trade and development as a strategy for addressing the climate change issue.
- Cereal production would decrease, particularly in the developing world, while prices and population at risk of hunger would increase due to climate change. The alternate assumptions of trade liberalization, economic development, and population growth made little difference with respect to the geopolitical patterns of relative effects of climate change.
- The magnitude of adverse climate impacts would be the least, however, under the conditions of low population growth. Low population growth would minimize the population at risk of hunger both in the presence and absence of climate change.

Overall, the study suggests that the worst situation would arise from a scenario of severe climate change, low economic growth, and little farm-level adaptation. In order to minimize possible adverse consequences – production losses, food price increases, and people at risk of hunger – the way forward is to encourage the agricultural sector to continue to develop crop breeding and management programs for heat and drought conditions (which would be useful even today in improving productivity in marginal environments), and for the nations of the world to take measures to slow the growth of the human population. The latter step would also be consistent with efforts to slow emissions of greenhouse gases, the source of the problem, and thus the rate and eventual magnitude of global climate change.

INTRODUCTION

In the coming decades, global agriculture faces the prospect of a changing climate (IPCC, 1990a, 1992), as well as the known challenge of feeding a world population that is projected to double its present level of 5 billion by about the year 2060 (International Bank for Reconstruction and Development/World Bank, 1990). The prospective climate change is global warming (with associated changes in hydrological regimes and other climatic variables) induced by the increasing concentration of radiatively active greenhouse gases (IPCC, 1990a, 1992). Climate change could have far-reaching effects on patterns of trade among nations, development, and food security. To help prepare for this uncertain but challenging future, this study examined the potential effects of climate change on crop yields, world food supply, and regions vulnerable to food deficits (see Box 1).

Despite such technological advances as improved crop varieties and irrigation systems, weather and climate are still key factors in agricultural productivity. For example, weak monsoon rains in 1987 caused large shortfalls in crop production in India, Bangladesh, and Pakistan, contributing to reversion to wheat importation by India and Pakistan (World Food Institute, 1988). The 1980s also saw the continuing deterioration of food production in Africa, caused in part by persistent drought and low production potential, and international relief efforts to prevent widespread famine. Moreover, the effects of climate on agriculture in individual countries cannot be considered in isolation. Agricultural trade has grown dramatically in recent decades and now provides significant increments of national food supplies to major importing nations and substantial income for major exporting nations (Table 1). These examples emphasize the close links between agriculture and climate, the international nature of food trade and food security, and the need to consider the impacts of climate change in a global context.

Recent research has focused on regional and national assessments of the potential effects of climate change on agriculture. These efforts have, for the most part, treated each region or nation in isolation, without considering the effects of changes in production in other places. At the same time, there has been a growing emphasis on understanding the interactions of climatic, environmental, and social factors in a wider context (Parry, 1990), leading to more integrated assessments in national agricultural impact

BOX 1. UNDERSTANDING VULNERABILITY TO HUNGER: HOW GREAT IS THE THREAT OF CLIMATE CHANGE?

Vulnerability to hunger is the prospect that the hierarchy of systems that provide an individual with enough food to meet his/her requirements for an active and healthy life may fail. In contrast to the immense driving forces of global change – population change, economic growth, trade liberalization, democratization – hunger is intimately connected with the local nuances of human ecology, exchange entitlement and political economy. The threat of climate change to individual vulnerability depends on household characteristics (for example, too little labor for both timely planting and off-farm employment), socio-economic class (for example, access to wealth), regional production systems (with varying sensitivity to climatic perturbations), and national activities (for example, that acquire and allocate food aid).

While global trends in nutrition and production are well-documented, information on vulnerability is less systematic. Projections into the future are equally uncertain, often depending on critical assumptions, such as the desired metabolic rate, distribution of income, or nature of governance (and the absence of civil strife). Given the local scale of hunger, the difficulty of compiling systematic databases for future projections, and the great potential for surprise in both natural and social systems, a useful approach is to look at climate change as a future risk to present vulnerability. This focus naturally complements research based on scenarios and projections of food systems using integrated or linked models. Research in progress in Zimbabwe aims to define the present and future social space of vulnerability.

Surveys of food security in Zimbabwe delineate livelihood groups that are presently vulnerable to food insecurity – urban (unemployed, informal workers) and rural (communal farmers, landless/farm workers/unemployed) (Table A-1). While the definition of vulnerable groups is robust and widely accepted, reference points for enumerating the extent of vulnerability, food insecurity, poverty, and malnutrition differ in time, region, and measurement. However, a reasonable estimate of the nature and extent of vulnerability in Zimbabwe emerges.

Some 6 percent of the total population are vulnerable to food insecurity in urban areas, comprised of the unemployed and households headed by informal workers. Among these 125,000 households, some 10 percent are headed by divorced, separated, or widowed women – a particularly vulnerable group with low employment, restricted social relations, and high numbers of dependents.

Most of the vulnerable are in rural areas. Agricultural land use in Zimbabwe spans a range from the productive specialized, intensive farming regions of central to eastern Zimbabwe (zones I and II), to the lowland, extensive farming zones of western and southern Zimbabwe (zones IV and V). Communal farmers are smallholder agriculturists, concentrated in the semi-arid, extensive farming zones. Even in years of average production, food-insecure communal farmers number almost 10 percent of the population. Rural households with no land, those dependent on agricultural labor or those with no employment, comprise 12.5 percent of the population. As in urban areas, female-headed households, 3.5 percent of the population, may be critically vulnerable.

Although food security is a priority for national policy, the causes of vulnerability are embedded in the human ecology, political economy, and entitlements relations of post-independence Zimbabwe. Maize, the staple food, has increased in production, more than doubling among communal farmers between the 1970s and 1980s. Yet, only 10–20 percent of communal farmers consistently produce a surplus. Weak macroeconomic performance, inequitable land distribution,

Table A-1. *Vulnerable Livelihood Groups in Zimbabwe, 1991*

Group	Number of Food-Insecure Households[a]	Percent of Population
Urban		
Unemployed	72,000	3.7
Informal Workers	53,000	2.7
Urban Total	125,000	6.4
Rural		
Communal Farmers:[b]		
Zones I and II	20,000–39,000	1–2
Zone III	22,500–98,000	1–5
Zones IV and V	137,000–450,500	7–23
Landless, Farm Workers and Unemployed	210,000	12.5
Rural Total[b]	389,500–797,500	21.5–42.5
Total[b]	514,500–922,500	27.9–48.9

Notes:

[a] Christensen and Sack report estimates of food insecurity based on the confluence of poverty, malnourishment, and variable incomes. This corresponds to our broader definition of vulnerability.

[b] The lower number is food security in years with average rainfall, while the higher number includes additional vulnerability due to exposure to crop failure in years with lower than average rainfall.

Source:

Christensen, G., and J. Stack. 1992. *The Dimensions of Household Food Insecurity in Zimbabwe, 1980–1991.* Working Paper No. 5. Oxford: Food Studies Group.

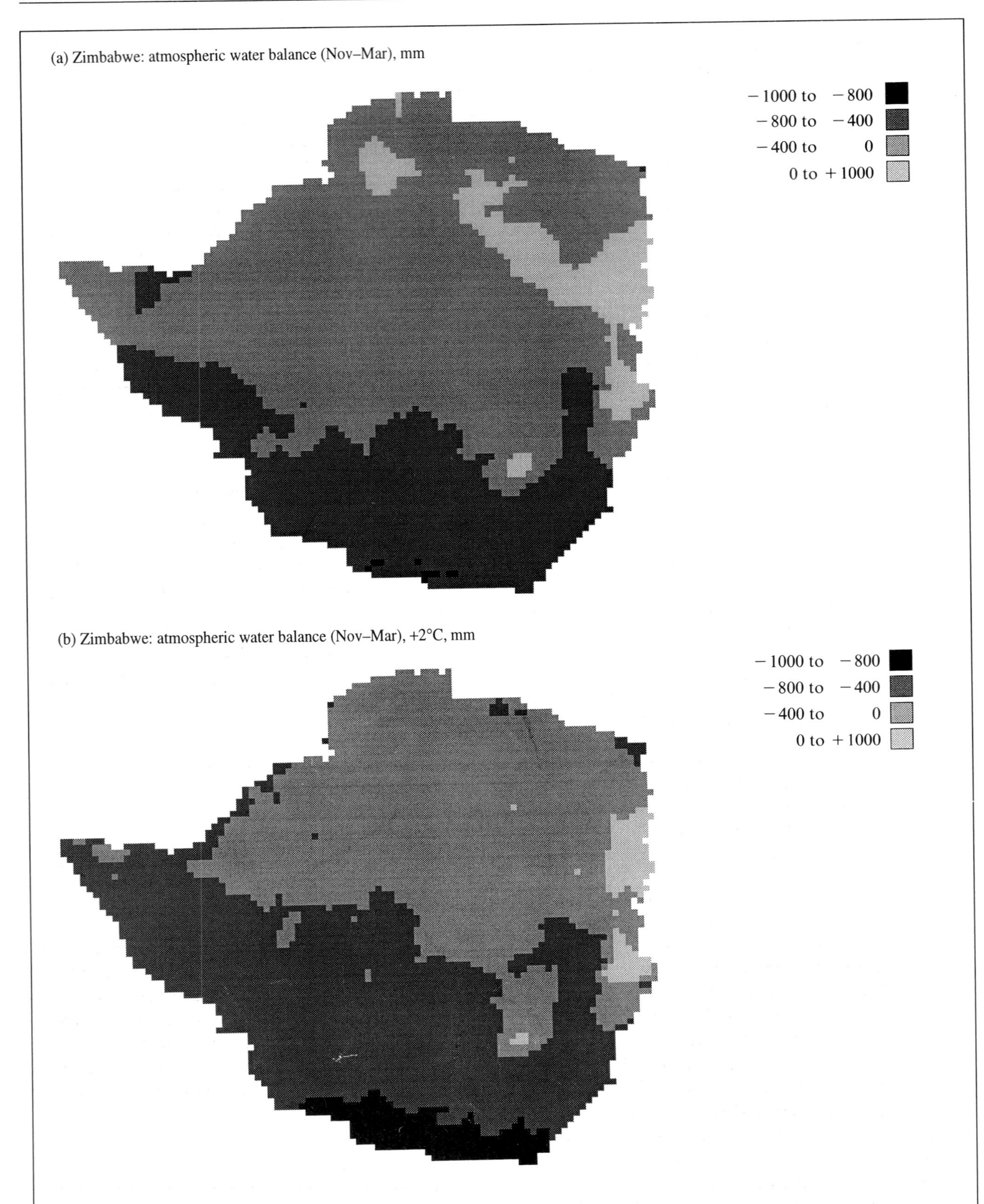

Figure A-1 Seasonal atmospheric water balance in Zimbabwe, precipitation–potential evaporation. (a) Current average conditions for November to March, and (b) with + 2°C warming. Source: Downing, 1992.

and misdirected social policy all contribute to vulnerability. External shocks of recurrent drought in the last ten years have further stressed vulnerable groups.

Among rural households, vulnerability doubles in times of drought, particularly if national institutions fail to provide timely support. Drought is recurrent in Zimbabwe, with serious impacts in 1991/92. Small changes in agroclimatic resources entail significant spatial shifts of land-use zones.

A simple index of the atmospheric water balance, precipitation minus potential evaporation, gauges the extent to which agricultural land use may be subject to changes in water resources – often the limiting factor in semi-arid regions of Zimbabwe. The present balance shows the wetter highlands (with surplus water) and the extreme water deficits of southern Zimbabwe (Figure A-1a). With a temperature increase of 2°C and no change in precipitation, the wet zones (with a water surplus) decrease by a third, from 9 percent of Zimbabwe to about 2.5 percent (Figure A-1b). The driest two zones double in area. A further increase in temperature, to +4°C, reduces the summer water-surplus zones to less than 2 percent of Zimbabwe, approximately corresponding to the 1991/92 drought.

In addition to a shrinkage of the agricultural area, crop yields in marginal zones would suffer. Simulations for one semi-arid area indicate that with 2°C warming, adequate yields currently expected 70 percent of the time would only be exceeded in less than 40 percent of the years.

Increased risk in food production directly affects vulnerable farmers. Surveys in Buhera, in the semi-extensive, semi-arid farming region near Chisumbanje show that household food security is affected by erratic rainfall, sandy and infertile soils, and low levels of crop technology. Maize yields are low – averaging 650 kg/ha. Farm sizes are relatively large, but households are also large with 10 people on average. Farmers on average fail to support their households. With climate change, household food security would deteriorate, possibly with a 10–20 percent decrease in food availability in vulnerable households.

The semi-extensive farming zone, on the margin of more intensive land uses, appears to be particularly sensitive to small changes in climate. Socioeconomic groups in this area, already vulnerable in terms of self-sufficiency and food security, would be further marginalized. Increased variations in rainfall and yields would alter the mix of appropriate response strategies. Successful farming systems would have to be responsive to both good and bad seasons, implying improved use of weather information, flexible markets for inputs and produce, and reliable drought responses.

Climate change threatens each vulnerable group through the multiple effects that diminish resource endowment, and will possibly increase resource conflicts and tension between agricultural and industrial/commercial sectors. Thus, broad-scale shifts in agricultural capability would affect not only rural populations, but the nation as a whole.

See the following for detailed references:

Downing, T. E. 1992. Climate change and vulnerable places: Global food security and country studies in Zimbabwe, Kenya, Senegal and Chile. Research Report #1. Oxford University, Environmental Change Unit. Oxford: Oxford University.

Bohle, H. G., T. E. Downing, and M. Watts. 1994. Climate change and social vulnerability: toward a sociology and geography of food insecurity. *Global Environmental Change* 4(1).

Thomas E. Downing
Programme Leader, Climate Impacts and Responses
Environmental Change Unit
University of Oxford
Oxford, U.K.

studies conducted in the United States (Adams *et al.*, 1990; Smith and Tirpak, 1989), Canada (Smit, 1989), Brazil (Magalhaes, 1992) and Indonesia, Malaysia, and Thailand (Parry *et al.*, 1992). Regional studies have been conducted in high-latitude and semi-arid agricultural areas (Parry *et al.*, 1988a, 1988b) and in the U.S. Midwest (Rosenberg and Crosson, 1991). The results of these and other agricultural impact studies have been summarized in the IPCC Working Group II Report (IPCC, 1990b). Studies of the sensitivity of world agriculture to potential climate changes have indicated that the effect of moderate climate change on world and domestic economies may be small as reduced production in some areas would be balanced by gains in others (Kane *et al.*, 1992). Potential changes in crop yield and distribution have been modeled (Leemans and Solomon, 1993). However, there has, to date, been no integrated (i.e., combined biophysical and economic) assessment of the potential effects of climate change on world agriculture.

The study reported here assessed the potential effects of climate change on world food supply. The research involved estimating the responses of crop yields to greenhouse gas-induced climate change scenarios and then simulating the economic consequences of these potential changes in crop yields. The analysis provides estimates of changes in terms of production and prices of major food crops and the number of people at risk of hunger.

APPROACH

World food supply study design

The structure and research methods for the world food

Table 1. *Major cereal importers and exporters*

A. Major net cereal importers, 1988 (mmt)		B. Major net cereal exporters, 1988 (mmt)	
USSR	34	USA	98
Japan	28	France	27
China	17	Canada	23
Korea, Rep.	9	Australia	15
Egypt	8	Argentina	10
Mexico	6	Thailand	6
Iran	5	Denmark	2
Italy	5	UK	1
Iraq	4	South Africa	1
Saudi Arabia	3	New Zealand	—

Source: FAO, 1988.

supply study are illustrated in Figure 1. There were two main components:

Estimation of potential changes in crop yield

Agricultural scientists in eighteen countries (see Appendices 1 and 2) simulated potential changes in grain yields using compatible crop models developed by the U.S. Agency for International Development's International Benchmark Sites Network for Agrotechnology Transfer (IBSNAT, 1989). The crops modeled were wheat, rice, maize, and soybeans. Wheat, rice, and maize account for approximately 85 percent of world cereal exports; soybean accounts for about 67 percent of world trade in protein cake equivalent. The crop models were run for current climate conditions, for arbitrary changes in climate (2°C and 4°C increases in temperature and +/−20 percent precipitation), and for climate conditions predicted by general circulation models (GCMs) for doubled atmospheric CO_2 levels. The direct effects of increasing levels of CO_2 on crop growth and water use were taken into account. For the GCM climate change scenarios, site-specific estimates of crop yield changes were aggregated by current regional production to estimate national crop yield changes at two levels of farmer adaptation. The national crop yield changes were then extrapolated to provide estimates of yield changes (for the three GCM scenarios) for other countries and crops included in the food trade analysis.

Estimation of world food trade responses

The national crop yield changes derived from the first component of the study were used as inputs for a world food trade model, the Basic Linked System (BLS), developed at the International Institute for Applied Systems Analysis (IIASA) (Fischer *et al.*, 1988). The BLS was run first for a reference scenario projecting the agricultural system to the year 2060 assuming no change in climate, and then with the

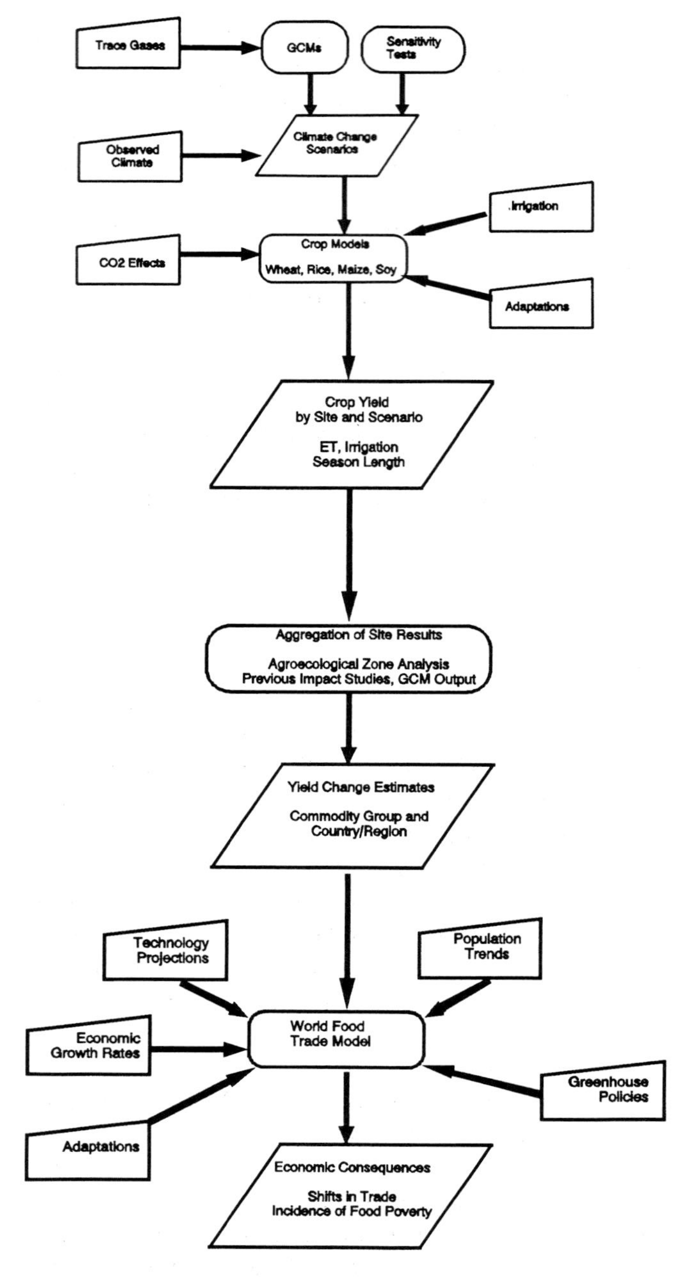

Figure 1 Key elements of the crop yield and world food trade supply.

three GCM climate change scenarios. Other BLS simulations included the effects of two levels of farmer adaptation and scenarios of different future trade liberalization policies and different economic and population growth rates. Outputs from the BLS simulations provided information on food production, food prices, and the number of people at risk of hunger (defined as those with an income insufficient to either produce or procure their food requirements) for these scenarios projected up to the year 2060.

Throughout the climate change study, a distinction was made between *farm-level adaptations*, which were tested by the crop models and result in yield changes, and *economic adjustments* to the yield changes, which were tested by the BLS world food trade model and result in national and regional production changes and price responses. Farm-level adaptations tested in the crop models included shifts in planting date, more climatically adapted crop varieties, changes in amounts of irrigation, and changes in fertilizer application. Economic adjustments tested by the BLS included increased agricultural investment, reallocation of agricultural resources according to economic returns (including crop switching), and reclamation of additional arable land as a response to higher cereal prices. These economic adjustments were assumed not to feed back to the yield levels predicted by the crop modeling study. The crop yield and economic modeling components are described in greater detail below.

Climate change scenarios

Sensitivity tests

Arbitrary climate sensitivity tests were conducted to test crop model responses to a range of temperature (+2°C and +4°C) and precipitation (+/−20 percent) changes. The sensitivity test results were not utilized in the economic analysis with the BLS world food trade model because of their lack of realism; for example, temperature change at high latitudes is predicted to be greater than the global mean in winter (IPCC, 1990a) rather than to increase by 2°C or 4°C in all regions of the world throughout the year. Farm-level adaptations were not tested in the sensitivity studies.

Scenarios based on general circulation model results

Scenarios of climate change were developed in order to estimate their effects on crop yields and food trade. The GCMs used were those from the Goddard Institute for Space Studies (GISS), the Geophysical Fluid Dynamics Laboratory (GFDL), and the United Kingdom Meteorological Office (UKMO). (See Chapter 1.) The predictions of these scenarios are presented in Table 2. See Box 2 for analysis of a scenario with lesser warming. Mean monthly changes in temperature, precipitation, and solar radiation from the appropriate GCM grid boxes were applied to observed daily climate records to create climate change scenarios for each site.

CO_2 level and timing

For the crop modeling part of this study, climate changes derived from the doubled CO_2 GCM simulations were utilized with an associated level of 555 ppm CO_2 (see Chapter 1). The 555 ppm CO_2 level is based on the GISS GCM trace gas scenario A (Hansen *et al.*, 1988), in which the simulated climate had warmed to the effective doubled CO_2 level of ~4°C by 2060. The level of CO_2 is important when estimating potential impacts on crops, because crop growth and water use have been shown to benefit from increased levels of CO_2 (Cure and Acock, 1986). For the BLS world food trade projections it was assumed that these conditions would occur in 2060. However, it is not known what the rates of future emissions of trace gases will be and when the full magnitude of their effects will be realized.

Table 2. *GCM climate change scenarios*

GCM	Year[1]	Resolution (lat × long)	CO_2 (ppm)	Change in average global: Temperature (°C)	Change in average global: Precipitation (%)
GISS[2]	1982	7.83° × 10°	630	4.2	11
GFDL[3]	1988	4.4° × 7.5°	600	4.0	8
UKMO[4]	1986	5.0° × 7.5°	640	5.2	15

Notes:
[1] When calculated.
[2] Hansen *et al.*, 1983.
[3] Manabe and Wetherald, 1987.
[4] Wilson and Mitchell, 1987.

Crop models and yield simulations

Crop models

The IBSNAT crop models were utilized by the participating agricultural scientists to estimate how climate change and increasing levels of carbon dioxide may alter yields of world crops at 112 sites in 18 countries. The sites represented both major and minor production areas at low, mid, and high latitudes (Figure 2). The crop models used were CERES-Wheat (Ritchie and Otter, 1985; Godwin *et al.*, 1989), CERES-Maize (Jones and Kiniry, 1986; Ritchie *et al.*, 1989), CERES-Rice (paddy and upland) (Godwin *et al.*, 1993), and SOYGRO (Jones *et al.*, 1989).

The IBSNAT models are composed of parameterizations of important physiological processes responsible for plant growth and development, evapotranspiration, and partitioning of photosynthate to produce economic yield. The simplified functions enable the prediction of the growth of crops as influenced by the major factors that affect yields, i.e., genetics, climate (daily solar radiation, maximum and minimum temperatures, and precipitation), soils, and management practices. This type of dynamic process crop growth model is considered to be a significant advance over traditional regression-based methods (see, e.g., Thompson, 1969) that were used to estimate crop yields from simple climate and management inputs with geographic and temporal specificity. The IBSNAT models include a soil moisture balance submodel so that they could be used to predict both rainfed

BOX 2. ANALYSIS OF GISS-A[1] CLIMATE CHANGE SCENARIO

In order to test a climate scenario with lower levels of projected warming, a fourth climate change scenario was included utilizing modified project methodology. The IBSNAT crop models were run at 70 sites with the climate predicted by the Transient Run A of the GISS GCM for the decade of the 2030s. Wheat, maize, soybean, and rice were simulated at 58, 31, 18, and 8 sites, respectively. Simulations were done with and without physiological CO_2 effects (555 ppm), but farm-level adaptations were not tested. Crop model results by site were then aggregated and extended to create estimates of country yield changes as described in the main body of the chapter and BLS simulations were made with these GISS-A yield change estimates.

As expected, crop yields were less negatively affected by the GISS-A climate change scenario (Figure B-1, Plate 4). However, crop modeling results showed that mean national crop yield changes were still mostly negative without the direct physiological effects of CO_2 on crop growth and yield. Yield changes without direct CO_2 effects ranged from −30 to +20 percent. With direct CO_2 effects, yield changes were mostly positive, ranging from −10 percent to +30 percent. In this case, the beneficial CO_2 effects more than compensated for detrimental climate effects in most locations. The geographical distribution of effects found with the three doubled CO_2 climate change scenarios remained the same, with more negative or less positive effects occurring in low latitudes and more positive effects occurring at high latitudes.

The economic adjustments simulated by the BLS reflected the more benign yield changes of the GISS-A scenario (Table B-1). Global cereal production declined about 4 percent when yield estimates did not include direct CO_2 effects and increased about 2.5 percent with crop yield changes that included direct CO_2 effects. These production effects were reflected in cereal price increases of about 65 percent in the case without CO_2 effects and price decreases of about 15 percent with CO_2 effects. People at risk of hunger in developing countries (see main body of chapter for description of this index) increased about 40 percent without CO_2 effects and decreased about 10 percent with CO_2 effects.

[1] The GISS-A scenario uses a lesser amount of warming (2.4°C) than UKMO, GISS, and GFDL. It was defined as the average change in the 2030s of the GISS-A transient run and was used as a scenario for doubled CO_2 conditions. Note that the IPCC's most likely warming range is 1.5 to 4.5°C for CO_2 doubling.

Table B-1. *Change in cereal production, cereal price index, and people at risk of hunger in 2060 for GISS-A climate change scenario*

	Reference 2060	LE* (330 ppm)	LE* (555 ppm)
GLOBAL			
Cereal Production (mmt)	3286	−145	82
Cereal Price Index	121	81	−21
People at Risk** (mil)	641	265	−84
DEVELOPED			
Cereal Production (mmt)	1449	26	153
DEVELOPING			
Cereal Production (mmt)	1836	−170	−71

Notes:
* Change relative to Reference Scenario in 2060.
** In developing countries.

Testing of a climate change scenario close to the middle of the IPCC range of 1.5 to 4.5°C (IPCC, 1990a; 1992) is important to characterize more fully potential impacts on the world agricultural system. Several points are worth emphasizing with the additional perspective provided by the GISS-A scenario. First, the magnitudes of impact on world agricultural production across all scenarios tested were generally low, whether direction of change was positive or negative. Only the UKMO scenario with a high level of predicted warming (5.2°C) produced a moderate decrease in world production. However, more negative consequences to global production were simulated than positive ones, even when minor and major farm-level adaptations and lower amounts of global warming (GISS-A climate change scenario) were included. Second, the regional imbalance of effects that suggest that developing countries may be more vulnerable to climate change than developed nations occurred across all scenarios. Finally, the direction of change in global production continued to depend on the inclusion of significant beneficial physiological effects of CO_2 on crop growth and yield even in the GISS-A climate change scenario. Further research is required to lessen the uncertainties of how these effects will be manifested in farmers' fields in both temperate and tropical regions in the presence of a changing climate.

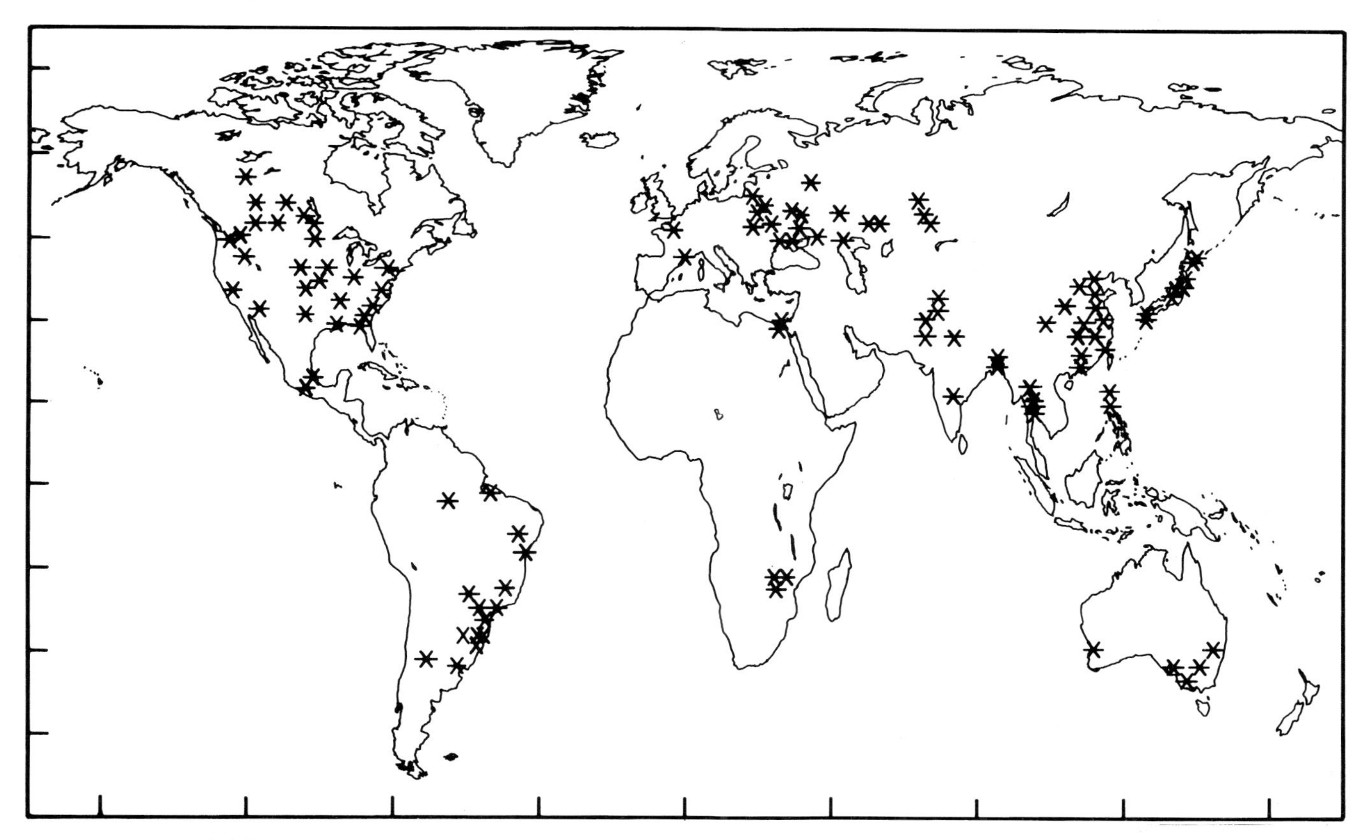

Figure 2 Crop model sites.

and irrigated crop yields. The cereal models simulate the effects of nitrogen fertilizer on crop growth, and these effects were studied in several countries in the context of climatic change. For the most part, however, the results of this study assume optimum nutrient levels.

The IBSNAT models were selected for use in this study because they have been validated over a wide range of environments (see, e.g., Otter-Nacke *et al.*, 1986) and are not specific to any particular location or soil type. They are better suited for large-area studies in which crop-growing and soil conditions differ greatly than are more detailed physiological models that have not been tested as widely. The validation of the crop models over different environments also improves the ability to estimate effects of changes in climate. Because the crop models have been tested over essentially the full range of temperature and precipitation regimes at which crops are grown in today's climate, and to the extent that future climate change is estimated to bring temperature and precipitation regimes within these ranges, the models may be considered useful tools for assessment of potential climate change impacts. Furthermore, because management practices, such as the choice of crop varieties, planting date, fertilizer application, and irrigation, may be varied in the models, they permit experiments that simulate adaptation by farmers to climatic change.

Physiological effects of CO_2

Most plants growing in experimental environments with increased levels of atmospheric CO_2 exhibit increased rates of net photosynthesis (i.e., total photosynthesis minus respiration) and reduced stomatal openings. (Experimental effects of CO_2 on crops have been reviewed by Acock and Allen [1985] and Cure [1985]). Partial stomatal closure leads to reduced transpiration per unit leaf area and, combined with enhanced photosynthesis, often improves water-use efficiency (the ratio of crop biomass accumulation or yield to the amount of water used in evapotranspiration). Recent field free-air release studies have found overall positive CO_2 effects under current climate conditions (Hendry, 1993). Thus, by itself, increased CO_2 can increase yield and reduce water use (per unit biomass).

The crop models used in this study accounted for the beneficial physiological effects of increased atmospheric CO_2 concentrations on crop growth and water use (Peart *et al.*, 1989). Ratios were calculated between measured daily photosynthesis and evapotranspiration rates for a canopy exposed to high CO_2 values, based on published experimental results (Allen *et al.*, 1987; Cure and Acock, 1986; and Kimball, 1983), and the ratios were applied to the appropriate variables in the crop models on a daily basis. The ratios (555 ppm CO_2:330 ppm CO_2) for soybean, wheat and rice, and maize were 1.21, 1.17, and 1.06, respectively. Changes in stomatal resistance were set at 49.7/34.4 s/m for C3 crops and 87.4/55.8 s/m for C4 crops, based on experimental results by Rogers *et al.* (1983). As simulated in this study, the direct effects of CO_2 may bias yield changes in a positive direction, since there is uncertainty regarding whether experimental results will be observed in the open field under conditions likely to be operative when farmers are managing crops. Plants growing in experimental settings are often subject to fewer environmental stresses and less competition from weeds and pests than are likely to be encountered in farmers' fields.

Table 3. *Current world crop yield, area, production, and percent world production aggregated for countries participating in study*

Crop	Yield (t/ha)	Area (ha × 1000)	Production (t × 1000)	Study countries (%)
Wheat	2.1	230,839	481,811	73
Rice	3.0	143,603	431,585	48
Maize	3.5	127,393	449,364	71
Soybeans	1.8	51,357	91,887	76

Source: FAO, 1988.

Limitations of crop growth models

The crop models embodied a number of simplifications. For example, weeds, diseases, and insect pests were assumed to be controlled; there were no problem soil conditions (e.g., salinity or acidity); and there were no extreme weather events, such as tornadoes. The models were calibrated to experimental field data, which often have yields higher than those currently typical under farming conditions. Thus, the absolute effects of climatic change on yields in farmers' fields may be different from those simulated by the crop models.

Although the crop models simulated the current range of agricultural technologies available around the world, including the use of high-yielding varieties that are responsive to technological inputs, agricultural technology is likely to be very different by the year 2060. The models may be used to test the effects of some potential improvements in agricultural production, such as the use of crop varieties with higher thermal requirements and the installation of irrigation systems, but they do not include possible future improvements. (The BLS economic model used in the study does include future trends in yield improvement, but it does not include technological developments induced by negative climate change impacts.) Finally, models for crops such as millet and cassava were not yet sufficiently tested for use in this study. Potential yield changes of such crops, which may respond differently to both climate change and increases in CO_2, are needed for better assessment of climate change impacts in tropical and subtropical regions.

Yield simulations

Crop modeling simulation experiments were performed for the baseline climate (1951–1980), values from the arbitrary sensitivity tests, and GCM doubled CO_2 climate change scenarios with and without taking into account the physiological effects of CO_2. The experiments involved the following tasks:

1. For the countries studied, geographic boundaries were defined for the major production regions; agricultural systems (e.g., rainfed and/or irrigated production, number of crops grown per year) were described; and data on regional and national rainfed and irrigated production of major crops were gathered.
2. Observed climate data for representative sites within these regions were obtained for the baseline period (1951–1980), or for as many years of daily data as were available, and the soil, crop variety, and management inputs necessary to run the crop models at the selected sites were specified.
3. The crop models were validated with experimental data from field trials, to the greatest extent possible.
4. The crop models were run with baseline data, values from the arbitrary sensitivity tests, and GCM climate change scenarios, with and without taking into account the direct effects of CO_2 on crop growth. Rainfed and/or irrigated simulations were carried out as appropriate to current growing practices.
5. Alterations in farm-level agricultural practices that would lessen any adverse consequences of climate change were identified and evaluated by simulating irrigated production and other adaptation responses (for example, shifts in planting date and substitution of crop varieties).

Deriving estimates of potential crop yield changes

Aggregation of site results

Table 3 shows the percentages of world production of wheat, rice, maize, and soybean for the countries in which simulations were conducted. Simulations were carried out in regions representing 70–75 percent of the current world production of wheat and maize. Even though model runs for soybean were conducted in only two countries (Brazil and the United States), these together account for 76 percent of world production. Rice production was less well represented in the model simulations than the other crops, because India, Indonesia, and Vietnam have significant production areas that were not included in the study. Further research is needed in these key countries in order to improve the reliability of the projections of climate change impacts on rice production.

Crop model results for wheat, rice, maize, and soybean from the 112 sites in the 18 countries were aggregated by weighting regional yield changes (based on current production) to estimate changes in national yields. The aggregations were either calculated by the participating scientists or developed jointly with them (see Rosenzweig and Iglesias, 1994). The scientists in each country selected sites representative of major agricultural regions, described the regional agricultural practices, and provided production data for the estimation of regional contributions to the national yield changes. Other crop production data sources included the United Nations Food and Agriculture Organization (FAO, 1988), the U.S. Department of Agriculture (USDA) Crop Production Statistical Division, and the USDA International Service. The regional yield estimates represented the current mix of rainfed and irrigated production and the current crop varieties, nitrogen management, and soil types.

Table 4. *Increments added to estimated yield change to account for direct effects of CO_2*

Crop	Percentage[1]
Wheat	22
Rice	19
Soybeans	34
Coarse grains[2]	7
Other C3 crops	25
Other C4 crops	7

Notes:

[1] Based on crop model simulations.

[2] Weighted by relative production of C3 and C4 crops constituting coarse grain production in the particular country or region.

Yield change estimates for crops and regions not simulated

Changes in national yields of other crops and commodity groups and for regions not simulated were estimated based on three criteria: (1) similarities to growing conditions for the modeled crops, (2) results from approximately 50 previously published and unpublished regional climate change impact studies (AIR Group unpublished manuscript), and (3) projected temperature and precipitation changes (and hence soil moisture availability for crop growth) from the GCM climate change scenarios.

Estimates of yield changes were made with and without taking into account the direct effects of CO_2. Increments added to the estimated crop yield changes to account for direct CO_2 effects were based on average responses of crop production to CO_2 and climate change scenarios in the crop model simulations (Table 4). These increments differ from the photosynthesis ratios employed in the crop models because they incorporate the combined responses of the simulated crops to changes in photosynthesis and evapotranspiration as well as to changes in climate. In the crop model simulations, the responses to CO_2 did not vary greatly across regions and climate change scenarios.

Limitations of crop yield change estimates

The primary source of uncertainty in the estimates lay in the sparseness of the crop modeling sites and the fact that they may not have adequately represented the variability of agricultural regions within countries, the variability of agricultural systems within similar agro-ecological zones, or dissimilar agricultural regions. However, since the study sites were similar to regions that account for about 70 percent of world grain production, the conclusions concerning world totals of cereal production contained in this study are believed to be adequately substantiated. Another source of uncertainty lay in the simulation of grain crops only, leading to the estimation of yield changes for other commodities, such as root crops and fruit, based primarily on previous estimates. The previous estimates tended to be less negative than the crop responses modeled in this study, and this introduced a bias in favor of these other crops in the world food trade model.

Farm-level adaptations

In each country, the agricultural scientists used the crop models to test possible responses to the worst climate change scenario (which was usually, but not always, the UKMO scenario). These adaptations included changes in planting date, cultivar, amounts of irrigation, fertilizer use, and crop variety. Irrigation simulations in the crop models assumed automatic irrigation to field capacity when plant available water dropped to 50 percent and 100 percent irrigation efficiency. All adaptation possibilities were not simulated at every site and country: the choice of adaptations to be tested was made by the participating scientists, based on their knowledge of current agricultural systems (Table 5).

For the economic analysis in the BLS, the crop model results reported by the participating scientists were grouped into two levels of adaptation. Adaptation Level 1 implied little change to existing agricultural systems, reflecting relatively easy farmer response to a changing climate. Adaptation Level 2 implied more substantial change to agricultural systems, possibly requiring resources beyond the farmer's means.

Adaptation Level 1 included:

1. Shifts in planting date (± 1 month)
2. Additional application of irrigation water to crops already under irrigation

Table 5. *Adaptations tested in crop modeling study*

Country	Crop tested[a]	Change of planting date	Change of cultivar/crop	Additional irrigation	Additional N fertilizer
Argentina	m	x	x[1,7]	x	
Australia	r,w	x[2]	x	x[2]	
Bangladesh	r		x		
Brazil	w,m,s	x[2]	x[1]	x,x[2,3]	x
Canada	w	x[6]		x,x[2]	
China	r	x	x,x[5,7]		
Egypt	m,w	x	x	x	
France	m,w	x,x[7]	x	x	
India	w			x	
Japan	r,w,m	x[2]		x[2]	
Mexico	m	x	x[1]	x[3]	x
Pakistan	w	x		x	
Philippines	r	x[5]	x[5]		
Thailand	r		x		
Uruguay	b	x	x	x	x,x[4]
USA	w,m,s	x	x	x	
USSR	w	x[6,7]	x		
Zimbabwe	m	x[2]		x,x[2]	x

Notes:
[a]w = wheat; m = maize; r = rice; s = soybean; b = barley
[1] Hypothetical new cultivar.
[2] Combination of irrigation and change in planting date.
[3] Combination of irrigation and increased nitrogen fertilizer.
[4] Combination of change in planting date and increased nitrogen fertilizer.
[5] Combination of new cultivar and change in planting date.
[6] Change to winter wheat.
[7] Suggested shift in the zone of crop production.

3. Changes in crop variety to currently available varieties more adapted to the altered climate

Adaptation Level 2 included:

1. Large shifts in planting date (> 1 month)
2. Increased fertilizer application (included here because of the implied costs for farmers in developing countries)
3. Installation of irrigation systems
4. Development of new crop varieties (tested by the manipulation of genetic coefficients in crop models)

Yield changes for both adaptation levels were based on crop model simulations where available and extended to other crops and regions using the estimation methods described above. For the crops and regions not simulated, the negative impact of climate change was halved if adaptations were estimated to partially compensate for them; if compensation was estimated to be full, yield changes were set to 0. If yield changes were positive in response to climate change and the direct effects of CO_2, adaptation to produce even greater yield increases was not included, with the assumption that farmers would lack incentive to adapt further. The adaptation estimates were developed only for the scenarios that included the direct effects of CO_2, as these were judged to be most realistic. Examples of the crop yield change estimates for Adaptation Levels 1 and 2 for the UKMO climate change scenario for several countries are shown in Table 6.

Limitations of the adaptation analysis

The adaptation simulations were not comprehensive, because all possible combinations of farmer responses were not tested at every site. Spatial analyses of crop, climatic, and soil resources are needed to fully test the possibilities for crop substitution. Neither the availability of water supplies for irrigation nor the costs of adaptation were considered in this study; these are both critical needs for further research. A related study on the integrated impacts of climate change in Egypt, which utilized the results of this work, does address future water availability for national agricultural production in that country (see Chapter 7).

At the local level, there may be social or technical reasons

Table 6. *Changes in wheat yield[1] estimated for the UKMO $2 \times CO_2$ climate change scenario, alone and with two levels of farm adaptation*

	UKMO (%)	AD1[2] (%)	AD2[3] (%)
Argentina	−30	−20	−10
USA	−14	−7	−3
Eastern Europe & former USSR	−20	−10	−5

Notes:
[1] With direct CO_2 effects.
[2] Adaptation Level 1 implies minor change to current agricultural systems.
[3] Adaptation Level 2 implies major change to current agricultural systems.

why farmers are reluctant to implement adaptation measures. For example, increased fertilizer application and improved seed stocks may be capital-intensive and not suited to indigenous agricultural strategies. Furthermore, such measures may not necessarily result in sustainable production increases. In the case of irrigation, initial benefits may eventually give way to soil salinization and lower crop yields. Thus, Adaptation Level 2 represents a fairly optimistic assessment of world agriculture's response to changed climate conditions as characterized by the GCMs tested in this study and may possibly require substantial changes in current agricultural systems, investment in regional and national agricultural infrastructure, and policy changes. However, estimation of the effect of changes in regional, national, and international agricultural policies relating to farm-level adaptation were beyond the scope of the analysis.

The world food trade model

The world food system is a complex, dynamic interaction of producers and consumers, mediated through global markets. Related activities include input production and acquisition, transportation, storage, and processing. Although there is a trend toward internationalization in the world food system, only about 15 percent of the total world agricultural production currently crosses national borders (Fischer *et al.*, 1990). National governments shape the system by imposing regulations and by making investments in agricultural research, infrastructure improvements, and education. The system functions to meet the demand for food, to produce food in increasingly efficient ways, and to trade food within and across national borders. Although the system does not guarantee stability, it has generated long-term real declines in prices of major food staples (Fischer *et al.*, 1990).

The Basic Linked System consists of linked national agricultural sector models. It was designed at IIASA for food policy studies, but it also can be used to evaluate the effect of climate-induced changes in crop yield on world food supply and agricultural prices. It consists of 16 national (including the European Community [EC]) models with a common structure, 4 models with country-specific structures, and 14 regional group models (Table 7). The 20 models in the first two groups cover about 80 percent of attributes of the world food system, such as demand, land, and agricultural production. The remaining 20 percent are covered by the 14 regional models for countries that have broadly similar attributes (e.g., African oil-exporting countries, Latin American high-income exporting countries, Asian low-income countries, etc.). The grouping is based on country characteristics, such as geographic location, income per capita, and the country's position with regard to net food trade.

The BLS is a general equilibrium model system, with representation of all economic sectors, empirically estimated parameters, and no unaccounted supply sources or demand sinks (see Fischer *et al.* [1988] for a complete description of the model). In the BLS, countries are linked through trade, world market prices, and financial flows (Figure 3). It is a recursively dynamic system: a first round of exports from all countries is calculated for an assumed set of world prices, and international market clearance is checked for each commodity. World prices are then revised, using an optimizing algorithm and are again transmitted to the national models. Next, new domestic equilibria are generated and net exports are adjusted. This process is repeated until the world markets for all commodities are cleared. At each stage of the iteration, domestic markets are in equilibrium. This process yields international prices as influenced by governmental and intergovernmental agreements. The system is solved in annual increments, simultaneously for all countries. Summary indicators of the sensitivity of the world system include world cereal production, world cereal prices, and prevalence of population in developing countries at risk of hunger.

The BLS does not incorporate any climate relationships per se. Effects of changes in climate were introduced to the model as changes in the average national or regional yield per commodity. Ten commodities were included in the model: wheat, rice, coarse grains, protein feed, bovine and ovine meat, dairy products, other animal products, other food, non-food agriculture, and non-agriculture. Yield change estimates for coarse grains were based on the percentage of maize grown in the country or region; soybean crop model results were used to estimate the protein feed category; and the estimates for the non-grain crops were based on the modeled grain crops and previous estimates of climate

Table 7. *Models in the Basic Linked System*

Models with a common structure	Models with country-specific structures	Regional group models
Argentina	Eastern Europe & former USSR	Africa Oil Exporters
Australia	China	Africa Medium-Income Exporters
Austria	India	Africa Medium-Income Importers
Brazil	United States	Africa Low-Income Exporters
Canada		Africa Low-Income Importers
Egypt		Latin American High-Income Exporters
Indonesia		Latin American High-Income Importers
Japan		Latin American Medium Income
Kenya		Southeast Asia High-Medium Exporters
Mexico		Southeast Asia High-Medium Importers
Nigeria		Asia Low Income
New Zealand		Southwest Asia Oil Exporters
Pakistan		Southwest Asia Medium-Low Income
Thailand		Rest of the World
Turkey		
European Community		

Note: See Appendix 3 for countries within regional groups.

change impacts as described above. A positive bias toward non-grain crops was introduced by this procedure, since the previous estimates of yield changes of the non-grain crops were less negative than the modeled results from this study.

Economic growth rates

Economic growth rates are a product of several BLS functions. Non-agricultural production utilizes a Cobb–Douglas production function, with labor and capital as production factors. Non-agricultural labor input depends primarily on population growth and somewhat on relative prices between agriculture and non-agriculture by means of a sector migration function. Capital accumulation depends on investment and depreciation, which in turn depend on rates of saving and depreciation. Depreciation rates and rates of saving were estimated from historical data and were kept constant after 1990. There was an exogenous assumption based on historical data for technical progress in the production function. For the lower growth scenario, the rate of saving was reduced, resulting in about 10 percent lower gross domestic product in 2060.

The economic growth rates predicted by the BLS in the reference case followed historical trends, as shown in Table 8. For the period 1980 to 2060, the BLS predicted a growth of 1.3 percent, 1.7 percent, and 2.4 percent annually for world, developed, and developing countries, respectively, as compared to average population growth rates of 1.1 percent, 0.3 percent, and 1.3 percent.

Yield trends

In general, the rate of exogenous technical progress started from historical values and for cereal crops approached 0.5 percent per annum by 2060. Representing improvement in agriculture productivity due to technological progress, the annual yield trends used in the BLS for the period 1980–2000 were 1.2 percent, 1.0 percent, and 1.7 percent for world, developed countries, and developing countries, respectively. According to FAO data, yields have been growing at an average of around 2 percent annually during the period 1961–1990, both for developed and developing (excluding China) countries (FAO, 1991). From 1965 to 1985, annual productivity for less-developed countries grew at about 1.5 percent/year. In the 1980s, however, yields grew globally by an average of only 1.3 percent, implying a falling trend in yield growth rates.

The falling growth rates utilized in the reference case of the BLS may be justified for several reasons. Historical trends suggest decreasing rates of increase in crop yields, and yield improvements from biotechnology have yet to be realized. Much of the large yield increases in developed countries in the 1950s and 1960s and in developing countries thereafter was due to the intensification of chemical inputs and mechanization. Apart from economic reasons and environmental concerns, which suggest that maximum input levels may have been reached in many developed countries, there are likely to be diminishing rates of return for further input increases. In some developing countries, especially in Africa, increase in input levels and intensification of production are likely to continue for some time but may also ultimately level off. Furthermore, since Africa has the lowest average cereal yields of all the regional groups combined with a high population growth rate, it will likely contribute an increasing share of cereal production, thereby reducing average global yield increases.

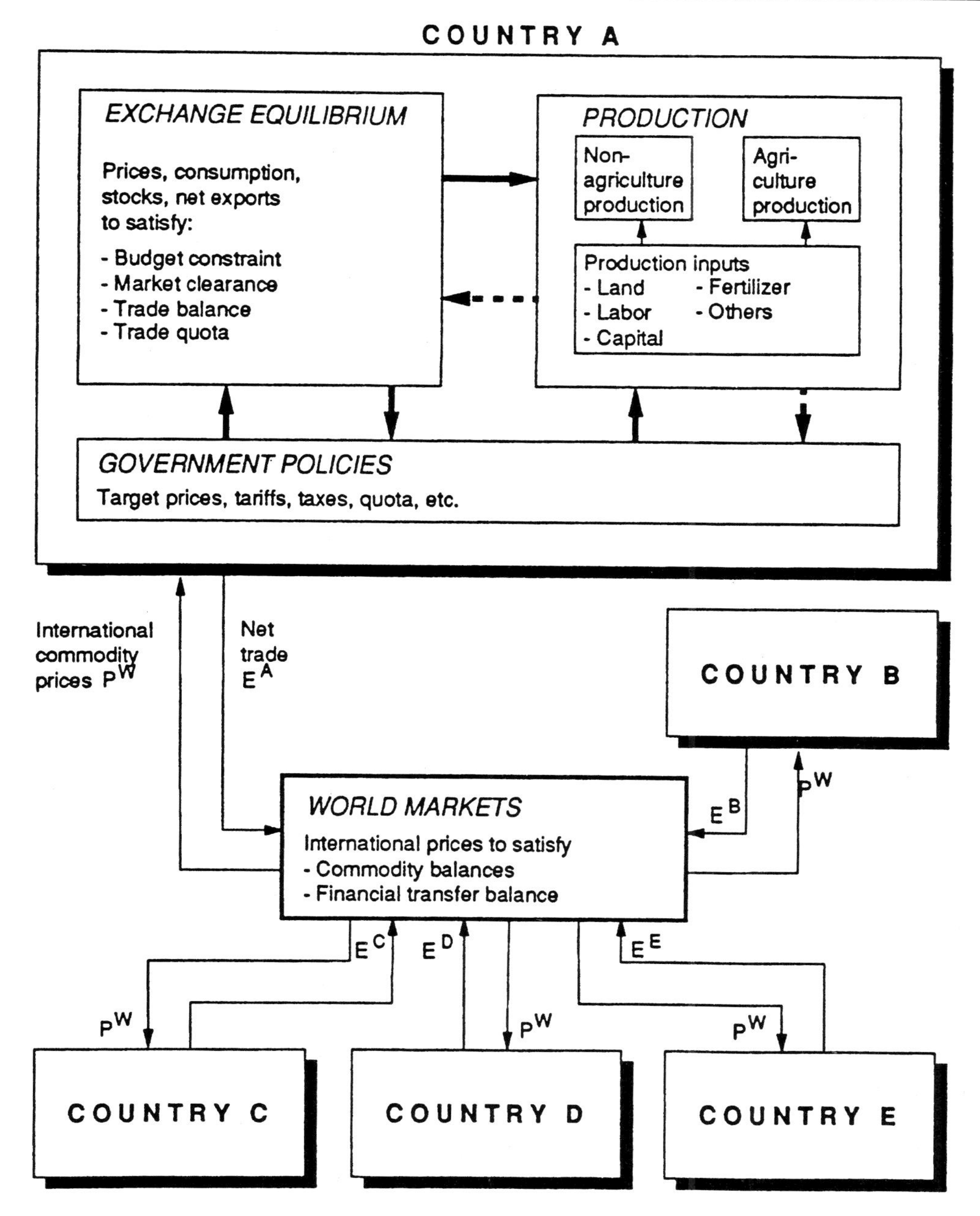

Figure 3 The Basic Linked System: relationships between country components and world markets. Arrows to countries represent international commodity prices; arrows to world markets represent net trade.

Arable land
Estimates of the availability of arable land for expansion of crop production were based on FAO data. In the BLS standard national models, a piece-wise linear time-trend function was used to impose upper bounds (inequality constraints) on land use. In addition, this time-trend function was modified with an elasticity term (usually 0.05 or less) which reacts to changes in shadow prices of land in comparison to 1980 levels. The upper limits imposed by the time-trend function utilized the FAO data on potential arable land. The arable land limits were not adjusted for climate change, even though they may be affected positively in some locations by an extension of season length or a drying of wet soils, or negatively by sea-level inundation or desertification.

Risk of hunger indicator
The indicator of the number of people at risk of hunger used in the BLS was defined as those people in developing countries (excluding China) with an income insufficient to either produce or procure their food requirements. The measure was derived from FAO estimates and methodology for developing market economies (FAO, 1984 and 1987). The FAO estimates were obtained by stipulating that calorie consumption distribution in a country is skewed and can be

Table 8. *Historical and simulated average annual growth rates of GDP (% per annum)*

	Historical				Basic Linked System		
	1960 to 1970	1970 to 1980	1980 to 1990	1980 to 2000	2000 to 2020	2020 to 2040	2040 to 2060
World	4.8	3.6	2.9	2.9	2.0	1.5	1.1
Developed countries	4.7	3.1	2.8	2.6	1.8	1.3	1.0
Developing countries	5.2	5.4	3.1	4.0	2.6	1.8	1.3

Source: FAO, 1991.

represented by a beta distribution. The parameters of these distributions were estimated by the FAO for each country based on country-specific data and cross-country comparisons. The estimate of the energy requirement of an individual was based on the basal metabolic rate (time in a fasting state and lying at complete rest in a warm environment). Body weight, age, and sex have an impact on this requirement. The FAO presented two estimates of the number of undernourished people, based on minimum maintenance requirements of 1.2 and 1.4 (the latter judged as more appropriate) basal metabolic rate. The BLS estimate for 1980, based on the 1.4 basal metabolic rate requirements, was 501 million undernourished people in the developing world, excluding China.

Limitations of the world food trade model

The economic adjustments simulated by the BLS were assumed not to alter the basic structure of the production functions. These relationships may be altered in a changing climatic regime and under conditions of elevated CO_2. For example, yield responses to nitrogen fertilization may be altered due to changing nutrient solubilities in warmer soils. Furthermore, in the analysis of BLS results, consideration was limited to the major cereal food crops, even though shifts in the balance of arable land and livestock agriculture are also likely under changed climatic regimes. Livestock production is a significant component of the global food system and is also potentially sensitive to climatic change. The non-agriculture sector was poorly modeled in the BLS, leading to simplifications in the simulation of economic responses to climatic change.

Finally, recent changes in global geopolitics and related changes in agricultural production were not well represented in the BLS. To account for these changes, prices in previously planned economies were made more responsive compared to earlier versions, 'plan targets' for allocation decisions were replaced, and some constraints were relaxed in the agricultural sector model. Better analysis depends on the development of new models for these emerging capitalist economies.

The set of model experiments

The estimates of climate-induced changes in food production potential were used as inputs to the BLS in order to assess possible impacts of climate change on future levels of food production, food prices, and the number of people at risk of hunger (see Figure 1). Impacts were assessed for the year 2060, with estimates of population growth, technology trends, and economic growth projected to that year. Assessments were first made for a reference scenario that assumed no climate change and were subsequently made for the GCM scenarios (see discussion above). The difference between the two assessments is the climate-induced effect. A further set of assessments examined the efficacy of two levels of farmer adaptation in mitigating climate change impacts and the effect on future production of different rates of economic and population growth, and of liberalizing the world trade system. Results for these scenarios are described in the following sections.

THE REFERENCE SCENARIO

The reference scenario projected the agricultural system to the year 2060, assuming no climate change and no major changes in the political or economic context of world food trade. It assumed:

- UN medium population estimates (10.2 billion people by 2060) (International Bank for Reconstruction Development/World Bank, 1990).
- 50 percent trade liberalization in agriculture (e.g., removal of import restrictions), introduced gradually by 2020.
- Moderate economic growth (ranging from 3.0 percent/year in 1980–2000 to 1.1 percent/year in 2040–2060).
- Technology projected to increase yields over time (1990–2060). Cereal yields for the world, developing countries, and developed countries were assumed to increase annually by 0.7 percent, 0.9 percent, and 0.6 percent, respectively.

CLIMATE CHANGE SCENARIOS

These are projections of the world system, including effects of climate change on agricultural yields under the GCM scenarios. The food trade simulations for these three scenarios began with 1990 and assumed a linear change in yields until the doubled CO_2 changes would be reached in 2060. Simulations were made both with and without taking into account the physiological effects of 555 ppm CO_2 on crop growth and yield for the equilibrium yield estimates. In these scenarios, internal economic adjustments in the model occur, such as increased agricultural investment, reallocation of agricultural resources according to economic returns (including crop switching), and reclamation of additional arable land as a response to higher cereal prices. These are based on shifts in supply and demand factors that alter the comparative advantage among countries and regions in the world food trade system. These economic adjustments are assumed not to feed back to the yield levels predicted by the crop modeling study.

SCENARIOS INCLUDING THE EFFECTS OF FARM-LEVEL ADAPTATIONS

The food trade model was first run with yield changes assuming no external farm-level adaptation to climate change and was then re-run with different climate-induced changes in yield projected from the two levels of adaptation described above. Policy, cost, and water resource availability were not studied explicitly and were assumed not to be barriers to adaptation. Switching from one enterprise to another based on production and demand factors was included in the BLS.

SCENARIOS OF DIFFERENT FUTURE TRADE POLICIES AND DIFFERENT LEVELS OF ECONOMIC AND POPULATION GROWTH

A final set of scenarios assumed changes to the world tariff structure and different rates of economic and population growth, yielding insight into alternate futures. As with the previous experiments, these were conducted both with and without taking climate change into account. These scenarios included:

- *Trade liberalization*, with full trade liberalization in agriculture being introduced gradually by 2020.
- *Slower rates of economic growth*, ranging from 2.7 percent/year in 1980–2000 to 1.0 percent in 2040–2060. Such low rates would result in a global GDP in 2060 that would be 10.3 percent lower than the reference scenario and would be 11.2 percent lower in developing countries and 9.8 percent lower in developed countries.
- *Low population growth*, following UN low population estimates (8.6 billion people by 2060).

The analysis of trade liberalization in this study was restricted to the removal of distortions between trade prices and domestic prices at the level of the raw materials of the agricultural commodities. Where applicable, trade and production quotas were released. Other types of domestic assistance, e.g., input subsidies, export credit, and insurance, were not included in the analysis. For a given world market price for an agricultural commodity, the domestic price under trade liberalization depends upon whether the country is a net exporter or net importer of the commodity, the differential being a margin for international freight and insurance.

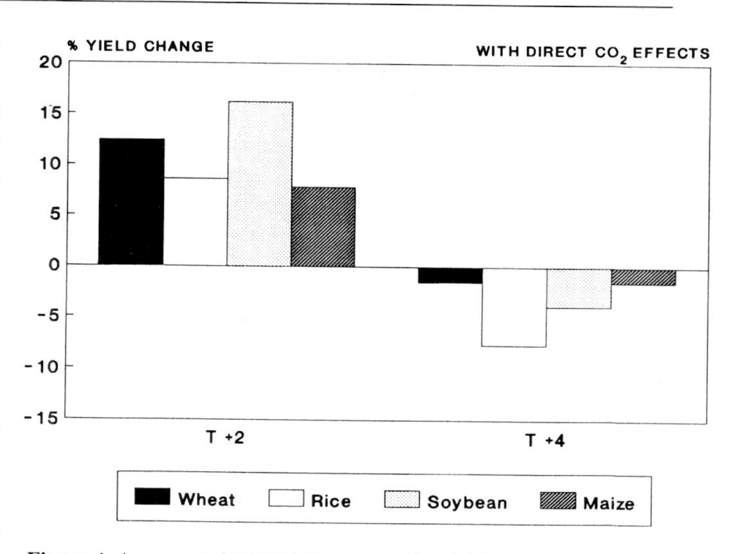

Figure 4 Aggregated IBSNAT crop model yield changes for +2°C and +4°C temperature increase. Country results are weighted by contribution of national production to world production. Direct effects of CO_2 on crop growth and water use are taken into account.

EFFECTS ON CROP YIELDS

Crop yields with arbitrary sensitivity tests

With the direct effects of CO_2 and precipitation held at current levels, average crop yields weighted by national production showed a positive response to +2°C warming and a negative response to +4°C (Figure 4). Wheat and soybean yields were estimated to increase 10–15 percent, and maize and rice yields were estimated to increase about 8 percent, with a +2°C temperature rise. Yields of all four crops turn negative at +4°C, indicating a threshold of the compensation of direct CO_2 effects for temperature increases between 2 and 4°C, as simulated in the IBSNAT crop models. Rice and soybean would be most negatively affected at +4°C. These averaged results, however, mask differences among countries. For example, the effects of latitude are such that in Canada, a +2°C temperature increase with no precipitation change would result in wheat yield increases (with direct effects of CO_2 taken into account), whereas the same changes in Pakistan would result in average wheat yield decreases of about 12 percent. In general, a 20 percent increase in precipitation would improve the simulated yields of the crops tested, and a 20 percent decrease would lower yields of all crops.

Crop yields without adaptation

Table 9 shows modeled wheat yield changes for the GCM doubled CO_2 climate change scenarios (the yield changes include results from both rainfed and irrigated simulations,

Table 9. *Current production and change in simulated wheat yield under GCM 2 × CO_2 climate change scenarios, with and without accounting for the direct effects of CO_2*[1]

	Current production				Change in simulated yields					
					Without CO_2			With CO_2		
Country	Yield (t/ha)	Area (ha × 1000)	Prod. (t × 1000)	% Total	GISS[2] (%)	GFDL[2] (%)	UKMO[2] (%)	GISS[3] (%)	GFDL[3] (%)	UKMO[3] (%)
Australia	1.38	11,546	15,574	3.2	−18	−16	−14	8	11	9
Brazil	1.31	2,788	3,625	0.8	−51	−38	−53	−33	−17	−34
Canada	1.88	11,365	21,412	4.4	−12	−10	−38	27	27	−7
China	2.53	29,092	73,527	15.3	−5	−12	−17	16	8	0
Egypt	3.79	572	2,166	0.4	−36	−28	−54	−31	−26	−51
France	5.93	4,636	27,485	5.7	−12	−28	−23	4	−15	−9
India	1.74	22,876	39,703	8.2	−32	−38	−56	3	−9	−33
Japan	3.25	237	772	0.2	−18	−21	−40	−1	−5	−27
Pakistan	1.73	7,478	12,918	2.7	−57	−29	−73	−19	31	−55
Uruguay	2.15	91	195	0.0	−41	−48	−50	−23	−31	−35
Former USSR										
winter	2.46	18,988	46,959	9.7	−3	−17	−22	29	9	0
spring	1.14	36,647	41,959	8.7	−12	−25	−48	21	3	−25
USA	2.72	26,595	64,390	13.4	−21	−23	−33	−2	−2	−14
WORLD[4]	2.09	231	482	72.7	−16	−22	−33	11	4	−13

Notes:

[1] Results for each country represent the site results weighted according to regional production. The world estimates represent the country results weighted by national production. Sources: see Appendix 2.

[2] GCM 2 × CO_2 climate change scenario alone.

[3] GCM 2 × CO_2 climate change scenario with direct CO_2 effects.

[4] World area and production × 1,000,000.

weighted by current percentage of the respective practice). When the direct physiological effects of CO_2 were not taken into account, climate changes caused decreases in simulated wheat yields in all cases; when accounted for, the direct effects of CO_2 mitigated the negative effects primarily in mid- and high latitudes.

The magnitudes of the estimated yield changes varied by crop (Table 10 and Figure 5, Plate 1). Global wheat yield changes weighted by national production were positive with the direct CO_2 effects in two out of three scenarios, while maize yield was most negatively affected, reflecting its greater production in low-latitude areas where simulated yield decreases were greater. Of all of the crops, maize production declined most with direct CO_2 effects, probably due to its lower response to the physiological effects of CO_2. Simulated soybean yields were most reduced without the direct effects of CO_2 but were least affected in the less severe GISS and GFDL climate change scenarios when direct CO_2 effects were simulated. Soybean responded positively to increased CO_2, but it is the crop most affected by the high temperatures of the UKMO scenario.

The differences among countries in simulated crop yield responses to climate change when the direct effects of CO_2

Table 10. *Changes in simulated wheat, rice, maize, and soybean yield*

	Change in simulated yields[1]					
	Without CO_2			With CO_2		
Crop	GISS[2] (%)	GFDL[2] (%)	UKMO[2] (%)	GISS[3] (%)	GFDL[3] (%)	UKMO[3] (%)
Wheat	−16	−22	−33	11	4	−13
Rice	−24	−25	−25	−2	−4	−5
Maize	−20	−26	−31	−15	−18	−24
Soy	−19	−25	−57	16	5	−33

Notes:

[1] Crop yield changes were obtained by weighting site results first by regional production within countries and then by national contribution to total production simulated in the study. Sources: see Appendix 2.

[2] GCM climate change scenario alone.

[3] GCM climate change scenario with direct effects of CO_2.

were not taken into account were primarily related to differences in current growing conditions. Higher temperatures tended to shorten the growing period at all locations tested. At low latitudes, however, crops are currently grown at higher temperatures, produce lower yields, and are nearer the limits of temperature tolerances for heat and water stress. Warming at low latitudes thus results in accelerated growing periods for crops, more severe heat and water stress, and greater yield decreases than at higher latitudes. In many mid- and high-latitude areas, where current temperature regimes are cooler, increased temperatures, although still shortening grain-filling periods, thus exerting a negative influence on yields, do not significantly increase stress levels. At some sites near the high-latitude boundaries of current agricultural production, increased temperatures can benefit crops otherwise limited by cold temperatures and short growing seasons.[1] The potential for expansion of cultivated land is embedded in the BLS world food trade model and is reflected in shifts in production calculated by that model.

Simulated yield increases in the mid- and high latitudes were caused primarily by:

1. *Positive physiological effects of CO_2.* At sites with cooler initial temperature regimes, increased photosynthesis more than compensated for the shortening of the growing period caused by warming.
2. *Lengthened growing season and amelioration of cold temperature effects on growth.* At some sites near the high-latitude boundaries of current agricultural production, increased temperatures extended the frost-free growing season and provided regimes more conducive to greater crop productivity.

The primary causes of decreases in simulated yields were:

1. *Shortening of the growing period.* Higher temperatures during the growing season speed annual crops through their development (especially at the grain-filling stage), allowing less grain to be produced. This was projected to occur in most crops and sites, one exception being those sites with the coolest growing season temperatures in Canada and the former USSR.
2. *Decrease in water availability.* This factor is the result of a combination of increases in evapotranspiration rates in the warmer climate, enhanced losses of soil moisture, and, in some cases, a projected decrease in precipitation in the climate change scenarios.
3. *Poor vernalization.* Vernalization is the requirement of some temperate cereal crops, e.g., winter wheat, for a period of low winter temperatures to initiate or accelerate the flowering process. Low vernalization results in low flower bud initiation and, ultimately, in reduced yields. Projected decreases in winter wheat yields at some sites in Canada and the former USSR were due to lack of vernalization.

[1] The extent of soil suitable for expanded agricultural production in these regions was not studied explicitly.

Figure 6 (Plate 2) shows estimated potential changes in average national grain crop yields for the GISS, GFDL, and UKMO doubled CO_2 climate change scenarios with and without allowing for the direct effects of CO_2 on plant growth. The maps were created from the nationally averaged yield changes for wheat, rice, coarse grains, and protein feed estimated for the BLS simulations for each country or group of countries in the world food trade model; regional variations within countries are not reflected. Latitudinal differences are apparent for all the scenarios. With direct CO_2 effects, high-latitude changes are less negative or even positive in some cases, whereas lower-latitude regions suffer more detrimental effects of climate change on agricultural yields.

The GISS and GFDL climate change scenarios, with CO_2 effects on crop growth and water use, produced yield changes ranging from +30 to −30 percent. Effects on crop yields under the GISS scenario were, in general, more adverse than under the GFDL scenario for parts of Asia and South America, while effects under the GFDL scenario resulted in more negative yields in the United States and Africa and less positive results in the former USSR. The UKMO climate change scenario, which had the greatest warming (5.2°C global surface air temperature increase), showed average national crop yields to decline almost everywhere (by up to 50 percent in Pakistan) even with beneficial CO_2 effects taken into account.

Crop yields with adaptation

The adaptation studies conducted by the scientists participating in the project suggest that ease of adaptation to climate change is likely to vary with crop, site, and adaptation technique (Table 11). For example, at present, many Mexican producers can afford to use only small doses of nitrogen fertilizer at planting; if more fertilizer becomes available to more farmers, some of the yield reductions under the climate change scenarios might be offset. However, given the current economic and environmental constraints in countries such as Mexico, a future with unlimited water and nutrients is unlikely (Liverman *et al.*, 1992). In contrast, switching from spring to winter wheat at the modeled sites in the former USSR would produce a favorable response (Menzhulin *et al.*, 1992), suggesting that agricultural productivity may be enhanced there with the relatively easy shift to winter wheat varieties.

Yield estimates for the two levels of adaptation developed for the BLS simulations for the GCM scenarios are shown in Figure 7 (Plate 3). As in Figure 6 (Plate 2), results shown are averages for countries and groups of countries, and regional variations within countries are not reflected. Direct CO_2 effects on crop growth and water use are taken into account. Adaptation Level 1, simulating minor changes to existing agricultural systems, compensated for the climate change

Table 11. *Adaptation tests in Mexico and the former USSR*

A. Effect of GCM 2 × CO_2 climate change scenarios on CERES-Maize yields at Tlaltizapan and Poza Rica, Mexico, with and without nutrient stress.

	Simulated yield[1] (t/ha)			
	Nutrient stress[2]		No nutrient stress	
Scenario	Tlaltizapan	Poza Rica	Tlaltizapan	Poza Rica
BASE	4.02	3.18	4.49	3.98
GISS	3.07	2.97	3.77	3.30
GFDL	3.20	2.70	3.47	3.18
UKMO	1.56	2.35	3.93	2.67

B. Effect of GCM 2 × CO_2 climate change scenarios on aggregated spring and winter CERES-Wheat yield in the former USSR.

	Simulated yield change[1] (%)	
Scenario	Spring wheat	Winter wheat
GISS	+21	+41
GFDL	−4	+12
UKMO	−18	+9

Notes:
[1] With direct CO_2 effects.
[2] Actual conditions.

scenarios incompletely, particularly in the developing countries. For the GISS and GFDL scenarios, adaptation implying major changes to current agricultural systems (Adaptation Level 2) compensated almost fully for the negative climate change impacts. With the high level of global warming from the UKMO climate change scenario, neither Level 1 nor Level 2 Adaptation fully overcame the negative climate change effects on crop yields in most countries, even when direct CO_2 effects were taken into account.

EFFECTS OF CLIMATE CHANGE ON FOOD PRODUCTION, FOOD PRICES, AND RISK OF HUNGER

The reference scenario (the future without climate change)

Assuming no effects of climate change on crop yields and current trends in economic and population growth rates, the models predicted world cereal production[2] to be 3,286 million metric tons (mmt) in 2060 (compared to 1,795 mmt in 1990). Per capita cereal production in developed countries would increase from 690 kg per capita in 1980 to 984 kg per capita in 2060. In developing countries (excluding China) cereal production would increase from 179 to 282 kg per capita. Aggregated world per capita cereal production would decrease from 327 kg per capita in 1980 to 319 kg per capita in 2060. The declining aggregate trend for the future is caused by the relatively large difference in per capita cereal production in the developed and developing countries and the demographic changes assumed by the model.

Table 12. *Index of world prices simulated by the Basic Linked System reference case (1970 = 100)*

	Year			
	1980	2000	2020	2060
Cereals	102	125	126	121
Other crops	110	118	110	94
All crops	108	120	115	102
Livestock	105	131	135	119
Agriculture	107	123	121	107

Cereal prices were estimated at an index of 121 (1970 = 100) for the year 2060, reversing the trend of falling real cereal prices over the last 100 years (Table 12). This increase occurred because the BLS standard reference scenario had two phases of price development. During 1980 to 2020, while trade barriers and protection are still in place but are being reduced, there would be increases in relative prices; price decreases would follow when trade barriers are removed. The number of hungry people was estimated at about 640 million, or about 6 percent of total population in 2060 (compared to 530 million in 1990, about 10 percent of the total current population).

Effects of climate change with and without adjustments in the economic system

The BLS included the ability to simulate adjustments that the world food system might make to changes of yield (e.g., reallocation of agricultural land use, change in fertilizer use, and application of irrigation water). Simulations of the effects of climate change without such internal adjustments were of theoretical interest only as these would unrealistically imply no economic or behavioral response of producers and consumers. However, these hypothetical impacts help to define the adjustments taking place in the system over time. Under these conditions the effects of climate change and increased atmospheric CO_2 on crop yields derived from the GCM scenarios imply a 5 percent to an almost 20 percent reduction in total cereal production (Table 13). These esti-

[2] The estimate for cereals includes wheat, rice, maize, millet, sorghum, and minor grains (FAO, 1991); rice is included as rice milled equivalent (a factor of 0.67 is used to convert from rice paddy to milled rice).

Table 13. *Resilience[1] of the current food system to GCM climate change scenarios*

Cereal production[2]	GISS (%)	GFDL (%)	UKMO (%)
Without adjustment	−5.3	−8.5	−18.5
With adjustment	−1.2	−2.8	−7.6

Notes:

[1] Effect on world cereal production (percent change from reference) assuming economic adjustments/no economic adjustments (e.g., in land area, fertilizer use, and irrigation) simulated by the Basic Linked System.

[2] With direct CO_2 effects.

mates are changes to production levels projected for 2060 without climate change.

Adjustments within the economic system without the adaptation discussed above would tend to counteract negative yield impacts as agricultural production shifts to regions of more favorable comparative advantage. Such economic adjustment includes expansion of production in favorable areas and reduction of output in areas that become less productive. The BLS offset 65–80 percent of the potential impact on yield in scenarios for impacts below 10 percent of global cereal production (the GISS and GFDL climate change scenarios). The offset decreased to 60 percent under a scenario of greater yield reduction (UKMO).

Table 14. *Changes in cereal production, cereal prices, and number of people at risk of hunger in 2060 under GCM $2 \times CO_2$ climate change scenarios*

A. Change in cereal production

Region	Reference scenario[1]	GISS (%)	GFDL (%)	UKMO (%)
Global	3286	−1.2	−2.8	−7.6
Developed	1449	11.3	5.2	−3.5
Developing[2]	1836	−11.0	−9.2	−10.8

B. Change in cereal price index (1970 = 100)

	Reference scenario[1]	GISS (%)	GFDL (%)	UKMO (%)
Cereal prices	121	24	33	145

C. Change in number of people at risk of hunger

Region	Reference scenario[1]	GISS (%)	GFDL (%)	UKMO (%)
Developing[2]	641	10	17	58

Notes:

[1] Reference scenario is for 2060 assuming no climate change.

[2] Estimates for developing countries do not include China.

Effects of climate change with economic adjustment, but without farm-level adaptation

Changes in cereal production, cereal prices, and number of people at risk of hunger estimated for the GCM doubled CO_2 climate change scenarios (with direct CO_2 effects taken into account) are given in Table 14. These estimations are based upon dynamic simulations by the BLS in which the world food system was allowed to respond to climate-induced supply shortfalls of cereals and consequently higher commodity prices through increases in production factors (cultivated land, labor, and capital) and inputs such as fertilizer. The testing of climate change impacts without farm-level adaptation is unrealistic but was done for the purpose of establishing a baseline with which to compare the effects of farmer response.

Under the GISS scenario (which provides lower temperature increases) cereal production was estimated to decrease by just over 1 percent, while under the UKMO scenario (with the highest temperature increases) global production was estimated to decrease by more than 7 percent. The largest negative changes would occur in developing regions which average −9 percent to −11 percent, though the extent of decreased production would vary greatly by country depending on the projected climate. By contrast, production in developed countries was estimated to increase under all but the UKMO scenario (+11 percent to −3 percent). Thus, disparities in crop production between developed and developing countries were estimated to increase.

Price increases resulting from climate-induced decreases in yield were estimated to range between ~25 and 150 percent. In the case of the GISS scenario, the disequilibrium caused by the 5.3 percent reduction in yields of the unadjusted scenario would be resolved via market mechanisms in the adjusted case. This would result in a −1.2 percent consumer response and about a +4 percent (relative) producer response and would lead to 24 percent higher relative prices for cereals. Although this price response seems to be high, cereal prices account for only a modest fraction, perhaps 1/3 or less, of retail food prices. Hence, a 24 percent increase in world cereal prices does not imply a 24 percent increase in food prices.

These increases in price would likely affect the number of people with insufficient resources to purchase adequate amounts of food. The estimated number of hungry people increased approximately 1 percent for each 2–2.5 percent increase in prices (depending on climate change scenario). The number of people at risk of hunger increased by 10 percent to almost 60 percent in the climate change scenarios

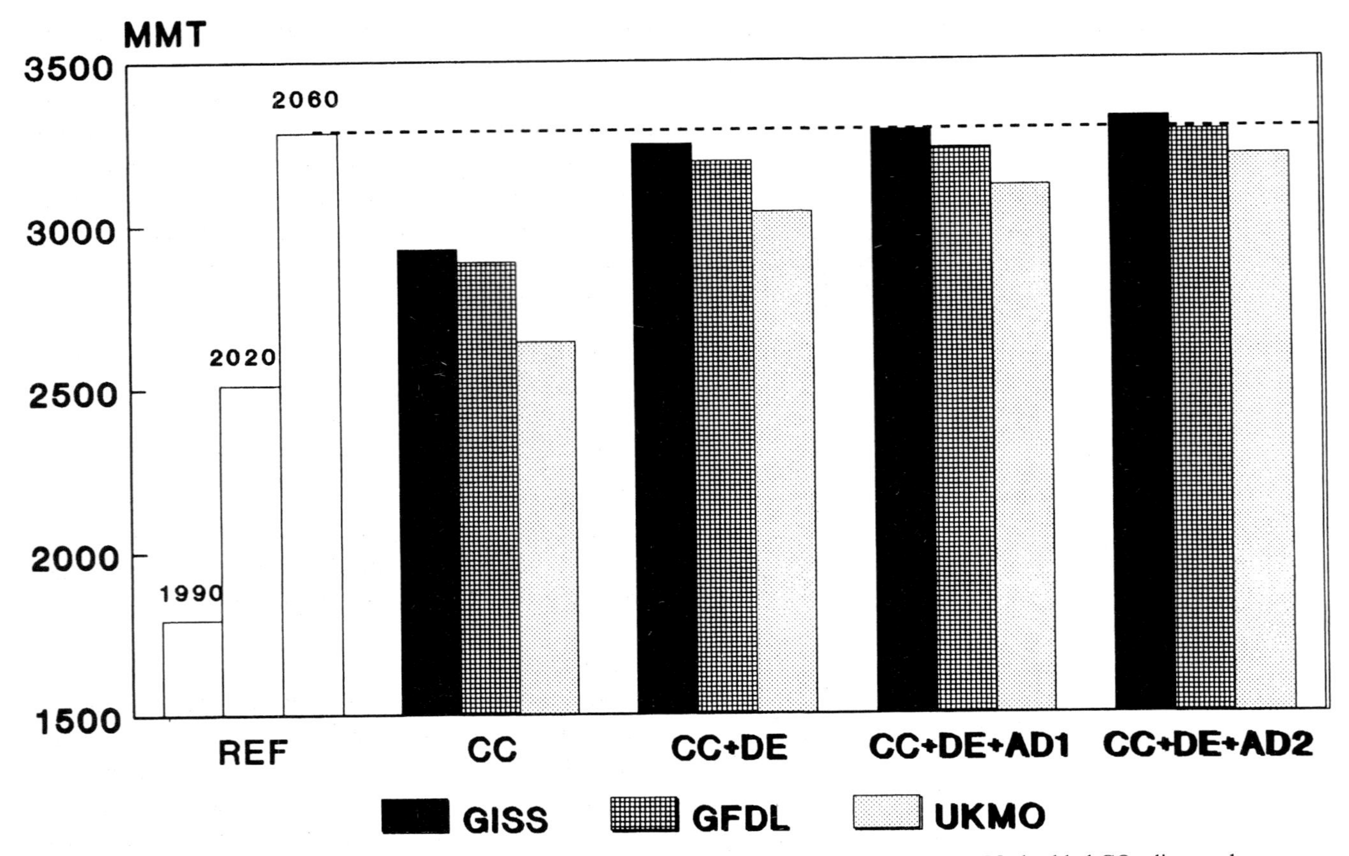

Figure 8 World cereal production projected by the BLS for the reference, GISS, GFDL, and UKMO doubled CO_2 climate change scenarios, with (CC + DE) and without (CC) direct CO_2 effects, and with Adaptation Levels 1 and 2 (AD1 and AD2). Adaptation Level 1 implies minor changes to existing agricultural systems; Adaptation Level 2 implies major changes.

tested, resulting in an estimated increase of between 60 and 350 million people in this condition (above the reference case of 640 million) by 2060.

Effects of climate change under different levels of farmer adaptation

Globally, both minor and major levels of adaptation were projected to help restore world production levels, when compared to the climate change scenarios with no adaptation (Figure 8). Growth in production under the reference scenario is far greater than average negative impacts of climate change. Averaged global cereal production would decrease by up to about 160 mmt (0 percent to − 5 percent) from the reference case of 3,286 mmt with Level 1 adaptations. These involved shifts in farm activities that are not very disruptive to regional agricultural systems. With adaptations implying major changes, global cereal production responses ranged from an additional 30 mmt, a slight increase, to a slight decrease of about 80 mmt (+ 1 percent to − 2.5 percent).

Level 1 adaptations would improve the comparative advantage of developed countries in world markets (Figure 9). In these regions cereal production was projected to increase by 4 percent to 14 percent over the reference case. However, the competitive positions of developing countries was estimated to benefit little from this level of adaptation. More extensive adaptation (Level 2) would virtually eliminate the global negative cereal yield impacts derived under the GISS and GFDL climate scenarios and would reduce impacts under the UKMO scenario to one third.

Figure 10 shows the estimated effects of climate change alone and of climate change with both levels of adaptation on cereal prices in 2060. As a consequence of climate change, world cereal prices were estimated to increase by about 25 percent to almost 150 percent. Under Adaptation Level 1, price increases range from 10 percent to 100 percent; under Adaptation Level 2, cereal price responses ranged from a decline of about 5 percent to an increase of 35 percent. When quantity of output declines, prices increase significantly. When quantity of output increases, prices drop.

As a consequence of climate change and Level 1 adaptations, the number of people at risk from hunger would increase by about 40 to 300 million (6 percent to 50 percent) from the reference case of 641 million (Figure 11). With more significant farmer adaptation (Level 2), the change in the

CHANGE IN CEREAL PRODUCTION IN 2060

% CHANGE WITH CO2 DIRECT EFFECTS

15 12 9 6 3 0 -3 -6 -9 -12 -15

GISS GFDL UKMO

GLOBAL DEVELOPED DEVELOPING

ADAPTATION 1

% CHANGE WITH CO2 DIRECT EFFECTS

15 12 9 6 3 0 -3 -6 -9 -12 -15

GISS GFDL UKMO

GLOBAL DEVELOPED DEVELOPING

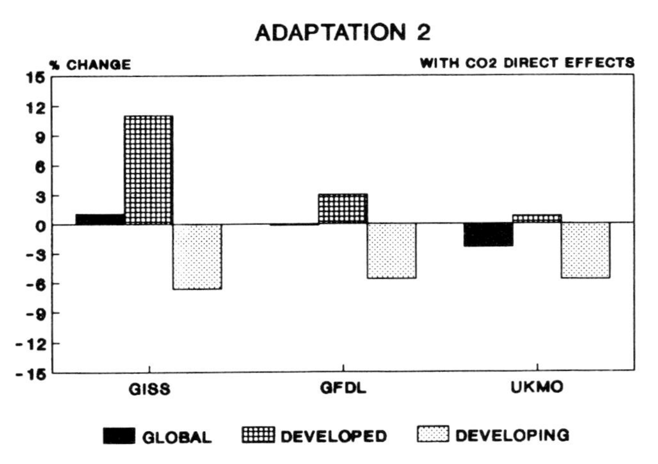

Figure 9 Change in world, aggregated developed country and developing country cereal production projected by the BLS under climate change scenarios in 2060 for no adaptation and Adaptation Levels 1 and 2. Reference scenario for 2060 assumed no climate change (global 3286 mmt, developed 1449 mmt, developing 1836 mmt).

number of people at risk from hunger ranged from −12 million for the GISS scenario to 120 million for the UKMO scenario (−2 percent and +20 percent). These results indicate that, except for the GISS scenario under Adaptation Level 2, the simulated farm-level adaptations did not entirely mitigate the negative effects of climate change on potential risk of hunger, even when economic adjustments, i.e., the production and price responses of the world food system, were taken into account.

Effects of climate change assuming full trade liberalization, lower economic growth rates, and lower population growth rates

For each of these alternate future assumptions, a new reference scenario was established with the BLS and then tested with the GCM climate change scenarios.

Full trade liberalization

Assuming full agricultural trade liberalization and no climate change by 2020 resulted in a projection of more efficient resource use. This led to a 3.2 percent higher value added in agriculture globally and 5.2 percent higher agricultural GDP in developing countries (excluding China) by 2060 compared to the original reference scenario. This policy change would result in almost 20 percent fewer people at risk of hunger. Global cereal production would increase by 70 mmt, with most of the production increases occurring in developing countries (Table 15).

Climate change impacts were then simulated under these new reference conditions. Under the same trade liberalization policies, global impacts due to climate change would be slightly reduced, with enhanced gains in production accruing to developed countries. Losses in production would be greater in developing countries. Price increases were reduced slightly from what would occur without full trade liberalization, and the number of people at risk of hunger was also reduced.

Reduced rate of economic growth

Estimates were also made of impacts under a lower economic growth scenario (10 percent lower than reference). These are indicated in Table 16. Lower economic growth results in a tighter supply situation, higher prices, and more people below the hunger threshold.

The effect of climate change on these trends is generally to reduce production, increase prices, and increase the number of people at risk from hunger. Developed countries increase cereal production in the GISS and GFDL scenarios even with the projected lower economic growth rates, but developing countries decrease production under all climate change scenarios.

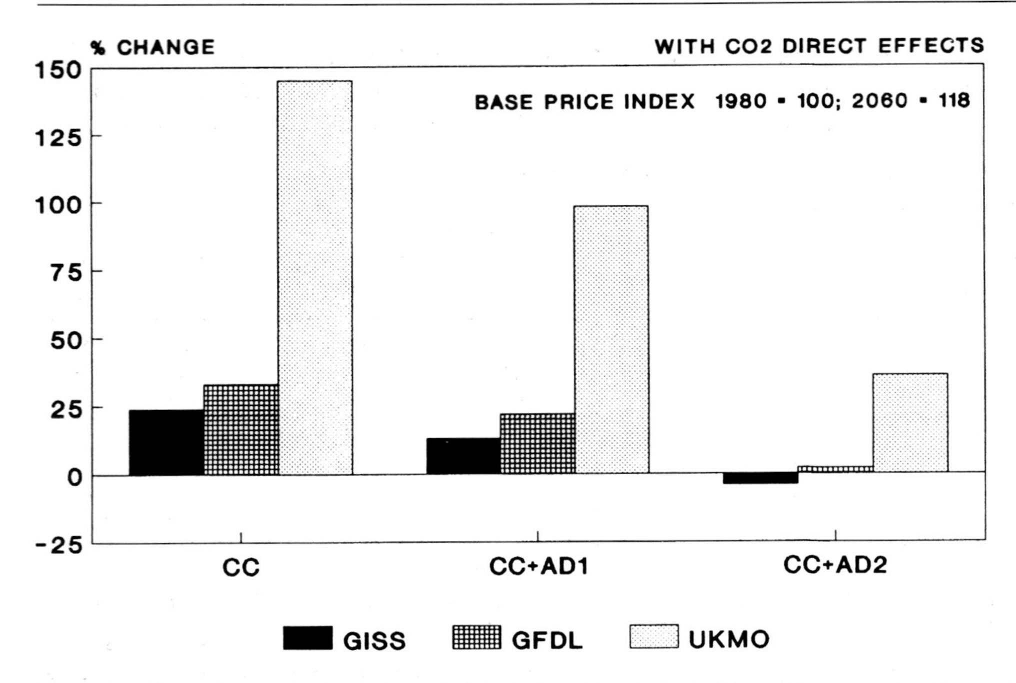

Figure 10 Change in cereal price index in 2060 calculated by the Basic Linked System under climate change scenarios for no adaptation and Adaptation Levels 1 and 2 (AD1 and AD2). Reference scenario for 2060 assumed no climate change (price index is 18 percent above 1980 levels).

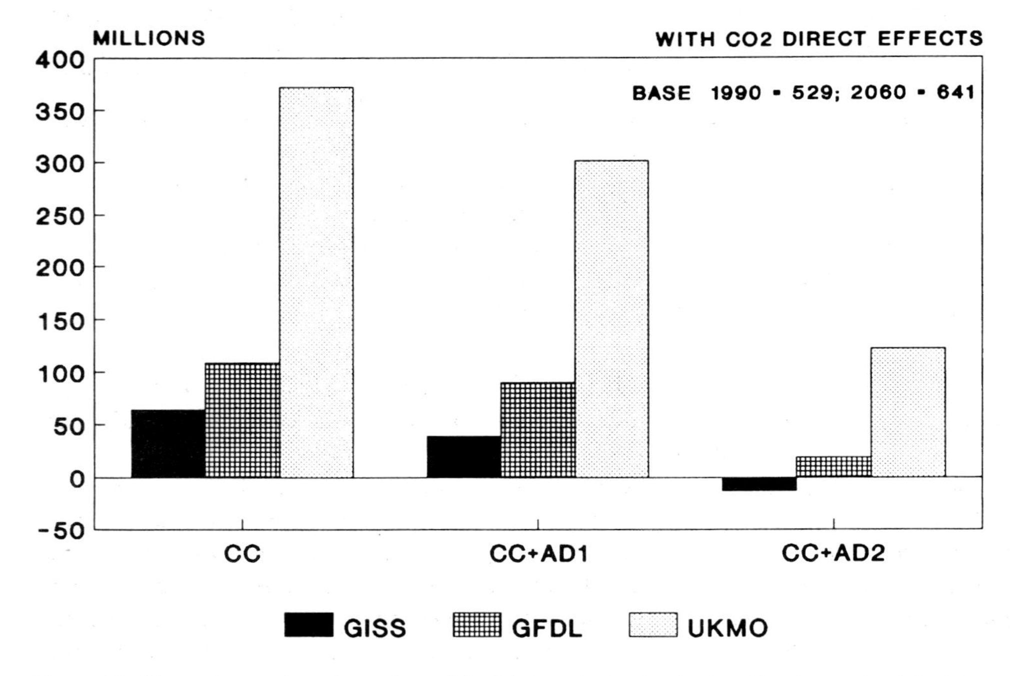

Figure 11 Change in number of people at risk of hunger in 2060 calculated by the Basic Linked System under climate change scenarios for no adaptation and Adaptation Levels 1 and 2 (AD1 and AD2). Reference scenario for 2060 assumes no climate change (529 million people are at risk of hunger in 1990; 641 million people are at risk of hunger in 2060).

Table 15. *Change in cereal production, price index, and the number of people at risk of hunger in 2060 assuming full trade liberalization and GCM climate change scenarios*

A. Cereal production (relative to Ref-FTL[1])

	Reference scenario[2] (mmt)	Ref-FTL (mmt)	GISS (mmt)	GFDL (mmt)	UKMO (mmt)
Global	3286	3356	−29	−87	−274
Developed	1449	1472	184	96	−55
Developing	1836	1884	−213	−183	−219

B. Change in cereal price index (% of Ref-FTL; 1970 = 100)

	Reference scenario	Ref-FTL	GISS	GFDL	UKMO
Cereal prices	121	153	19	30	135

C. Change in number of people at risk of hunger in 2060 (relative to Ref-FTL)

	Reference scenario (m)	Ref-FTL (m)	GISS (m)	GFDL (m)	UKMO (m)
Global[3]	641	532	53	98	378

Notes:

[1] Relative to Ref-FTL, reference scenario with full trade liberalization and no climate change.

[2] Reference scenario assumes no climate change.

[3] Entire increase in number of people at risk of hunger is in developing countries (excluding China).

Table 16. *Change in cereal production, price index, and the number of people at risk of hunger in 2060 assuming a low rate of economic growth and GCM climate change scenarios*

A. Cereal production (relative to Ref-E[1])

	Reference scenario[2] (mmt)	Ref-E (mmt)	GISS (mmt)	GFDL (mmt)	UKMO (mmt)
Global	3286	3212	−31	−87	−253
Developed	1449	1428	177	86	−51
Developing	1836	1786	−208	−173	−202

B. Change in cereal price index (% of Ref-E; 1970 = 100)

	Reference scenario	Ref-E	GISS	GFDL	UKMO
Cereal prices	121	137	21	30	139

C. Change in number of people at risk of hunger in 2060 (relative to Ref-E)

	Reference scenario (m)	Ref-E (m)	GISS (m)	GFDL (m)	UKMO (m)
Global[3]	641	757	63	119	412

Notes:

[1] Relative to Ref-E, reference scenario with a low rate of economic growth and no climate change.

[2] Reference scenario assumes no climate change.

[3] Entire increase in number of people at risk of hunger is in developing countries.

Altered rates of population growth

Lower population growth was shown to have a significant effect on cereal production, food prices, and number of hungry people (Table 17). Simulations based on rates of population growth according to UN low estimates resulted in a world population about 17 percent lower in year 2060 when compared to UN mid-estimates used in the reference run. The corresponding reduction in the developing countries (excluding China) would be about 19.5 percent, from 7.3 billion to 5.9 billion. The combination of higher GDP/capita (about 10 percent) and lower world population produced an estimated 40 percent fewer hungry people in the year 2060 compared to the reference scenario.

Even under the most adverse of the three climate scenarios (UKMO) the estimated number of hungry people was some 10 percent lower than the number estimated under the reference scenario without any climate change. Increases in world prices for agricultural products – in particular, cereals – under the climate change scenarios employing the low population projection were around 75 percent of those projected using the UN mid-estimate.

Figure 12 summarizes the generalized relative effects of different policies of trade liberalization, economic growth, and population growth on the production of cereals and people at risk of hunger. Alternative development assumptions made little difference with respect to the geopolitical patterns of the relative effects of climate change. In all cases, cereal production decreased, particularly in the developing world, while prices and population at risk from hunger increased due to climate change. The beneficial effects of trade liberalization and low population growth were of the same or an even greater (in the case of population) order of magnitude as the adverse effects of climate change. This suggests that there may be much to be gained from altering the conditions of trade and development as a strategy for addressing the climate change issue. The magnitude of

Table 17. *Changes in cereal production, price index, and the number of people at risk of hunger in 2060 assuming UN Low Estimate of Population growth and GCM climate change scenarios*

A. Cereal production (relative to Ref-P[1])

	Reference scenario[2] (mmt)	Ref-P (mmt)	GISS (mmt)	GFDL (mmt)	UKMO (mmt)
Global	3286	2929	−20	−76	−208
Developed	1449	1349	139	65	−52
Developing	1836	1582	−159	−141	−157

B. Change in cereal price index (% of Ref-P; 1970 = 100)

	Reference scenario	Ref-P	GISS	GFDL	UKMO
Cereal prices	121	92	19	28	116

C. Change in number of people at risk of hunger in 2060 (relative to Ref-P)

	Reference scenario (m)	Ref-P (m)	GISS (m)	GFDL (m)	UKMO (m)
Global[3]	641	395	18	50	183

Notes:
[1] Relative to Ref-P, reference scenario with UN Low Estimate of Population growth and no climate change.
[2] Reference scenario assumes no climate change.
[3] Entire increase in number of people at risk of hunger is in developing countries.

adverse climate impacts was lowest, however, under the conditions of low population growth. An assumption of low population growth rate minimized the population at risk of hunger in both the presence and absence of climate change in the BLS simulations.

CONCLUSIONS

Climate change induced by increasing greenhouse gases is likely to affect crop yields differently from region to region across the globe. Under the climate change scenarios adopted in this study, the effects on crop yields in mid- and high-latitude regions appeared to be less adverse than those in low-latitude regions. However, the more favorable effects on yield in temperate regions depended to a large extent on full realization of the potentially beneficial direct effects of CO_2 on crop growth. Decreases in potential crop yields are likely to be caused by shortening of the crop growing period, decrease in water availability due to higher rates of evapotranspiration, and poor vernalization of temperate cereal crops. When adaptations at the farm level were tested (e.g., change in planting date, switch of crop variety, changes in fertilizer application and irrigation), compensation for the detrimental effects of climate change was found to be more successful in developed countries.

When the economic implications of these changes in crop yields were explored in a world food trade model, the relative ability of the world food system to absorb impacts decreased with the magnitude of the impact. Regional differences in effects remained noticeable, and developed countries are expected to be less affected by climate change than developing economies. Dynamic economic adjustments can compensate for lower impact scenarios, such as the GISS and GFDL climate scenarios, but not higher impact ones, such as the UKMO scenario. Prices of agricultural products were found to be related to the magnitude of the climate change impact, and incidence of food poverty increased in all but one of the climate change scenarios tested.

When the effects of lower future population, lower economic growth rates, and partial trade liberalization were tested in the food trade model, reduced population growth rates were projected to have the largest effect on minimizing the impact of climate change. Lower economic growth resulted in tighter food supplies, and consequently resulted in higher rates of food poverty. Full trade liberalization in agriculture provided for more efficient resource use and reduced the number of people at risk of hunger by about 100 million (from the reference case of about 640 million in 2060). However, all of the scenarios of future climate adopted in this part of the study exacerbated estimates of the number of people at risk of hunger.

It should be emphasized that the results reported here are not a forecast of the future. There are very large uncertainties that preclude making forecasts, particularly the lack of information on possible climate change at the regional level, on the effects of technological change on agricultural productivity, on trends in demand (including population growth), and on the wide array of possible adaptations. The adoption of efficient adaptation techniques is far from certain. In developing countries, there may be social or technical constraints, and adaptive measures may not necessarily result in sustainable production over long time-frames. The availability of water supplies for irrigation and the costs of adaptation are both critical needs for further research.

Future trace gas emission rates, as well as when the full magnitude of their effects will be realized, are not certain, and only a limited range of GCM climate change scenarios, representing the upper end of the projected warming, was tested. However, it can be argued that the use of scenarios from the higher GCM projections provides perspective on

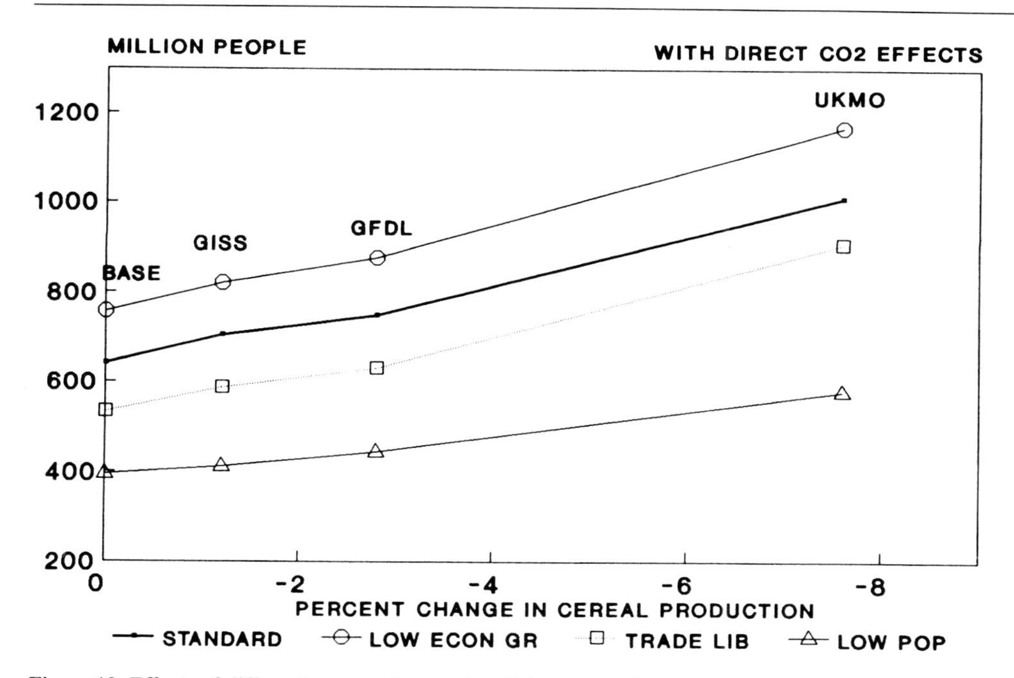

Figure 12 Effects of different assumptions and policies on number of people at risk from hunger calculated by the Basic Linked System under the climate change scenarios.

the downside risk of global warming projections. Because of these uncertainties, the study should be considered as an exploratory assessment of the sensitivity of the world food system to a limited number of what is, in effect, a much wider array of possible futures.

Determining how countries, particularly developing countries, can and will respond to reduced yields and increased costs of food is a critical research need arising from this study. Will such countries be able to import large amounts of food? From a political and social standpoint, these results show a decrease in food security in developing countries. The study suggests that the worst situation arises from a scenario of severe climate change, low economic growth, and little farm-level adaptation. In order to minimize possible adverse consequences – production losses, food price increases, and more people at risk of hunger – the way forward is to encourage the agricultural sector to continue to develop crop breeding and management programs for heat and drought conditions (measures that will be useful even today in improving productivity in marginal environments) and to encourage the nations of the world to take measures to slow population growth. The latter step would also be consistent with efforts to slow emissions of greenhouse gases, the source of the problem, and thus the rate and eventual magnitude of global climate change.

ACKNOWLEDGMENTS

This project was funded by the U.S. Environmental Protection Agency, Climate Change Division, with additional support from the U.S. Agency for International Development. Dr. Roy Jenne provided the GCM climate change scenarios. Ana Iglesias developed crop model aggregation techniques. Richard Goldberg provided technical assistance, and Christopher Shashkin provided project support. We are grateful to Drs. Richard Adams, Timothy Carter, William Cline, John Hayes, John Reilly, and Richard Warrick for their constructive reviews.

REFERENCES

Acock, B., and L. H. Allen, Jr. 1985. Crop responses to elevated carbon dioxide concentrations. In *Direct Effects of Increasing Carbon Dioxide on Vegetation*, eds. B. R. Strain and J. D. Cure, 33–97. DOE/ER-0238. Washington, DC: U.S. Department of Energy.

Adams, R. M., C. Rosenzweig, R. M. Peart, J. T. Ritchie, B. A. McCarl, J. D. Glyer, R. B. Curry, J. W. Jones, K. J. Boote, and L. H. Allen, Jr. 1990. Global climate change and U.S. agriculture. *Nature* 345(6272):219–22.

Allen, L. H., Jr., K. J. Boote, J. W. Jones, P. H. Jones, R. R. Valle, B. Acock, H. H. Rogers, and R. C. Dahlman. 1987. Response of vegetation to rising carbon dioxide: Photosynthesis, biomass and seed yield of soybean. *Global Biogeochemical Cycles* 1:1–14.

Cure, J. D. 1985. Carbon dioxide doubling responses: A crop survey. In *Direct Effects of Increasing Carbon Dioxide on Vegetation*, eds. B. R. Strain and J. D. Cure, 99–116. DOE/ER-0238. Washington, DC: U.S. Department of Energy.

Cure, J. D., and B. Acock. 1986. Crop responses to carbon dioxide doubling: A literature survey. *Ag. and For. Meteor.* 38:127–45.

Fischer, G., K. Frohberg, M. A. Keyzer, and K. S. Parikh. 1988. *Linked National Models: A Tool for International Food Policy Analysis*. Dordrecht, Netherlands: Kluwer.

Fischer, G., K. Frohberg, M. A. Keyzer, K. S. Parikh, and W. Tims. 1990. *Hunger – Beyond the Reach of the Invisible Hand*. Laxenburg, Austria: International Institute for Applied Systems Analysis, Food and Agriculture Project.

Food and Agriculture Organization (FAO). 1984. *Fourth World Food Survey*. Rome: United Nations FAO.

Food and Agriculture Organization (FAO). 1987. *Fifth World Food Survey*. Rome: United Nations FAO.
Food and Agriculture Organization (FAO). 1988. *1987 Production Yearbook*. Statistics Series No. 82. Rome: United Nations FAO.
Food and Agriculture Organization (FAO). 1991. *AGROSTAT/PC*. Rome: United Nations FAO.
Godwin, D., J. T. Ritchie, U. Singh, and L. Hunt. 1989. *A User's Guide to CERES-Rice – V2.10*. Muscle Shoals, Alabama: International Fertilizer Development Center.
Godwin, D., U. Singh, J. T. Ritchie, and E. C. Alocilja. 1993. *A User's Guide to CERES-Rice*. Muscle Shoals, Alabama: International Fertilizer Development Center.
Hansen, J., G. Russell, D. Rind, P. Stone, A. Lacis, S. Lebedeff, R. Ruedy, and L. Travis. 1983. Efficient three-dimensional global models for climate studies: Models I and II. *Monthly Weather Review* 111(4):609–62.
Hansen, J., I. Fung, A. Lacis, D. Rind, G. Russell, S. Lebedeff, R. Ruedy, and P. Stone. 1988. Global climate changes as forecast by the GISS 3-D model. *Journal of Geophysical Research* 93(D8):9341–64.
Hendry, G. R. 1993. *FACE: Free-Air CO_2 Enrichment for Plant Research in the Field*, ed. C. K. Smoley. Boca Raton, Florida: CRC Press.
International Bank for Reconstruction and Development/World Bank. 1990. *World Population Projections*. Baltimore, Maryland: Johns Hopkins University Press.
International Benchmark Sites Network for Agrotechnology Transfer (IBSNAT). 1989. *Decision Support System for Agrotechnology Transfer Version 2.1 (DSSAT V2.1)*. Honolulu: Dept. of Agronomy and Soil Science, College of Tropical Agriculture and Human Resources, University of Hawaii.
International Panel on Climate Change (IPCC). 1990a. *Climate Change: The IPCC Scientific Assessment*, eds. J. T. Houghton, G. J. Jenkins, and J. J. Ephraums. Cambridge: Cambridge University Press.
International Panel on Climate Change (IPCC). 1990b. *Climate Change: The IPCC Impacts Assessment*, eds. W. J. McG. Tegart, G. W. Sheldon, and D. C. Griffiths. Canberra: Australian Government Publishing Service.
International Panel on Climate Change (IPCC). 1992. *Climate Change 1992*. The Supplementary Report to the IPCC Scientific Assessment, eds. J. T. Houghton, B. A. Callander, and S. K. Varney. Cambridge: Cambridge University Press.
Jones, C. A., and J. R. Kiniry. 1986. *CERES-Maize: A Simulation Model of Maize Growth and Development*. College Station: Texas A&M Press.
Jones, J. W., K. J. Boote, G. Hoogenboom, S. S. Jagtap, and G. G. Wilkerson. 1989. *SOYGRO V5.42: Soybean Crop Growth Simulation Model. Users' Guide*. Gainesville: Department of Agricultural Engineering and Department of Agronomy, University of Florida.
Kane, S., J. Reilly, and J. Tobey. 1992. An empirical study of the economic effects of climate change on world agriculture. *Climatic Change* 21:17–35.
Kimball, B. A. 1983. Carbon dioxide and agricultural yield. An assemblage and analysis of 430 prior observations. *Agronomy Journal* 75:779–88.
Leemans, R., and A. M. Solomon. 1993. Modeling the potential change in yield and distribution of the earth's crops under a warmed climate. *Climate Research* 3:79–96.
Liverman, D., M. Dilley, K. O'Brien, and L. Menchaca. 1994. Possible impacts of climate change on maize yields in Mexico, in Rosenzweig, C. and A. Iglesias (eds.). *Implications of Climate Change for International Agriculture: Crop Modeling Study*. U.S. Environmental Protection Agency. EPA 230-B-94-003. Washington, DC.
Magalhaes, A. R. 1992. *Impacts of Climatic Variations and Sustainable Development in Semi-arid Regions*. Proceedings of International Conference. Fortaleza, Brazil: ICID.
Manabe, S., and R. T. Wetherald. 1987. Large-scale changes in soil wetness induced by an increase in CO_2. *Journal of Atmospheric Science* 44:1211–35.
Menzhulin, G. V., L. A. Koval, and A. L. Badenko, 1994. Potential effects of global warming and carbon dioxide on wheat production in the former Soviet Union, in Rosenzweig, C. and A. Iglesias (eds.). *Implications of Climate Change for International Agriculture: Crop Modeling Study*. U.S. Environmental Protection Agency. EPA 230-B-94-003. Washington, DC.
Otter-Nacke, S., D. C. Godwin, and J. T. Ritchie. 1986. *Testing and Validating the CERES-Wheat Model in Diverse Environments*. AgGRISTARS YM-15-00407. Houston: Johnson Space Center No. 20244.
Parry, M. L. 1990. *Climate Change and World Agriculture*. London: Earthscan.
Parry, M. L., T. R. Carter, and N. T. Konijn, eds. 1988a. *The Impact of Climatic Variations on Agriculture, Volume 1: Assessments in Cool Temperate and Cold Regions*. Dordrecht, Netherlands: Kluwer.
Parry, M. L., T. R. Carter, and N. T. Konijn, eds. 1988b. *The Impact of Climatic Variations on Agriculture, Volume 2: Assessments in Semi-Arid Regions*. Dordrecht, Netherlands: Kluwer.
Parry, M. L., M. B. de Rozari, A. L. Chong, and S. Panich, eds. 1992. *The Potential Socio-Economic Effects of Climate Change in South-East Asia*. Nairobi: UN Environment Programme.
Peart, R. M., J. W. Jones, R. B. Curry, K. Boote, and L. H. Allen, Jr. 1989. Impact of climate change on crop yield in the Southeastern U.S.A. *The Potential Effects of Global Climate Change on the United States*, eds. J. B. Smith and D. A. Tirpak. EPA-230-05-89-050. Washington, DC: U.S. Environmental Protection Agency.
Ritchie, J. T., and S. Otter. 1985. Description and performance of CERES-Wheat: A user-oriented wheat yield model. In *ARS Wheat Yield Project*, ed. W. O. Willis, 159–75. ARS-38. Washington, DC: Department of Agriculture, Agricultural Research Service.
Ritchie, J. T., U. Singh, D. Godwin, and L. Hunt. 1989. *A User's Guide to CERES-Maize – V2.10*. Muscle Shoals, Alabama: International Fertilizer Development Center.
Rogers, H. H., G. E. Bingham, J. D. Cure, J. M. Smith, and K. A. Surano. 1983. Responses of selected plant species to elevated carbon dioxide in the field. *Journal of Environmental Quality* 12:569–74.
Rosenberg, N. J., and P. R. Crosson. 1991. *Processes for Identifying Regional Influences of and Responses to Increasing Atmospheric CO_2 and Climate Change: The MINK Project. An Overview*. Resources for the Future. DOE/RL/01830T-H5. Washington, DC: Dept. of Energy.
Rosenzweig, C., and A. Iglesias, eds. 1994. *Implications of Climate Change for International Agriculture: Crop Modeling Study*. EPA-230-B-94-003. Washington, DC: U.S. Environmental Protection Agency.
Rosenzweig, C., and M. L. Parry. 1994. Potential impact of climate change on world food supply. *Nature* 367:133–8.
Smit, B. 1989. Climatic warming and Canada's comparative position in agricultural production and trade. In *Climate Change Digest*, 1–9. CCD 89–01. Environment Canada.
Smith, J. B., and D. A. Tirpak, eds. 1989. *The Potential Effects of Global Climate Change on the United States*. Report to Congress. EPA-230-05-89-050. Washington, DC: U.S. Environmental Protection Agency.
Thompson, L. M. 1969. Weather and technology in the production of corn in the U.S. corn belt. *Agronomy Journal* 61:453–6.
Wilson, C. A., and J. F. B. Mitchell. 1987. A doubled CO_2 climate sensitivity experiment with a global climate model including a simple ocean. *Journal of Geophysical Research* 92(13):315–43.
World Food Institute. 1988. *World Food Trade and U.S. Agriculture, 1960–1987*. Ames: Iowa State University.

APPENDIX 1. PROJECT PARTICIPANTS —

The study involved a large number of scientists in a broad range of countries. Key participants are listed below.

Argentina
Dr. Osvaldo Sala
University of Buenos Aires

Australia
Dr. Wayne S. Meyer
CSIRO
Dr. Doug Godwin
'Alton Park'

Austria
Dr. Günther Fischer
IIASA

Bangladesh
Dr. Z. Karim
Mr. M. Ahmed
Bangladesh Agricultural Research Council

Brazil
Dr. O. J. F. de Siqueira
CNPT/EMBRAPA

Canada
Dr. Michael Brklacich
Agriculture Canada

China
Dr. Zhiqing Jin
Jiangsu Academy of Agricultural Sciences

Egypt
Dr. H. M. Eid
Ministry of Agriculture

France
Dr. Richard Delécolle
INRA

Germany
Dr. Klaus Frohberg
University of Bonn

India
Dr. D. Gangadhar Rao
Central Research Inst. for Dryland Agriculture

Japan
Dr. Hiroshi Seino
National Inst. of Agro-Environmental Sciences

Mexico
Dr. Diana Liverman
Pennsylvania State University
Dr. Leticia Menchaca
National University of Mexico

Pakistan
Dr. Ata Qureshi
Climate Institute, Washington, DC

Philippines
Dr. Crisanto R. Escaño
Mr. Leandro Buendia
Council for Agr., For. and Natural Resources

Russia
Dr. Gennady V. Menzhulin
Dr. Larisa A. Koval
State Hydrological Institute

Spain
Dr. Ana Iglesias
CIT-INIA

Taiwan
Dr. Tien-Yin Chou
Feng Chai University

Thailand
Dr. M. L. C. Tongyai
Ministry of Agriculture & Co-operatives

UK
Dr. Martin L. Parry
University of Oxford

Uruguay
Dr. Walter E. Baethgen
International Fertilizer Development Center

USA
Dr. Cynthia Rosenzweig
Columbia University/GISS
Mr. Brian Baer
Michigan State University
Dr. Bruce Curry
University of Florida
Dr. Gerrit Hoogenboom
University of Georgia
Dr. Roy Jenne
National Center for Atmospheric Research
Dr. James W. Jones
University of Florida
Dr. Joe T. Ritchie
Michigan State University
Dr. Upendra Singh
International Rice Research Institute
Dr. Gordon Tsuji
IBSNAT Project, University of Hawaii

Zimbabwe
Dr. Thomas E. Downing
Dr. Paul Muchena
Plant Protection Research Institute

APPENDIX 2. CROP MODELING STUDIES (ROSENZWEIG AND IGLESIAS [EDS.], 1994).

Argentina
O. E. Sala and J. M. Paruelo. Implications of Climate Change for International Agriculture: Global Food Production, Trade, and Vulnerable Regions: The Argentinean Case.

Australia
B. D. Baer, W. S. Meyer, and D. Erskine. Global Climate Change: Possible Effects on Australian Agriculture.

Bangladesh
Z. Karim, M. Ahmed, S. G. Hussain, and Kh. B. Rashid. Impact of Climate Change on the Production of Modern Rice in Bangladesh.

Brazil
O. J. F. de Siqueira, J. R. B. Farias, and L. M. A. Sans. Potential Effects of Global Climate Change for Brazilian

Agriculture: Simulation Studies Applied for Wheat, Maize, and Soybean.

Canada
M. Brklacich, R. Stewart, V. Kirkwood, and R. Muma. Effects of Global Climatic Change on Wheat Yields in the Canadian Prairie: Summary Report.

China
Z. Jin, D. Ge, H. Chen, J. Fang, and X. Zheng. Effects of Climate Change on Rice Production and Strategies for Adaptation in Southern China.

Egypt
H. M. Eid. Climate Change and Crop Modeling Study in Egypt.

France
R. Delécolle, D. Ripoche, F. Ruget, and G. Gosse. Possible Effects of Increasing CO_2 Concentration on Wheat and Maize Crops in North and Southeast France.

India
D. G. Rao. Impact of Climate Change on Simulated Wheat Production in India.

Japan
H. Seino. Indications of Climate Change for Japanese Crop Production.

Mexico
D. Liverman, M. Dilley, K. O'Brien, and L. Menchaca. The Impacts of Climate Change on Maize in Mexico.

Pakistan
A. Qureshi and A. Iglesias. Effects of Climate Change on Simulated Wheat Production in Pakistan.

Philippines
C. R. Escaño and L. Buendia. Climate Impact Assessment for Agriculture in the Philippines.

Thailand
M. L. C. Tongyai. Effects of Climate Change on Thai Agriculture.

Russia
G. Menzhulin, L. Koval, and A. Badenko. Potential Effects of Global Warming and Atmospheric Carbon Dioxide on Wheat Production in the Commonwealth of Independent States.

USA
C. Rosenzweig, B. Curry, T. Y. Chou, J. Ritchie, J. Jones, and R. Peart. Simulated Wheat, Corn, and Soybean Response to Predicted Climate Change in the US.

Uruguay
W. E. Baethgen. Climate Change and Crop Modeling Study in Uruguay.

Zimbabwe
P. Muchena. Implications of Climate Change for International Agriculture: Global Food Production, Trade, and Vulnerable Regions.

APPENDIX 3. COUNTRIES IN BASIC LINKED SYSTEM REGIONAL GROUPS

EC
Belgium, Denmark, France, Federal Republic of Germany (pre-unification), Italy, Ireland, Luxembourg, Netherlands, and United Kingdom

Eastern Europe and former USSR
Bulgaria, former Czechoslovakia, former Democratic Republic of Germany, Hungary, Poland, Romania, and former USSR

Africa
Oil exporters
Algeria, Angola, Congo, Gabon
Medium-income/calorie exporters
Ghana, Ivory Coast, Senegal, Cameroon, Mauritius, Zimbabwe
Medium-income/calorie importers
Morocco, Tunisia, Liberia, Mauritania, Zambia
Low-income/calorie exporters
Benin, Gambia, Togo, Ethiopia, Malawi, Mozambique, Uganda, Sudan
Low-income/calorie importers
Guinea, Mali, Niger, Sierra Leone, Upper Volta, Central African Republic, Chad, Zaire, Burundi, Madagascar, Rwanda, Somalia, Tanzania

Latin America
High-income/calorie exporters
Costa Rica, Panama, Cuba, Dominican Republic, Ecuador, Surinam, Uruguay
High-income/calorie importers
Jamaica, Trinidad and Tobago, Chile, Peru, Venezuela
Medium to low income
El Salvador, Guatemala, Honduras, Nicaragua, Columbia, Guyana, Paraguay, Haiti, Bolivia

Far East Asia
High to medium income/calorie exporters
Malaysia, Philippines
High to medium income/calorie importers
Republic of Korea, Laos, Vietnam, Korea DPR, Cambodia
Low income
Burma, Sri Lanka, Bangladesh
Oil exporters/high income
Libya, Iran, Iraq, Saudi Arabia, Cyprus, Lebanon, Syria

Near East Asia
Medium to low income
Yemen, Afghanistan

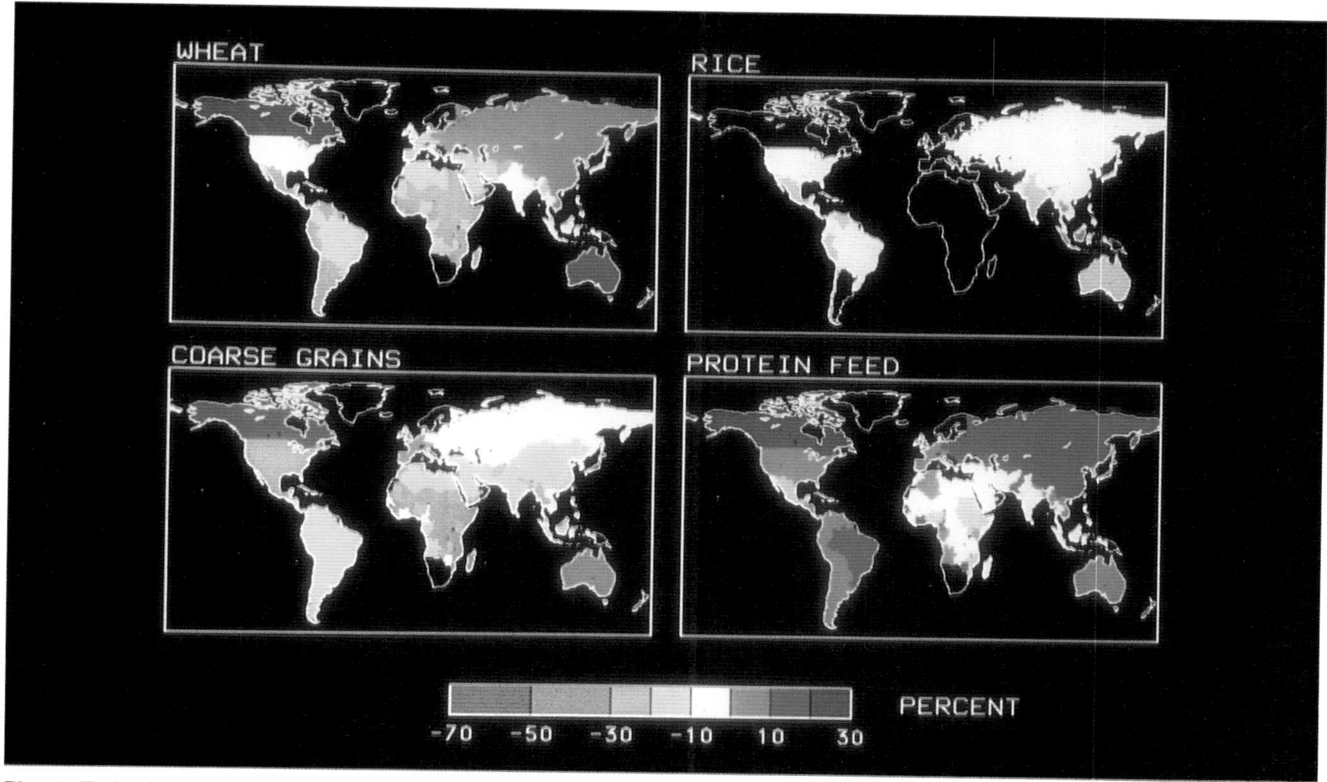

Plate 1 Estimated change in average wheat, rice, coarse grains, and protein feed yield for the GFDL climate change scenario with direct CO_2 effects. Results shown are averages for countries and groups of countries in the BLS world food trade model; regional variations within countries are not reflected.

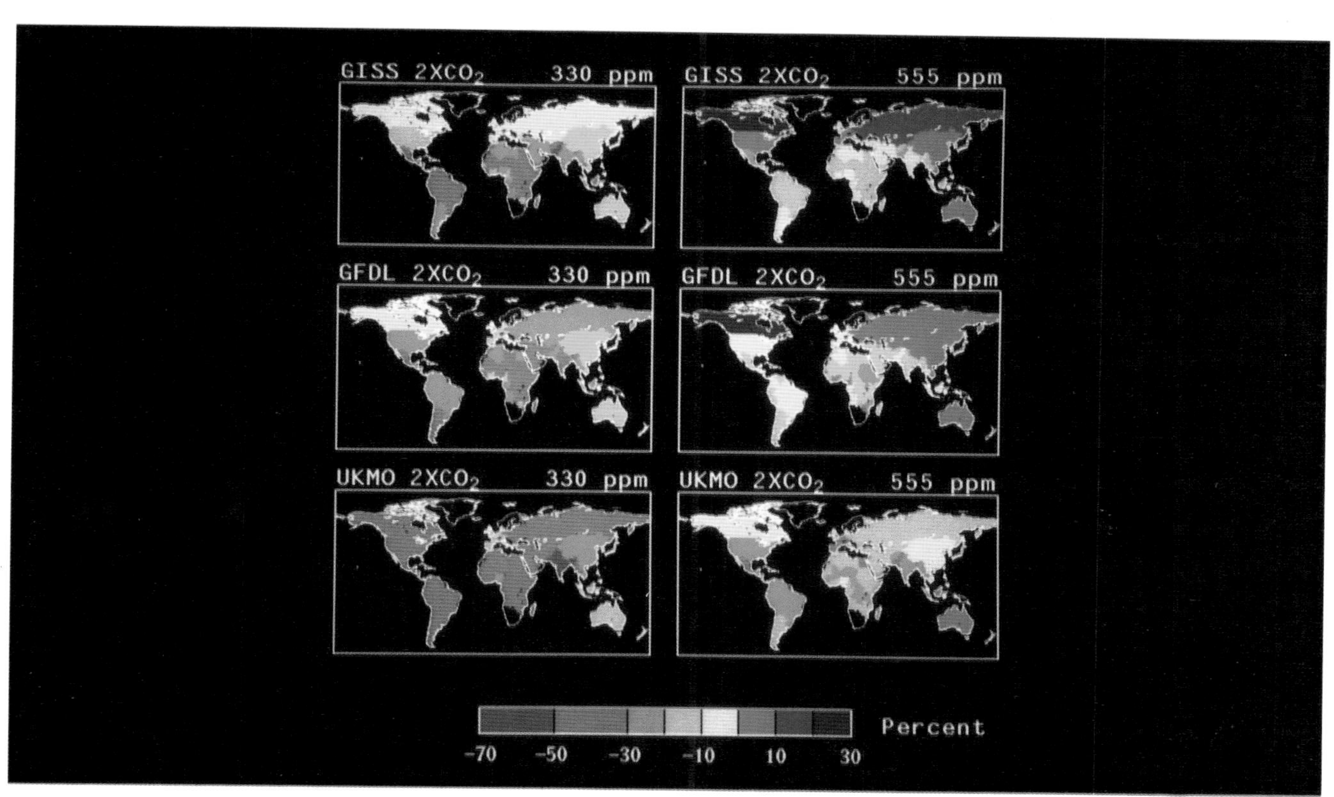

Plate 2 Estimated change in average national grain yield (wheat, rice, coarse grains, and protein feed) for the GISS, GFDL, and UKMO climate change scenarios. The left-hand column shows effects of climate change alone (330 ppm CO_2); the right-hand column shows the combined effects of climate change and direct CO_2 effects (555 ppm CO_2).

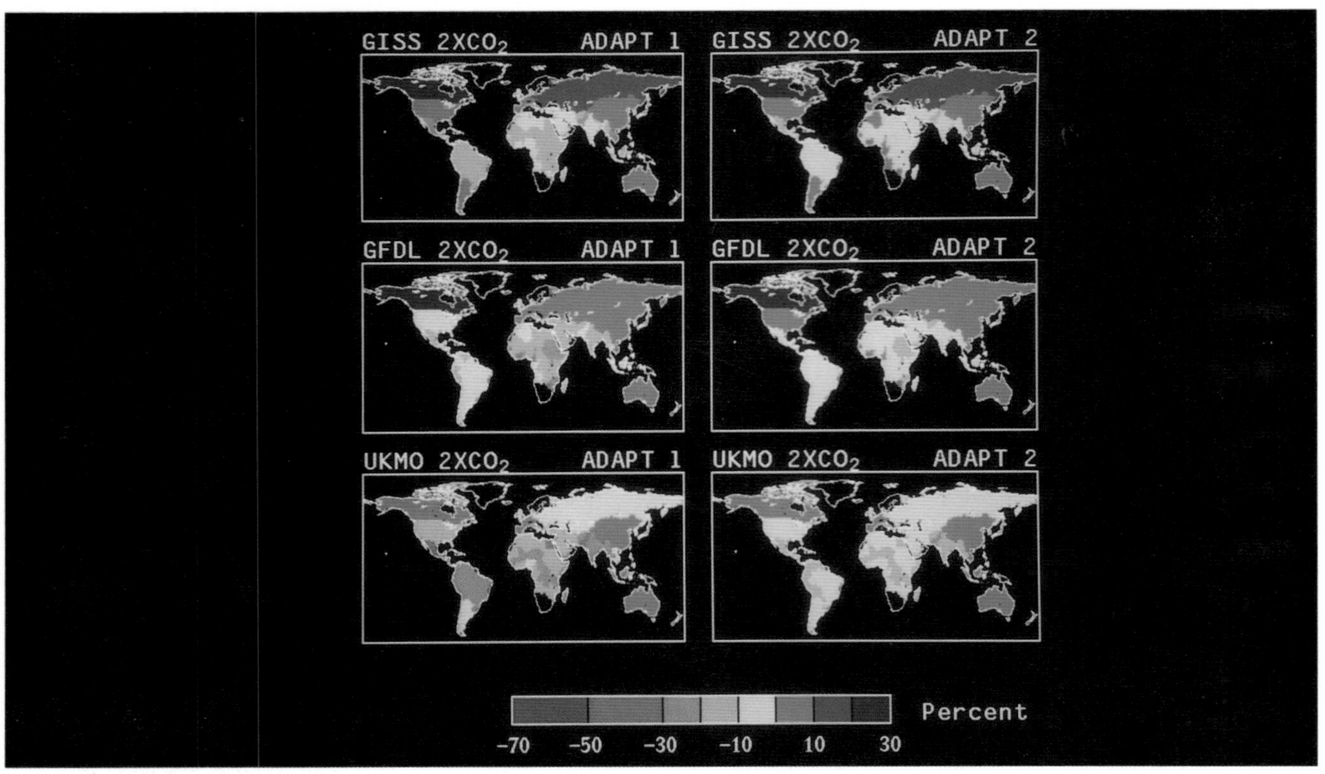

Plate 3 Estimated change in average national grain yield (wheat, rice, coarse grains, and protein feed with direct 555 ppm CO_2 effects) under two levels of adaptation for the GISS, GFDL, and UKMO climate change scenarios. Adaptation Level 1 signifies minor changes to existing agricultural systems; Adaptation Level 2 signifies major changes. Results shown are averages for countries and groups of countries in the BLS world food trade model; regional variations within countries are not reflected. (Rosenzweig and Parry, 1994).

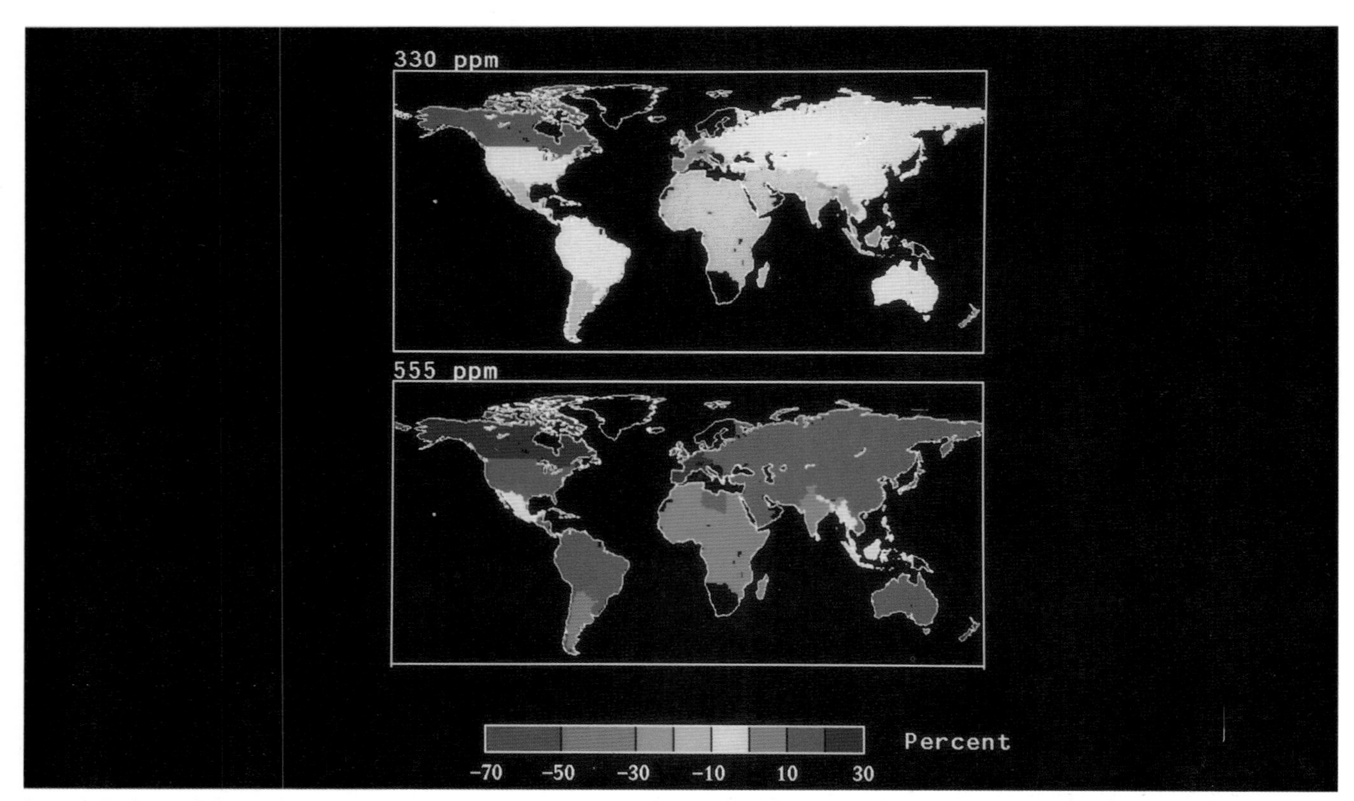

Plate 4 Estimated change in average national grain yield (wheat, rice, coarse grains, and protein feed) for the GISS-A (GISS 2030s) climate change scenario. The upper map shows effects of climate change alone (330 ppm CO_2); the lower map shows the combined effects of climate change and direct CO_2 effects (555 ppm CO_2). Results shown are averages for countries and groups of countries in the BLS world food trade model; regional variations within countries are not reflected.

Plate 5 An example of an undeveloped buffer of land adjacent to the coast in Uruguay, Parque del Plata, Department of Canelones. In this case the buffer is about 100 meters wide, preserving the natural sand dunes which provide protection against storms. The undeveloped land also allows some erosion to occur in the future with little or no adverse impacts for the adjacent properties. (Photograph by C.R. Volonté)

Plate 6 Erosion is already causing adverse impacts in Venezuela and many homes have had to be abandoned south of Boca de Aroa, State of Falcon. (Photograph by C.R. Volonté)

Plate 7 A typical tourist resort south of Dakar in Senegal. These facilities are particularly vulnerable to erosion as they were often built very close to the water's edge. (Photograph by K.C. Dennis)

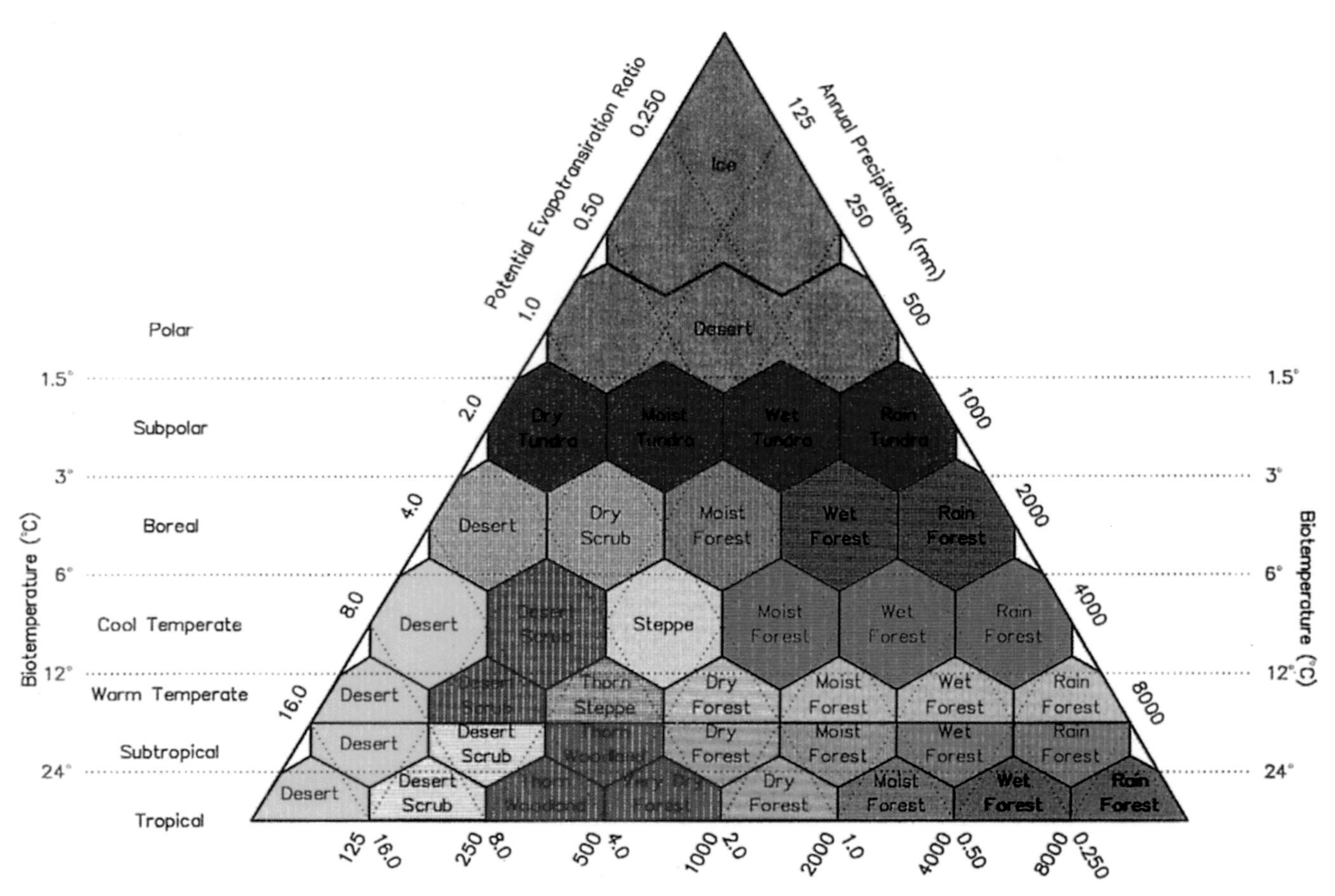

Plate 8 Holdridge climate–vegetation classification scheme (from Holdridge 1967).

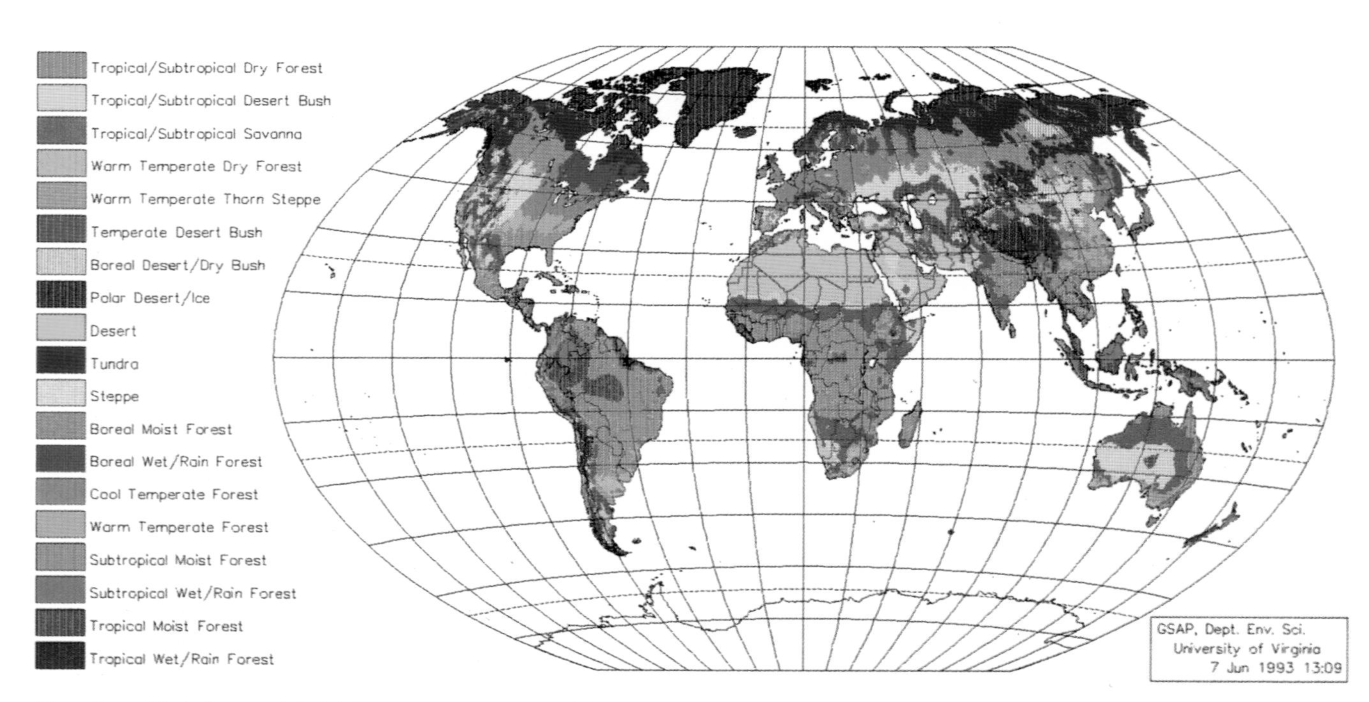

Plate 9a–e Global map of Holdridge Life Zones under (a) current climate conditions, and the climate change scenarios based on (b) GFDL, (c) GISS, (d) OSU, and (e) UKMO General Circulation Models (see Chapter 1). The resolution is 0.5° latitude × 0.5° longitude. Key relating aggregated classes to life zone is shown in Plate 8.

9a

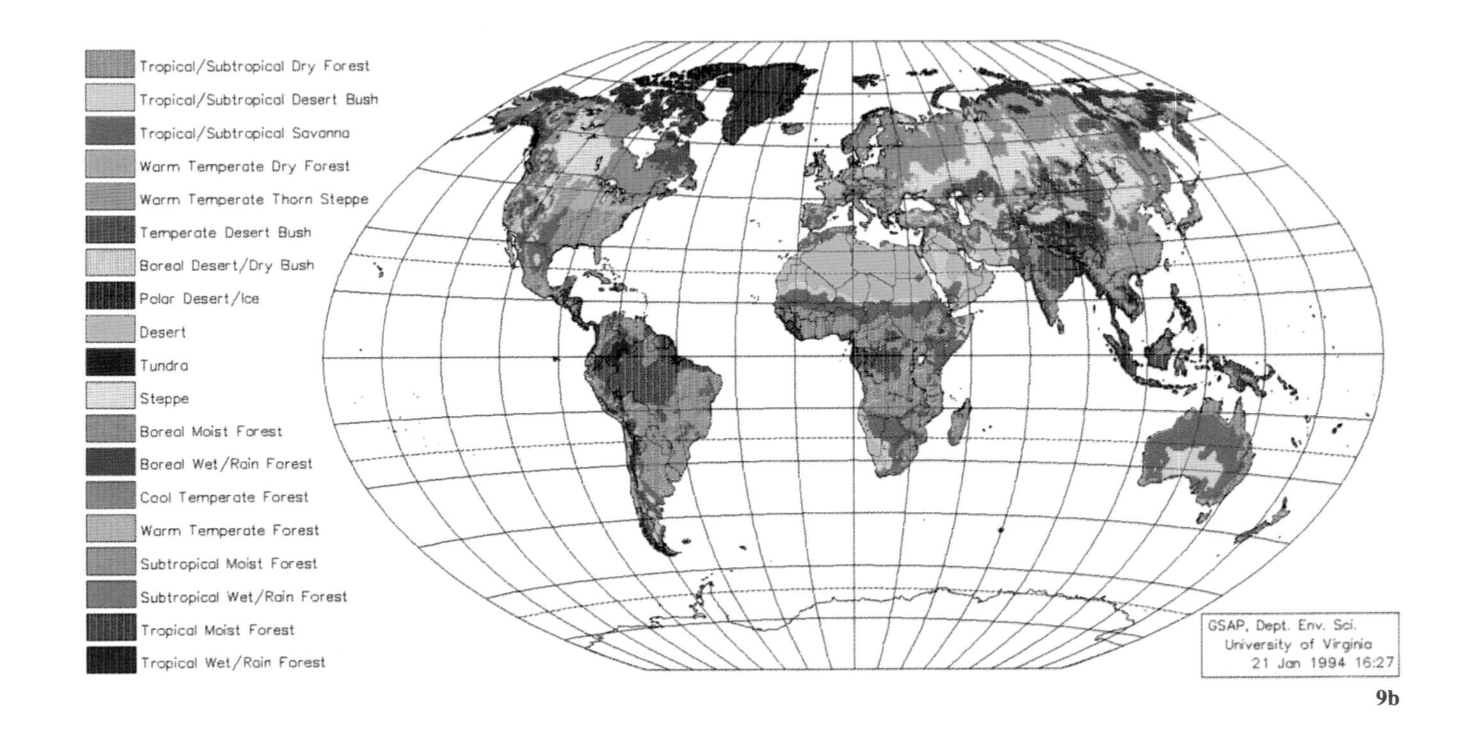
Tropical/Subtropical Dry Forest
Tropical/Subtropical Desert Bush
Tropical/Subtropical Savanna
Warm Temperate Dry Forest
Warm Temperate Thorn Steppe
Temperate Desert Bush
Boreal Desert/Dry Bush
Polar Desert/Ice
Desert
Tundra
Steppe
Boreal Moist Forest
Boreal Wet/Rain Forest
Cool Temperate Forest
Warm Temperate Forest
Subtropical Moist Forest
Subtropical Wet/Rain Forest
Tropical Moist Forest
Tropical Wet/Rain Forest
GSAP, Dept. Env. Sci.
University of Virginia
21 Jan 1994 16:27

9b

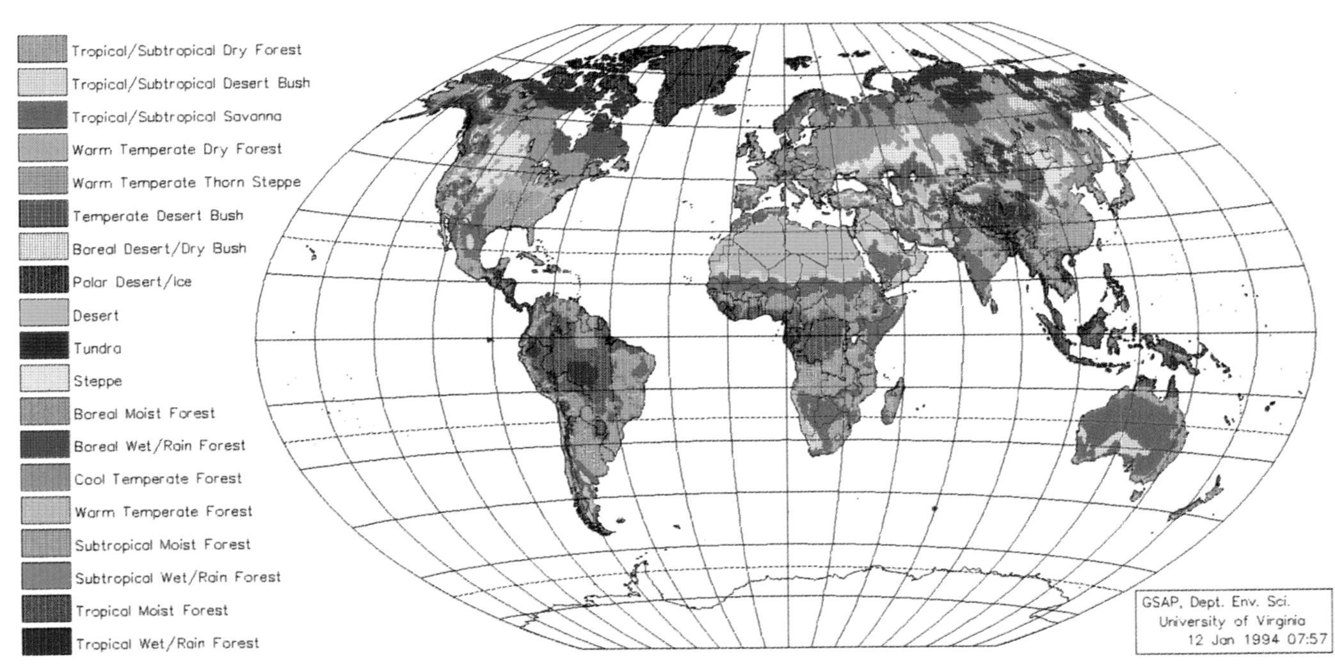
Tropical/Subtropical Dry Forest
Tropical/Subtropical Desert Bush
Tropical/Subtropical Savanna
Warm Temperate Dry Forest
Warm Temperate Thorn Steppe
Temperate Desert Bush
Boreal Desert/Dry Bush
Polar Desert/Ice
Desert
Tundra
Steppe
Boreal Moist Forest
Boreal Wet/Rain Forest
Cool Temperate Forest
Warm Temperate Forest
Subtropical Moist Forest
Subtropical Wet/Rain Forest
Tropical Moist Forest
Tropical Wet/Rain Forest
GSAP, Dept. Env. Sci.
University of Virginia
12 Jan 1994 07:57

9c

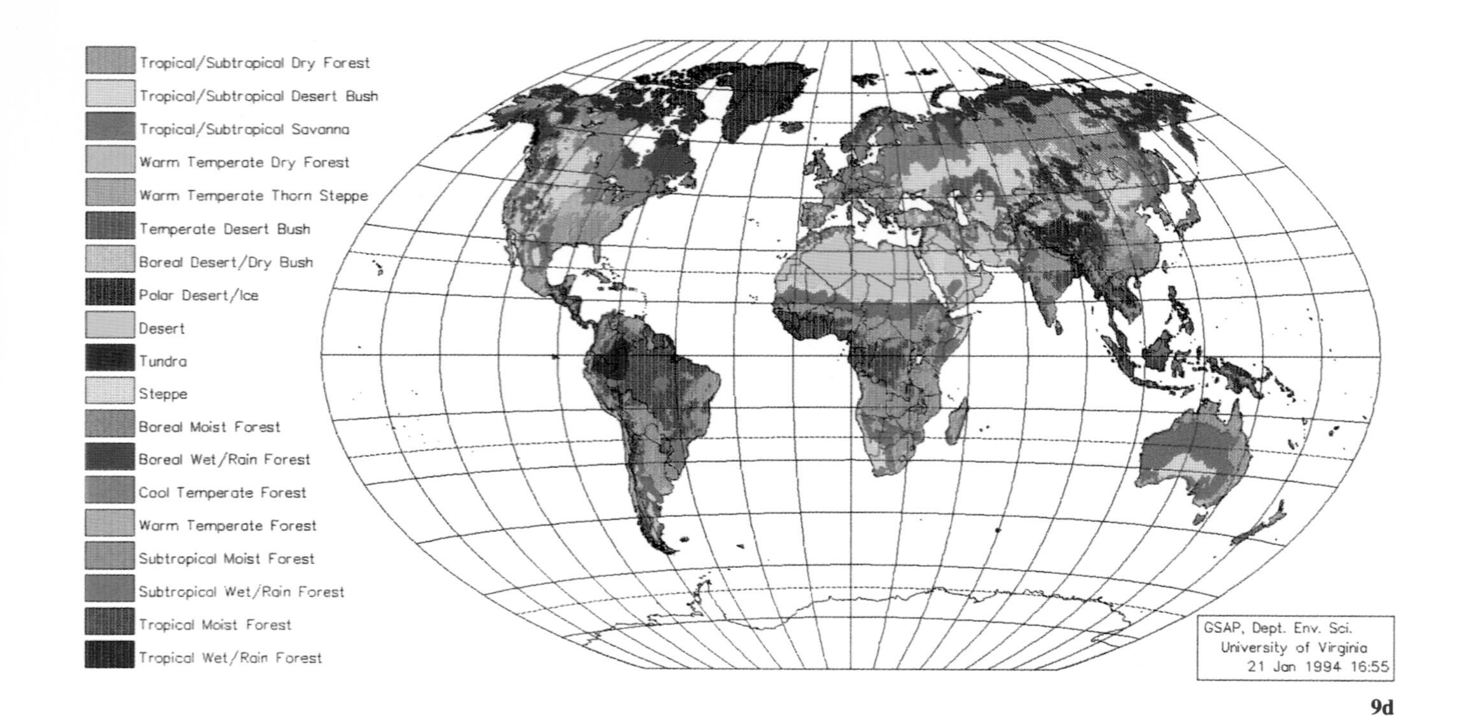
Tropical/Subtropical Dry Forest
Tropical/Subtropical Desert Bush
Tropical/Subtropical Savanna
Warm Temperate Dry Forest
Warm Temperate Thorn Steppe
Temperate Desert Bush
Boreal Desert/Dry Bush
Polar Desert/Ice
Desert
Tundra
Steppe
Boreal Moist Forest
Boreal Wet/Rain Forest
Cool Temperate Forest
Warm Temperate Forest
Subtropical Moist Forest
Subtropical Wet/Rain Forest
Tropical Moist Forest
Tropical Wet/Rain Forest
GSAP, Dept. Env. Sci.
University of Virginia
21 Jan 1994 16:55

9d

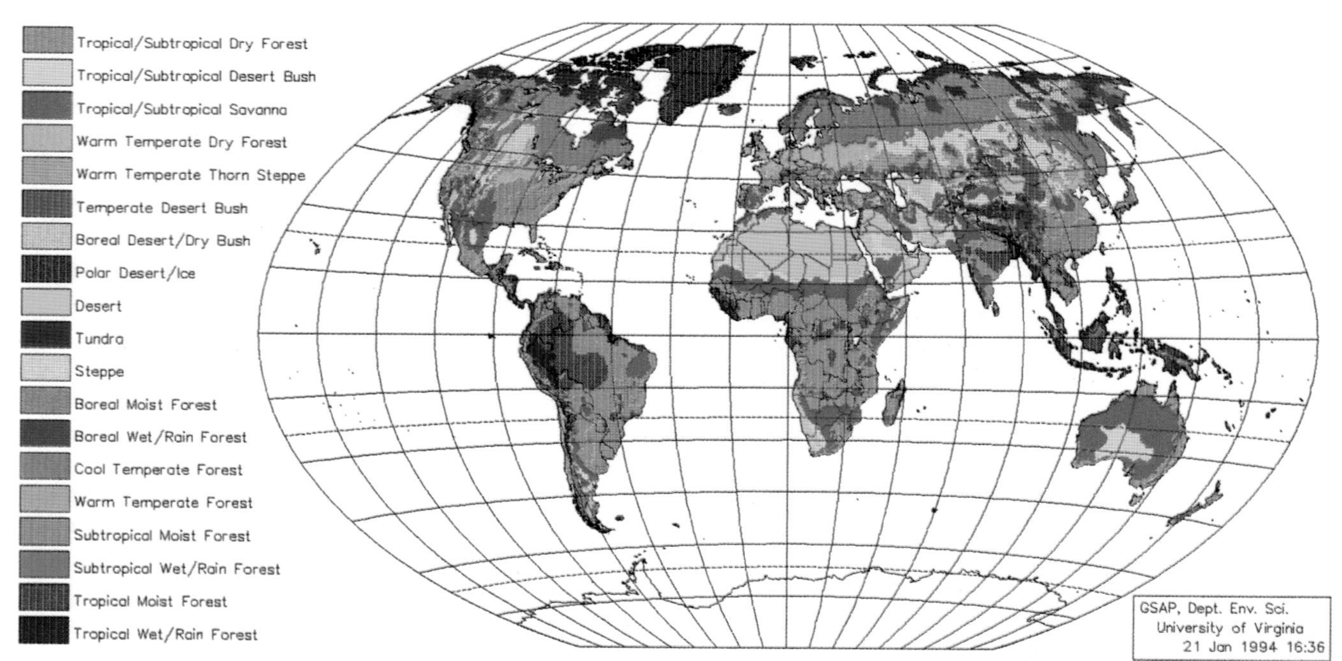
Tropical/Subtropical Dry Forest
Tropical/Subtropical Desert Bush
Tropical/Subtropical Savanna
Warm Temperate Dry Forest
Warm Temperate Thorn Steppe
Temperate Desert Bush
Boreal Desert/Dry Bush
Polar Desert/Ice
Desert
Tundra
Steppe
Boreal Moist Forest
Boreal Wet/Rain Forest
Cool Temperate Forest
Warm Temperate Forest
Subtropical Moist Forest
Subtropical Wet/Rain Forest
Tropical Moist Forest
Tropical Wet/Rain Forest
GSAP, Dept. Env. Sci.
University of Virginia
21 Jan 1994 16:36

9e

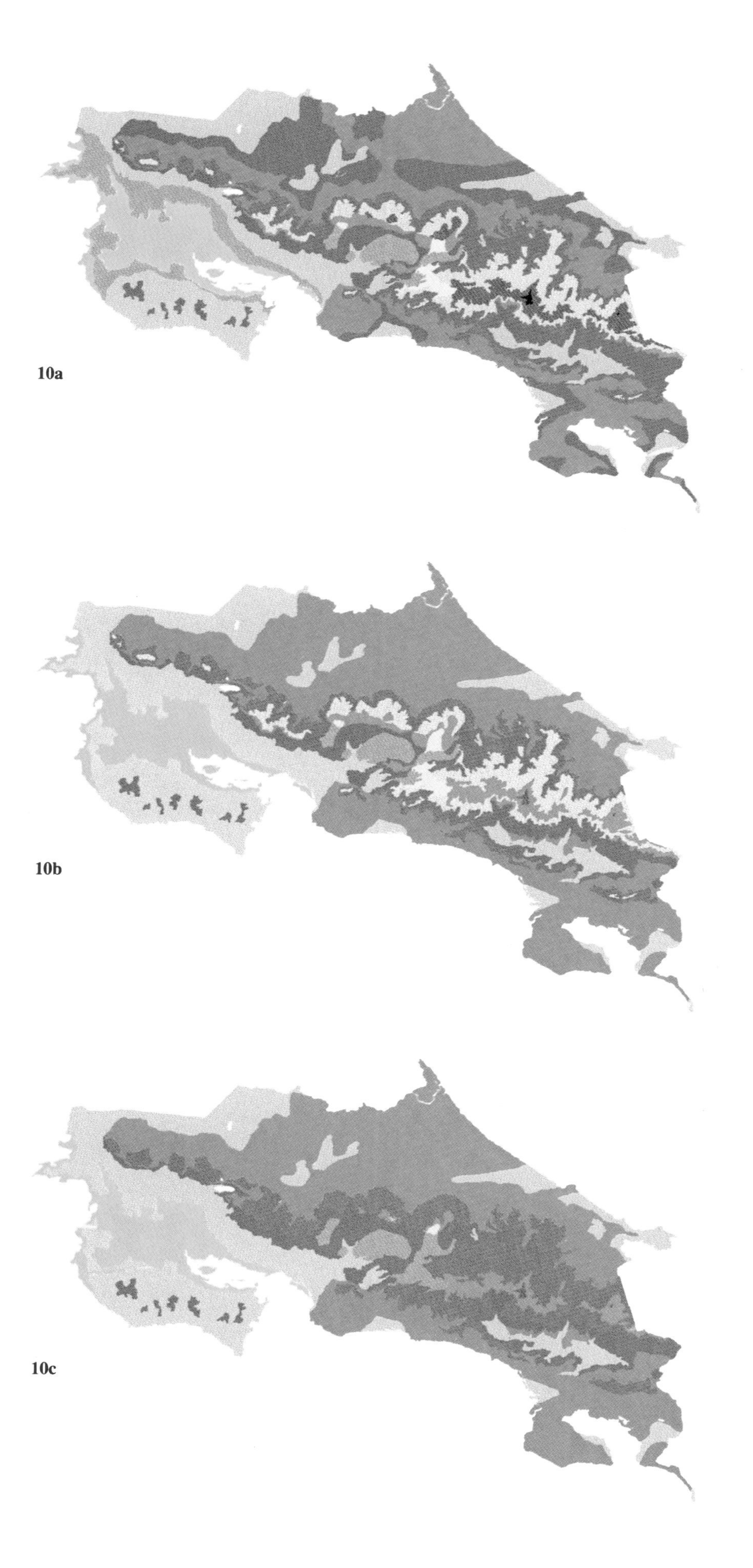

Plate 10a–c Eco-climatic zones of Costa Rica: (a) current climate; (b) under a +2.5°C, +10 percent precipitation scenario; and (c) under a +3.6°C, +10 percent precipitation scenario.

Tropical Dry Forest
Tropical Moist Forest
Tropical Wet Forest
Premontane Moist Forest
Premontane Wet Forest
Premontane Rain Forest
Lower Montane Moist Forest
Lower Montane Wet Forest
Lower Montane Rain Forest
Montane Wet Forest
Montane Rain Forest
Subalpine Rain Paramo

Key

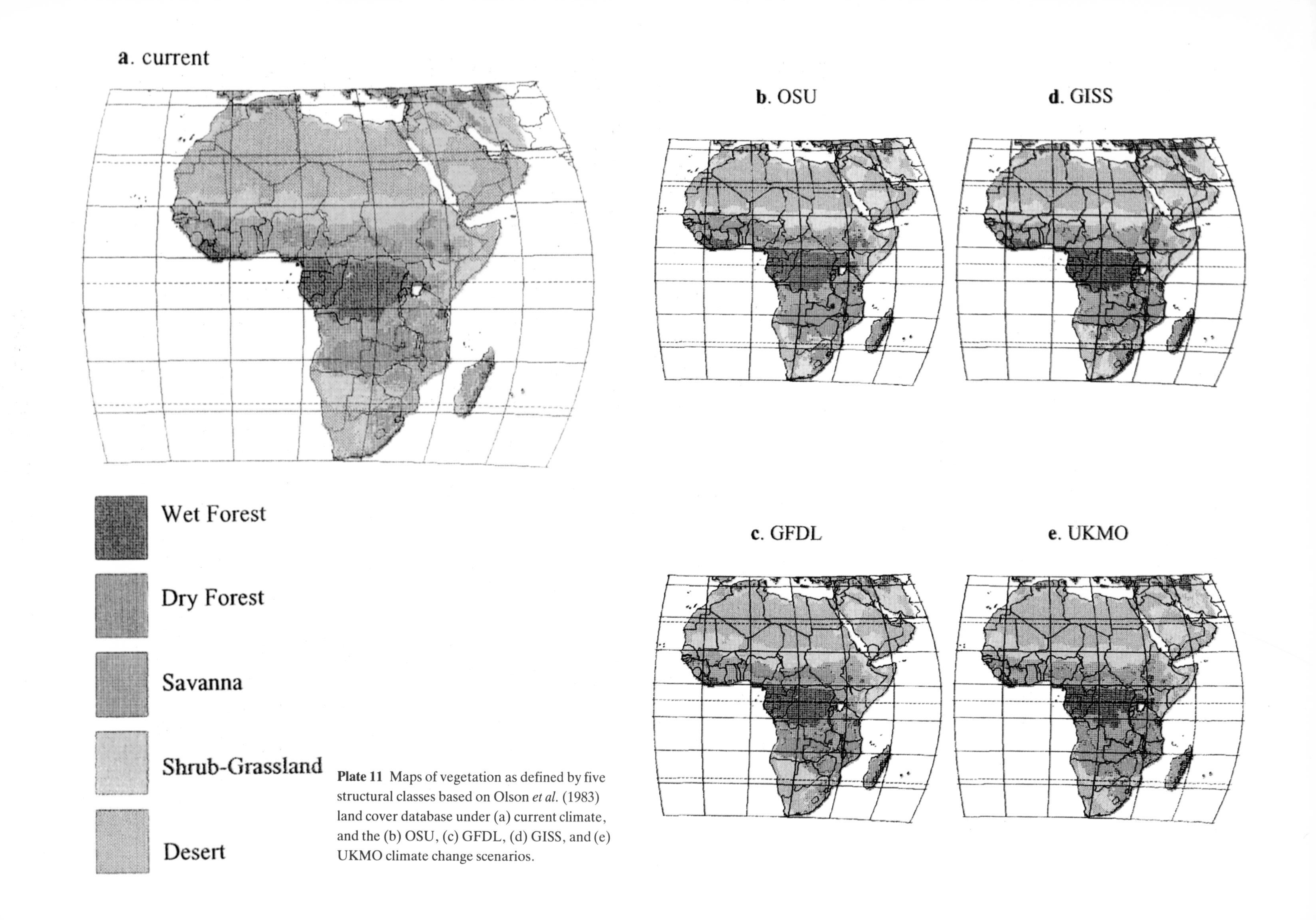

Plate 11 Maps of vegetation as defined by five structural classes based on Olson *et al.* (1983) land cover database under (a) current climate, and the (b) OSU, (c) GFDL, (d) GISS, and (e) UKMO climate change scenarios.

3 Complex River Basins

W. E. RIEBSAME, K. M. STRZEPEK, J. L. WESCOAT, Jr., R. PERRITT, G. L. GAILE, J. JACOBS, R. LEICHENKO, C. MAGADZA, H. PHIEN, B. J. URBIZTONDO, P. RESTREPO, W. R. ROSE, M. SALEH, L. H. TI, C. TUCCI, and D. YATES

Institute of Behavioral Science

University of Colorado

SUMMARY

River basin development is the foundation on which many developing countries seek to build future economic and social advancement. Because climate is intertwined with river basin characteristics and processes, the potential for global climate change raises serious implications for river basin management. To assess how climate change might affect basin development, climate scenarios associated with greenhouse warming were translated into potential effects and responses for five international rivers: the Nile, Zambezi, Indus, Mekong, and Uruguay. Collaborative teams in each river basin first assessed potential effects of climate change on runoff and other hydrological parameters and then evaluated how these changes would affect water resources management, including reservoir operations, hydropower production, irrigation, urban water supply, flood control, and environmental protection. The wide range of climate conditions projected by global climate models (GCMs) and arbitrary scenarios yielded a broad range of potential basin impacts. However, some important patterns emerged. For example, the drier basins (e.g., the Nile and Zambezi) accrued greater hydrological impacts from a given climate change. In terms of management, actual and planned water systems tuned to 'normal' climate tended to lose efficiency or produce less desirable outputs from any significant climate change, whether it resulted in increased *or* decreased river flow. Because water management in most of the basins incorporates schemes to deal with both floods and droughts (low flows), few unambiguously positive effects of climate change emerged, though increased flows did boost hydropower production in most basins. Despite this pattern of negative effects, basin managers felt that a mixture of traditional tactics, like reservoir construction or enlargement, and nonstructural approaches, such as watershed protection, would allow them to cope with climate change over the foreseeable future. Overall, they were optimistic that the types of changes postulated in this study could be accommodated by good planning and management.

CLIMATE CHANGE AND RIVER BASIN DEVELOPMENT

Water resources often form the nucleus of regional development plans, especially in the less developed countries. Because water resources are inextricably linked with climate, the prospect of global climate change has serious implications for regional development, and several research questions arise: how robust are the water resource management systems in the world's rapidly developing river basins? Which components of those systems are most likely to be affected by climate change? What options exist for adjusting water management to accommodate, perhaps even benefit from, future climate change?

Previous river basin impact assessments have tended to focus on developed regions (Smith and Tirpak, 1989; Waggoner, 1990). This study assessed potential climate impact and management adaptation in five international river basins shared by developing countries: the Indus, Mekong, Uruguay, Zambezi, and Nile (Figure 1). The basins have several characteristics in common: They are international, complex and rapidly developing, and possess the requisite scientific and modeling capabilities to undertake climate impact assessment. Scenarios of climate change for each basin were translated into runoff and water resource management impacts with simulation and optimization models used for water resources planning and management. In some cases,

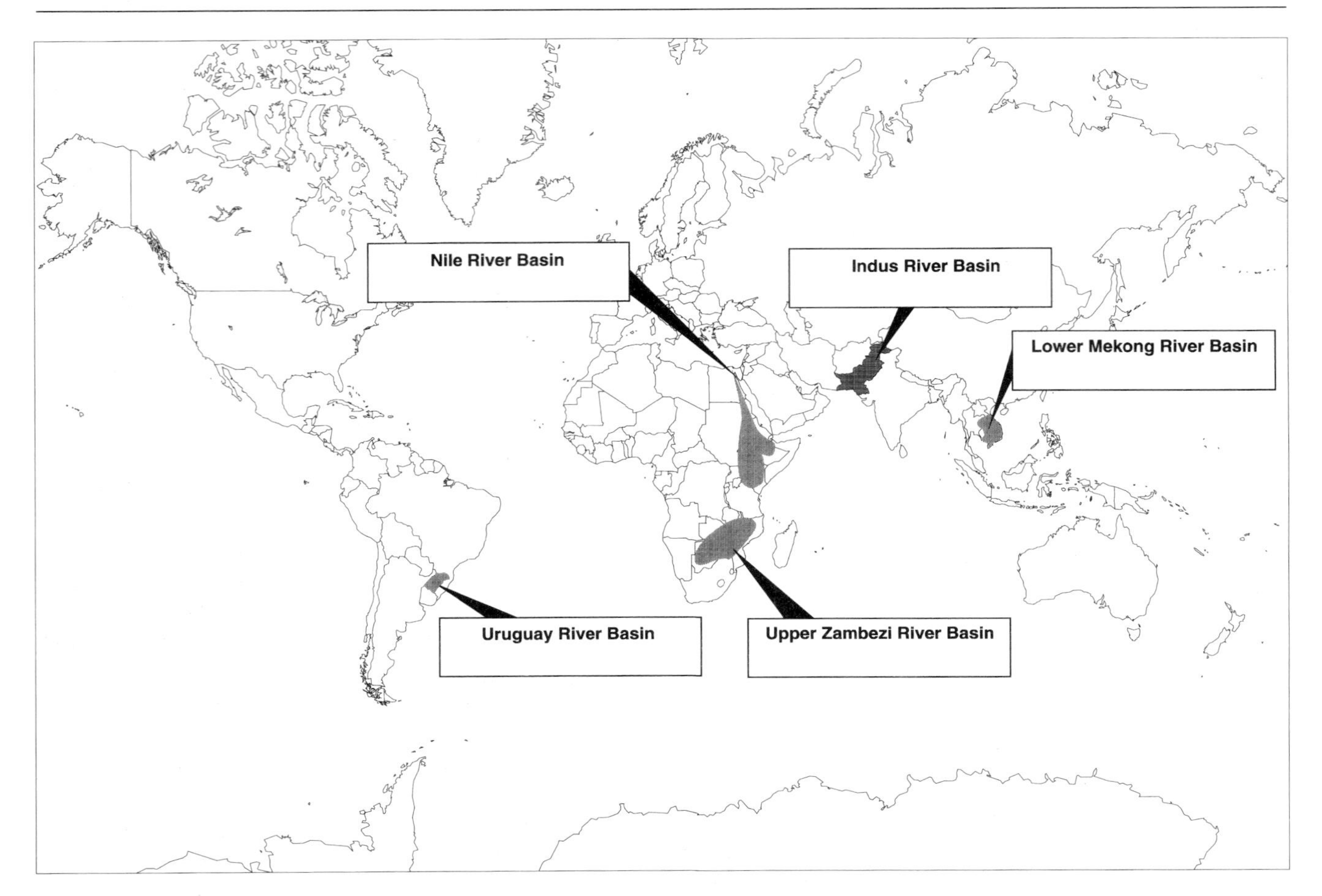

Figure 1 General location of the case study river basins.

models were already available; in others, new models were developed. A goal throughout the project was to develop collaboration among basin scientists and planners, thus creating the foundation for future impact assessments. This collaboration was especially important in the final step: evaluating regional capacity for adapting to environmental changes.

Our goals were not only to specify the effects of given scenarios of future climate, but to assess underlying basin vulnerability and adaptability. The Intergovernmental Panel on Climate Change (IPCC) concluded that even relatively small changes in climate can significantly worsen water resources problems (Tegart *et al.*, 1990). But river basins vary markedly in hydrological sensitivity, level of development, management capacity, and potential for further development that could mitigate the negative effects, or even capitalize on the positive effects, of climate change. We found great differences in the climate sensitivity and level of development among the five basins, differences that caution against universal conclusions about water resource impacts of climate change.

This chapter provides an overview of the river basin assessments, including a description of basin characteristics and the key findings from each case study. The discussion then turns to a comparative analysis meant to elicit cross-cutting insights and lessons for river basin development in a changing climate. Our approaches and foci differed among the basins; in all cases we began by focusing on traditional elements like water storage and hydropower systems, but in some cases we were able to expand to such areas as urban development, hazards, and ecological effects. Moreover, in some cases we were able to compare the relative effects of climate change with the effects of other factors influencing basin development, such as policy, institutional capacity, and potential internation conflicts.

RIVER BASIN DEVELOPMENT

River basins have long been attractive spatial units for regional resource development. Following World War II, integrated river basin planning was seen as the keystone of development schemes in many developing regions (United Nations Economic and Social Council, 1970). Much of this optimism stemmed from results of the Tennessee Valley

Authority experiment in the United States, which sought to combine water management and economic development in a depressed region (United Nations, 1950). In the 1950s, a number of complex projects were promoted in basins such as the Mekong and La Plata. The Mekong project, in particular, was seen as an integrated plan for unlocking the vast development potential of Southeast Asia (White, 1964). The Indus Basin Treaty of 1960 was followed by massive water resources investment in India and Pakistan aimed at developing South Asia along the same lines (Michel, 1967).

In his seminal work, White (1957, 1964) identified the major elements of integrated basin development as follows: (1) multiple-purpose storage reservoirs; (2) basin-wide planning; and (3) comprehensive regional development. Another goal was the integration of land and water resource planning under a unified river basin administration (Wescoat, 1992). Some or all of these elements were found and analyzed in the case study basins.

The integrated river basin development approach has received considerable attention in Africa and Asia, some of it critical. Perritt's (1989) review of African basin projects argues that negative effects, including urban development at the expense of rural areas, lack of local participation, and environmental degradation, have frustrated integrated development. Anticipated benefits 'had not materialized to affect positively the lives of large numbers of people in African countries' (Perritt, 1989, p. 205). A thorough assessment of the strengths and weaknesses of river basin development theory and practice is beyond the scope of this study, but the gap between the ideals and accomplishments is important for assessing regional vulnerability to climate change and for evaluating the prospects for 'sustainable development' in a changing climate. After discussing the main impacts of climate change projected in the basin case studies, we return to the broader issue of vulnerability and adjustment.

ASSESSMENT METHODS

We applied an assortment of methods, referred to as 'climate impact assessment' (see, e.g., Kates *et al.*, 1985), to evaluate basin sensitivities, impacts, and potential adjustments. Basin sensitivities were compiled by canvassing regional experts and the staffs of basin institutions, who then took part in the study. Standard hydrological and management modeling techniques were employed to project hydrological and water resource management effects of climate change (Table 1). The general strategy was to use climate–runoff models to project the hydrological effects of climate scenarios and then to employ systems management models to translate changed hydrology into effects on typical activities, such as hydropower production, irrigation, and flood control. Three GCMs were used: Goddard Institute of Space Science (GISS), Geophysical Fluid Dynamics Laboratory (GFDL) at Princeton University, and the United Kingdom Meteorological Office (UKMO) model (see Chapter 1 for details).

Table 1. *Modeling approaches in case study river basins*

Basin	Modeling: Hydrological	Modeling: Management
Uruguay	Type: physical Time scale: ave. year Time step: daily Spatial scale: partial w/ analog for full basin	Type: water resource planning tools Time step: monthly Spatial scale: partial project and regional
Mekong	Type: statistical Time scale: ave. year Time step: monthly Spatial scale: full basin	Type: river basin simulation MITSIM Time step: monthly Spatial scale: full basin
Indus	Type: physical Time scale: ave. year Time step: daily Spatial scale: partial w/ analog for full basin	Type: river basin optimization World Bank IBMR Time step: monthly Spatial scale: full basin
Zambezi	Type: physical/conceptual Time scale: time series Time step: monthly Spatial scale: full basin	Type: river basin simulation MITSIM Time step: monthly Spatial scale: full basin
Nile	Type: water balance Time scale: ave. year Time step: monthly Spatial scale: full basin	Type: river basin simulation DSSEWM Time step: monthly Spatial scale: full basin

It is worth noting here the variety of approaches dictated by different models, data availability, and institutional constraints in the five basins. With some modifications, each impact assessment was guided by a set of GCMs and sensitivity scenarios, but these were not applied in the same manner in each case; for example, temperature data were not used in the study of the Mekong basin, where the existing flow-forecast model neglects evaporation because the model is used in short time steps to predict floods. The basin areas assessed in each case study also differed: the entire catchments of the Zambezi, Nile, and Mekong were modeled, whereas the Uruguay and Indus studies focused on subbasin analysis that was extended to the full catchment by analogy. The Massachusetts Institute of Technology (MIT) River

Table 2. *Hydrological characteristics of case study river basins*

Parameters	Uruguay	Mekong	Indus	Indus (Hill)	Zambezi	Nile	Blue Nile
Length (km)	1,612	4,350	2,900	—	2,600	6,500	1,000
Area ($km^2 \times 10^3$)	239	810	960	450	1,330	2,880	313
Flow (m^3/s)	4,660	14,800	6,700	6,700	4,990	2,832	1,666
Flow ($10^9 m^3/yr$)	147	467	211	207	157	89	53
Spec Q ($l/s\text{-}km^2$)	19.5	18.3	7.0	14.9	3.8	1.0	5.3
Runoff (R) (mm)	615	576	220	460	118	31	168
Precip. (P) (mm)	1,200	1,380	458	612	990	730	784
R/P	0.51	0.42	0.48	0.75	0.12	0.04	0.21
PET/P	0.75	0.90	0.80	0.40	2.50	5.50	1.80

Notes:
Spec Q = specific discharge
PET = potential evapotranspiration

Basin Simulation model (MITSIM) was used to assess management impacts in the Zambezi and Mekong studies; the Decision Support System for Environmental and Water Management (DSSEWM), a derivative of MITSIM, was used in the Nile study. Traditional water resource regional planning tools, such as flow duration curves and storage yield relationships, were employed in the Uruguay study. The Indus study employed the World Bank's Indus basin investment planning/optimization model (IBMR).

Although the techniques varied, the models provided some comparative insights and represent the best approaches that could be taken within the logic of large-basin climate impact assessment. Indeed, the climate scenarios themselves contain at least as much uncertainty and regional difference as exists among the hydrological and management modeling approaches used in the various basins.

CLIMATE CHANGE AND THE CASE STUDY BASINS

Our focus in these studies was on the direct effects of climate change on basin hydrology and on the impacts of altered hydrology on water management. In each analysis we concentrated on a limited set of variables – chiefly, runoff amount and seasonality, reservoir management, hydropower generation, and irrigation – and on key impact areas, such as flood hazards, fisheries, and water allocations. The main issues varied among the basins based on our initial sensitivity assessments.

Hydrological characteristics of the basins

Key hydrological characteristics of the case study basins, expressed in aggregate parameters, are listed in Table 2. These are basin-wide averages that, in most cases, are good indicators of overall hydrology, but they conceal other factors that affect how a basin responds to climate change. The basins are not spatially homogenous, and the regional pattern of climate change will affect spatially averaged outcomes. For example, large parts of the Indus and Nile basins contribute little runoff due to very arid conditions; thus, parameters for the major contributing portions of the basins (the 'hill catchment' above the rim stations for the Indus and the Blue Nile basin for the Nile) are included in the comparative statistics. This example illustrates one of the problems of conceptualizing and characterizing climate impacts on such a large, differentiated landscape as a major river basin.

Figure 2 illustrates how the case study river basins fit into the global range of what might be termed 'runoff efficiency.' The graph shows the proportion of basin precipitation that actually runs off plotted against the evaporation climate of the basin – i.e., the runoff coefficient (runoff/precipitation) versus the dryness index (potential evapotranspiration/ precipitation). The figure makes apparent the wide range of hydrological responses among the case studies. It may surprise some readers to find the Indus so high on the curve, but its position is explained by the large proportion of the basin affected by low temperatures (low PET) and the large amount of runoff that starts as snow and ice in the upper basin. The lower Indus is, like the Nile, quite arid.

Basin sensitivity

Although GCM scenarios provided the main inputs for the impact assessments, basin-wide sensitivity analyses associated with arbitrary, homogenous changes in temperature and precipitation were also carried out in most cases. From these analyses we can assess rough hydrological elasticities of

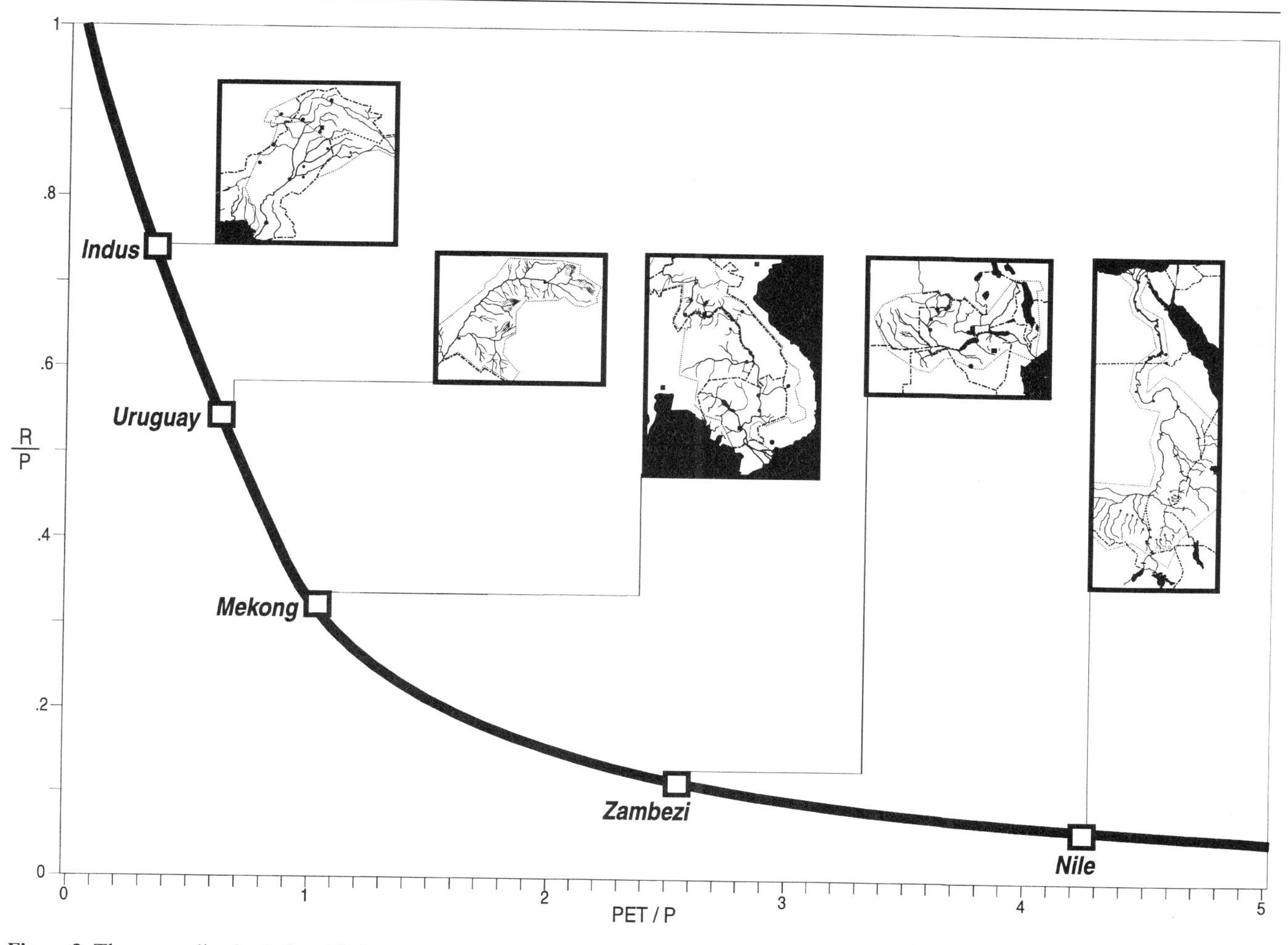

Figure 2 The generalized relationship between runoff and precipitation (R/P) and potential evapotranspiration and precipitation (PET/P) showing the relative dryness of the basins. Overall, sensitivity to climate change increases with lower R/P ratios and larger PET/P ratios.

the basins. The concept of hydrological elasticity, introduced for climate impacts on hydrology by Schaake (1990) and borrowed from economics, provides special insights into basin response to large-scale forces like climate. It is defined as the ratio of percentage change in runoff to percentage change in precipitation. Table 3 summarizes the hydrological elasticities for the case study basins. Although we have not been able to calculate the indices for the Uruguay or Mekong, both of which are relatively wet and thus less sensitive basins, Table 3 shows that the Nile stands out markedly as very sensitive. The Zambezi is also fairly sensitive. Figure 2 and Table 3 show that these latter basins have very low runoff coefficients and very high dryness indices, which empirically suggest high runoff elasticities. Such elasticity statistics probably provide more meaningful measures of sensitivity than do the results of GCM scenarios because the GCMs alter the temporal distribution of temperature and precipitation whereas the sensitivity studies provide for proportional changes to existing distributions.

General hydrological and management impacts

Summary statistics for selected climate change projections and hydrological impacts in the basins (Table 4) show that although the severity of the impacts of climate change on water resources depends primarily on the magnitude of the change, the different hydrological sensitivities of the basins are also important. The Zambezi and Nile are especially sensitive to climate warming: runoff decreases in those basins even when precipitation increases, due to the large hydrological role played by evaporation. The Mekong, Uruguay, and Indus show less hydrological sensitivity to temperature change but greater sensitivity to precipitation change.

Hydrological changes in the basins interact with a large set of management and development factors, especially relationships among the supply and demand for hydropower and irrigation, the yield and sustainability of important fisheries, flood hazard reduction, and the many social and institutional issues that arise in complex basin management.

Table 3. *Elasticities for the case study river basins*

Parameter	Uruguay	Mekong	Indus	Zambezi	Nile
Precipitation	n/a	n/a	0.950	1.88	3.6
Temperature	n/a	n/a	−0.125*	−1.68†	−4.5‡

Notes:
* Based on a 2°C warming on a 25°C basin average.
† Based on a 2°C warming on a 21°C basin average.
‡ Based on a 4°C warming on a 20°C basin average.

Table 4. *Summary of climate and hydrological changes for case study basins*

GCM		Uruguay	Mekong	Indus	Zambezi	Nile
GISS	Temp. (°C)	+4.5	n/a	+4.7	+4.0	+3.4
	Precip. (%)	97	97	110	109	131
	Runoff (%)	88	100	111	89	130
GFDL	Temp.	+3.6	n/a	+4.5	+3.7	+3.1
	Precip.	119	97	120	102	105
	Runoff	122	100	116	83	23
UKMO	Temp.	+5.6	n/a	n/a	+5.3	+4.7
	Precip.	104	107	n/a	119	122
	Runoff	106	115	n/a	118	88

Note: Modified scenarios were used in the Indus, as described in the Indus section in the text. Note also that selection and averaging of grid cells and runoff stations, and annual averaging of some marked seasonal changes, for this summary table obscure the complex effect of temperature changes, seasonality shifts, and subbasin contributions in the final discharge changes. For example, runoff increases proportionally greater than precipitation in the Uruguay under GFDL because much of the rainfall increase occurs in a single month.

Management impacts stemming from hydrological changes are a function of the extent of basin resources currently developed. One measure of resource development useful for explaining the management impacts of hydrological changes is the ratio of available storage to annual runoff (Table 5). In the Colorado River basin, for example, the storage to runoff ratio is 4.0, which suggests a highly developed and managed basin that may be able to mitigate a wide range of hydrological impacts. From Table 5, it can be seen that the Indus, Mekong, and Uruguay are more sensitive than the other basins in terms of water control and storage, although these sensitivities are mitigated by a more humid climate in the Mekong and by natural storage in snow and ice in the Indus.

The different basin sensitivities, levels of development, and hydrological changes associated with the climate scenarios all result in a wide range of water resource sector impacts, which are discussed now in more detail for each case study.

Table 5. *Storage to runoff ratio (S/R) for the case study basins*

	Uruguay	Mekong	Indus	Zambezi	Nile
Current S/R	0.10	0.10	0.25	3.00	2.50

THE URUGUAY BASIN

The Uruguay River basin (Figure 3) is located in South America's most developed region in terms of population, urbanization, agriculture, and industrial infrastructure. It is the most developed basin in the set of cases. The basin's commercial services, economic marketing, and industrial and agricultural production extend their influences throughout Latin America and the world, far beyond the hydrological watershed. Increasing demands from population and economic growth are shaping water resource planning in all countries that share the basin. For example, energy needs for increasing industrial and population growth, especially in southern Brazil, have raised the priority of hydroelectric planning in the Uruguay Basin such that it dominates other basin development goals.

Physical and social setting

The Uruguay River, with a main stem length of 1,600 km draining a total area of 239,000 km^2, begins in Brazil's southern highland plateau and flows along Brazil's border with Argentina; farther downstream it marks the boundary between Argentina and Uruguay. Precipitation in the Uruguay basin diminishes from an annual average of 2,000 mm in the upper basin to 1,200 mm downstream in Argentina and Uruguay at the confluence of the Uruguay River with the Parana and Plata rivers. Seasonal distribution of precipitation is characterized by two peaks, in winter and spring, which are also periods of minimum evapotranspiration. Droughts have occurred because of abnormally short rainy seasons, and floods occur frequently in the winter and spring, when typical precipitation is enhanced by mid-latitude depressions. Flooding is especially problematic in the large cities occupying flood plains in the middle and lower basin. Mechanized medium- to large-scale farms have taken advantage of this rainfall seasonality by alternate double cropping of winter wheat and spring soybeans. Soybean farming is very productive in the basin, but this development has

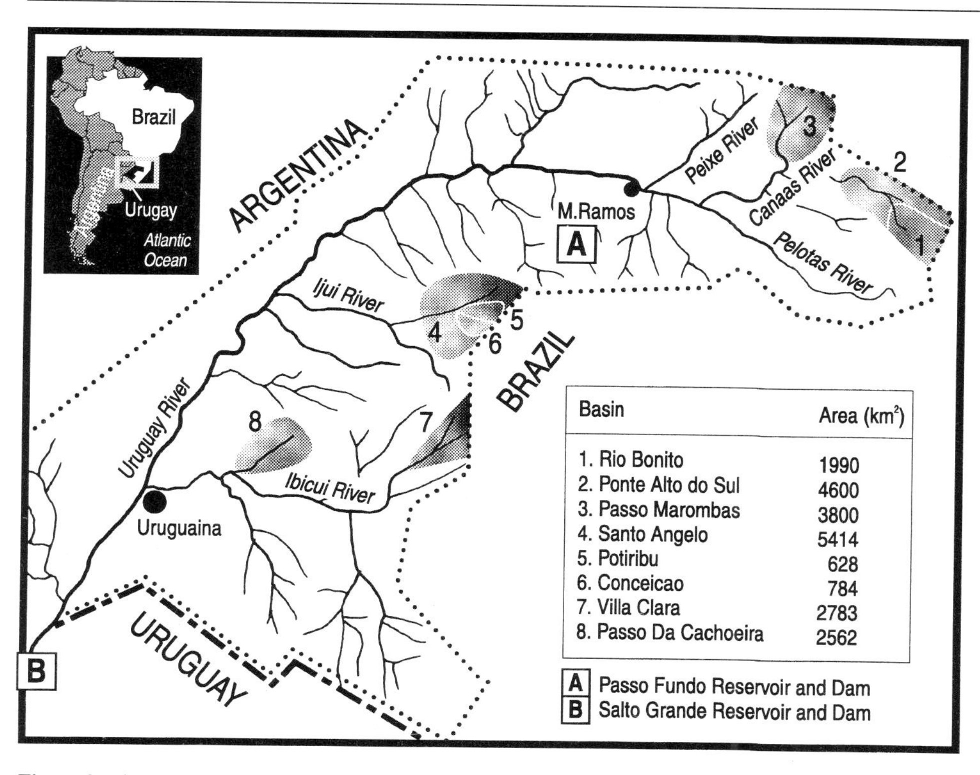

Figure 3 The Uruguay River Basin. Subbasins used to develop the basin-wide hydrological model are indicated by shading.

brought with it environmental and social costs, such as soil erosion and stream and reservoir siltation.

The upper Uruguay is the focus of most development. Here the river runs rapidly along a deep and rocky streambed with steep, vegetated banks. Brazil has increasingly viewed this portion of the basin as a tremendous hydroelectric resource and has plans to construct approximately 30 dams here. However, small farming communities in the upper basin have reacted strongly against this hydropower planning, protesting the inundation of farmlands and the threat of resettlement. Currently, only one upper basin hydropower dam exists, operated by Rio Grande do Sul's State Electric Utility Company at Passo Fundo. A second hydropower dam at Ita is under construction.

The middle Uruguay basin comprises flat terrain and soils conducive to irrigated rice farming, which is expanding in importance on the Brazilian side. A key feature of the middle basin is the city and dam of Salto Grande in Uruguay. Salto Grande Dam, the largest in the basin, was the world's first successful binational hydro project, joining Argentina and Uruguay in cooperative water resources development. The lower Uruguay becomes a characteristic plains river, running for 348 km below the Salto Grande Dam to the confluence of the Plata River at the border of Argentina and Uruguay.

In summary, the Uruguay basin is a rich and diversified agricultural and industrial region that supports a large population in three nations. It has a very high potential for hydroelectric production, and a precedent has already been set for international cooperation in this sector, although some plans have met with stiff opposition from local farmers. Other basin development goals, such as flood control, water supply, navigation, irrigation, and environmental quality, are becoming increasingly important but tend to be delegated and managed locally by sectoral interests of municipal, state, and national governments. Thus, although there is some comprehensive planning for hydropower, other goals are typically not pursued via an international, basin-wide planning approach.

Hydrological impacts

We used a rainfall/runoff model developed by Brazil's Institute of Hydraulic Research (IPH) to translate climate scenarios into runoff estimates for the Uruguay. The IPH III model, described in Motta and Tucci (1984), uses climate inputs to calculate runoff based on physical characteristics of the basin. A key parameter in IPH III, potential evapotranspiration (PET), was calculated by applying the standard Penman equation method to the GCM and arbitrary temperature scenarios. The results indicate an increase in evapotranspiration due to the warming common to all the scenarios.

IPH III has been used successfully for flood forecasting in various subbasins of the Uruguay. Eight key subbasins were

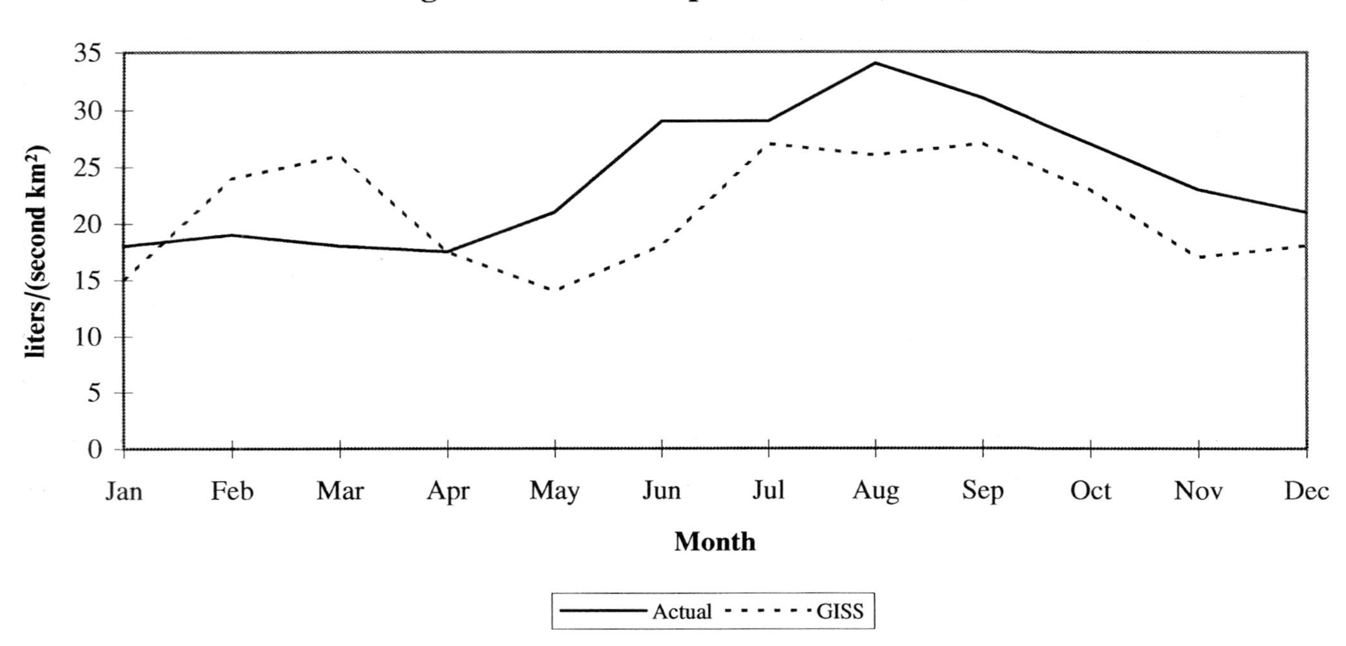

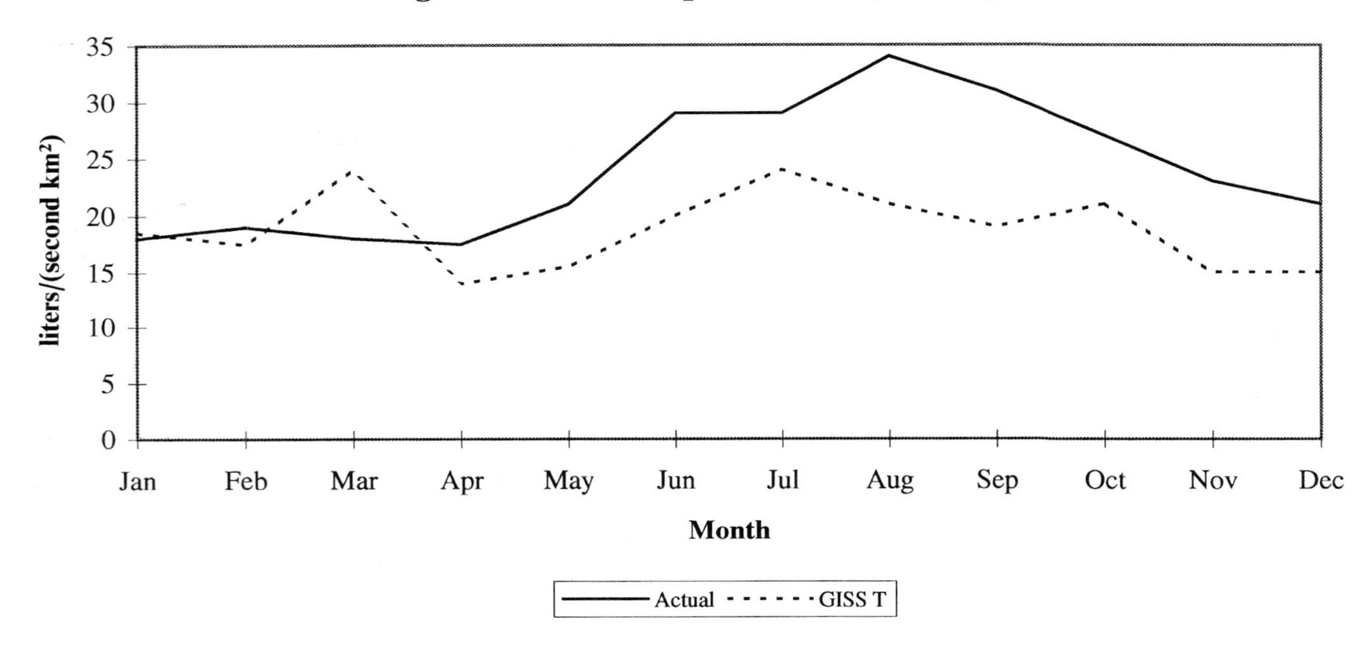

Figure 4 Actual vs. projected discharge in the Uruguay Basin for four global climate model scenarios.

selected for model runs based on actual data and the climate change scenarios, and the model presented a good fit to the recorded data. These subbasins were then used as analogs to other subbasins so that a discharge could be estimated for the full basin above Salto Grande Dam. The individual flows of the subbasins were regionalized by empirical standard coefficients, developed by Uruguay basin water managers, to obtain an estimate of water resource effects for the basin as a whole. This regionalization of hydrological functions enables recorded data and modeled hydrological functions or parameters from the eight subbasins to be transferred to others where such information does not exist or has been available only for short periods.

The basin was included in one GCM grid cell except in the case of the UKMO model, where two cells overlapped the basin. These were denoted as UKMO1 and UKMO2, corres-

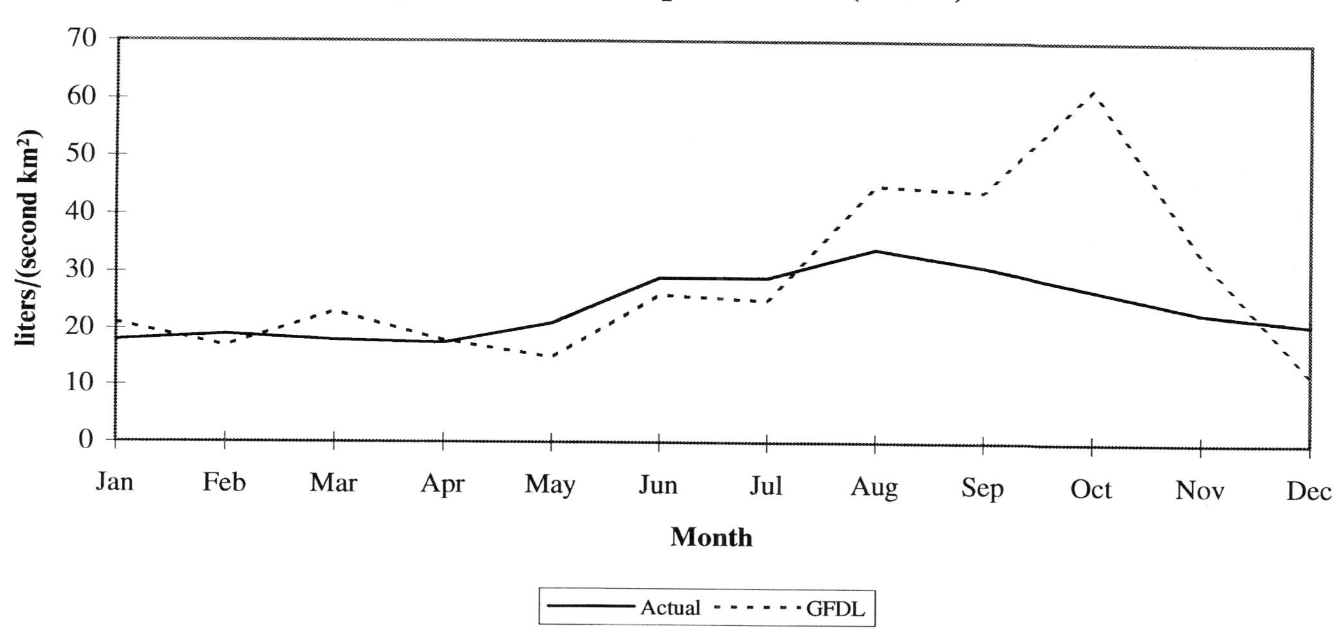

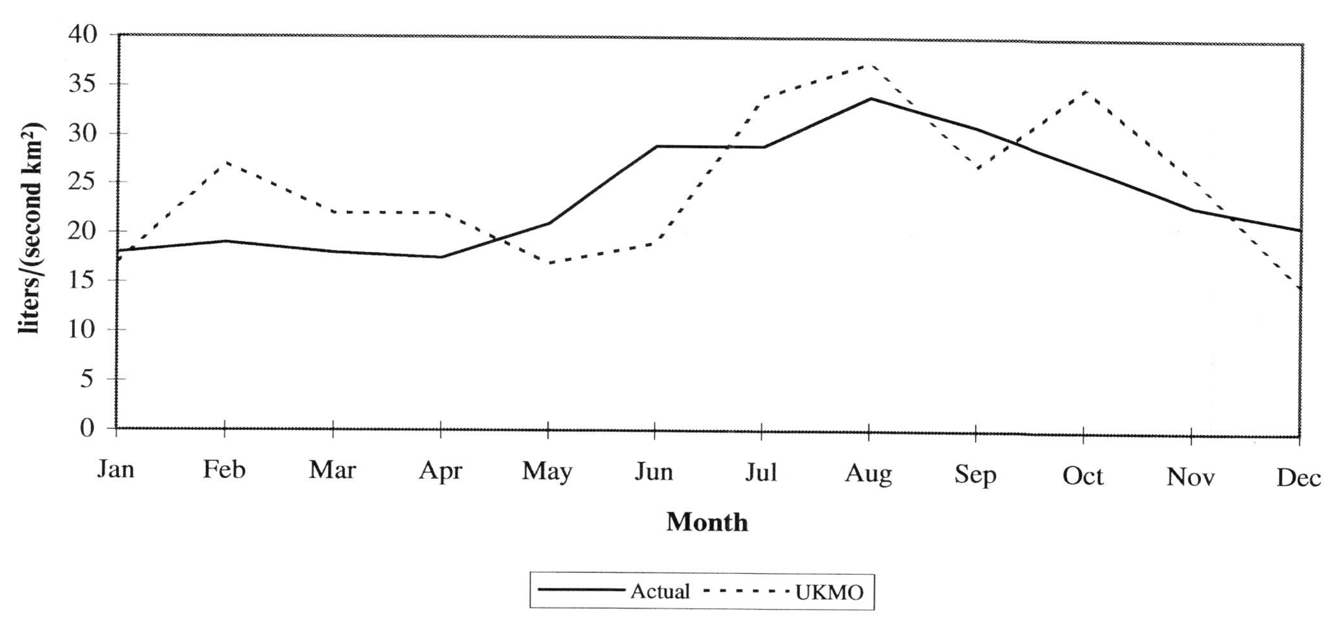

Figure 4 (*cont.*)

ponding to the upper and middle Uruguay, and were run separately.

The GISS and GISS Transient A scenarios project an average annual runoff reduction of 11.7 percent. The GFDL scenario projects an increase of 21.5 percent, and the UKMO simulation results in an average increase of 6.4 percent (Figure 4). All the GCM scenarios result in some shifts of seasonality of high and low flow periods and all the scenarios project a decrease in runoff during low flow periods of the year (less pronounced for UKMO1 than for the other scenarios). The GFDL scenario presents extreme variations in rainfall for two months in contrast to the other GCMs, with a 154 percent increase in rainfall for October and a significant decrease (46 percent) in December. Because of the October results, the GFDL scenario shows a significant increase in average annual rainfall and subsequent runoff.

Management impacts

Management functions projected for the climate change scenarios included mean specific flow; mean annual, flood, and low-flow probability curve functions; flow duration curve; and storage yield curve. Flood probability curves are used to design hydraulic structures and evaluate flood risks. Mean flows are used to assess water supply availability in the basin for domestic and industrial uses and for irrigation. Low flow probability curves are used to assess projects and problems related to water and environmental quality and navigation. Storage–yield curves are common in evaluations of storage reservoir volume for hydropower dams and flood control.

HYDROPOWER IMPACTS

It is likely that the critical factors relating climate changes to impacts on water resource planning in the Uruguay basin are changes in seasonality and intensity of rainfall during low flow months. These factors are important because of the basin's dependence on hydropower and its history of flood problems associated with intense rain events.

Hydropower is a key, sensitive sector in the region. The basin presents a large hydropower potential, with a yield of 40.5 kilowatts per square kilometer (kW/km^2). The total energy potential for the basin is 16,500 megawatts (MW); the total energy developed is 2,680 MW, and 13,500 MW are planned to be developed in the next 20 years. Existing projects and planned developments most likely to be built in the foreseeable future are identified in Table 6. Brazil has been the leader in hydroelectric development in the Uruguay basin, and hydropower accounts for over 90 percent of that country's energy consumption. Further development of the Uruguay basin is planned in order to keep pace with increasing Brazilian energy demands, especially to service the larger urban/industrial areas of Rio Grande do Sul. Moreover, because Brazil's energy grid is operated as a completely interlinked network, hydroelectricity from Uruguay basin projects can be transferred to the populous central portions of the country to supply huge industrial complexes and large cities, such as São Paulo.

Currently, Brazil operates one dam on the Uruguay and is constructing another, and plans call ultimately for 22 hydropower projects on the Brazilian reaches of the river. But not all of the basin is managed unilaterally. Three bilateral projects have been proposed by Brazil and Argentina on the international reach, and Argentina and Uruguay manage jointly the most important existing Uruguay River reservoir at Salto Grande. Project storage–yield curves calculated in engineering studies, regional storage ratios based on the GCMs, and descriptive data on actual and planned developments were used to assess the impacts of climate change on basin hydropower. For cascade developments, relative project conditions were maintained. The results should be considered relative, not absolute, since they come from different planning scenarios.

Table 6. *Uruguay hydropower reservoir characteristics*

Development	Operation year	Area (km^2)	Average flow (m^3/s)	Useful volume (km^3)	Power (MW/yr)
Passo Fundo	1972	2,300	51	1,560	220
Campos Novos	1997	14,200	267	527	880
Machadinho	2001	35,800	664	4,510	1,200
Garabi	1999	4,640	n/a	8,300	1,800
Monjolinho	2000	4,510	115	9	72
Barra Grande	2001	12,860	256	3,865	920
Garibaldi	2001	13,200	250	1,945	228
Pai-Quere	2001	6,250	129	1,742	288
Ita	1997	44,500	883	3,590	1,620
Salto Grande	1979	90,000	4,640	5,000	1,890

Table 7. *Uruguay power production impacts*

Development	Power (GWh/year)			
	Present/Planned	GISS	GFDL	UKMO
Passo Fundo	1,061	993	1,145	1,068
Salto Grande	9,112	8,852	11,032	8,563
Campos Novos	4,243	3,881	4,543	4,118
Machadinho	5,786	5,465	6,984	5,742
Garabi	8,678	8,639	10,386	8,493
Monjolinho	347	330	382	356
Barra Grande	4,436	4,140	4,990	4,432
Garibaldi	1,099	1,016	1,245	1,109
Pai-Quere	1,389	1,287	1,538	1,404
Ita	7,810	7,266	9,338	7,565

Storage–yield curves did not take evaporation into account. The increased evaporation associated with all of the scenarios would tend to reduce energy production. Though the typical response to such an impact is to compensate with increased reservoir capacity, there is a trend against developments that flood large areas due to the pressure from environmental interests and local land owners. The emerging trend toward smaller reservoirs could reduce the impact of evaporation on energy production. The existing developments do not flood large areas and are thus less sensitive to evaporation loss.

Planned (or actual in the case of Passo Fundo and Salto Grande) power production capacity is compared in Table 7 to predicted capacity for the GCM scenarios. The GISS scenario yields an average 4.8 percent reduction of energy

Table 8. *Summary of energy production impacts and associated costs*

	GISS		GFDL		UKMO	
	Energy (GWh/yr)	Cost (million U.S. $)	Energy (GWh/yr)	Cost (million U.S. $)	Energy (GWh/yr)	Cost (million U.S. $)
Total basin	−2,090	433	7,622	1,580	−1,108	229
Designed	−1,760	365	5,618	1,165	−576	117
Operation	−330	68	2,004	415	−541	112
Salto Grande	−260	53	1,920	398	−184	38
Garabi	−40	8	1,708	354	−548	113
Brazil	−1,810	375	4,848	1,005	−468	97

produced. The GFDL scenario forecasts an average increase of 17.3 percent and the UKMO scenario projects a reduction of only 2.5 percent. Table 8 shows the costs (gains or losses) associated with each scenario. With GISS, a reduction of 1,810 gigawatt hours per year (GWh/yr) in Brazil's total designed power would be possible, equivalent to a 375 MW power plant. To recover this energy it would be necessary to invest approximately U.S. $375 million (the current cost of constructing such a plant). In contrast, the GFDL scenario projects a 4,848 GWh/yr increase in power production, corresponding to a capacity of 1,005 MW, achieved without any increase in physical plant capacity and resulting in a windfall of over $1 billion in benefits. Because we did not conduct detailed systems simulations, we cannot estimate whether some of this extra power may be lost due to reservoir spillage. The UKMO scenario indicates a small change from present conditions: a 468 GWh/yr reduction, the equivalent of a 97 MW power plant. Overall, the GISS and UKMO scenarios predict reduction in hydropower generation from existing and planned projects by an average of 3 to 5 percent, a potential loss of roughly 1,810 GWh/yr. In contrast, the GFDL scenario predicts an increase in Brazilian power production by 4,848 GWh/yr (Table 8).

A small-scale and less centralized energy system also operates in the Uruguay basin to serve local municipalities and rural cooperatives. Some 30 small or micro hydro projects (10 MW or less, with little or no storage) currently operate, and the number continues to expand because of the basin's enormous potential for micro hydro development. The GISS and UKMO scenarios, however, would seriously undermine the expansion of small hydro projects in the basin by reducing expected benefits by 26 and 16 percent, respectively.

FLOOD IMPACTS

The impact of climate change on the flood-prone areas of two cities was evaluated by utilizing the maximum flow probability curves and the discharge curve for the lower and upper reaches of the basin. The results for Uruguaiana (Table 9) indicate a significantly reduced flood hazard under the GISS climate change scenario. However, GFDL projects a considerable increase in the risk of flooding. The GFDL projection is influenced especially by the predicted change in October rainfall which would cause very high maximum flows. The UKMO model gave results similar to those for GISS.

Table 9. *Impacts on flood-prone areas of Uruguaiana*

Return period (years)	Probability (annual)	Change in flood stage (m) GISS	GFDL	UKMO	GISS T
10	0.10	−0.66	4.18	−0.64	−1.60
30	0.03	−0.80	4.92	−0.78	−1.85
50	0.02	−0.84	5.28	−0.80	−2.25
100	0.01	−0.87	5.70	−0.82	−2.53

In theory, the hydropower reservoirs are operated so that they maintain an empty volume that is available for controlling floods downstream and for dam safety. However, emphasis on hydropower production has detracted from efforts to provide flood control, and no basin-level planning to reduce flood risks has been undertaken. The flood control volume is calculated based on historical series and on operational parameters, and, unfortunately, the volume of *existing* reservoirs is yet too small to have significant effects on downstream floods. The Salto reservoir may contribute to flood control downstream, but for large upstream floods the reservoir volume is small compared with the volume of the potential flood hydrograph.

The potential for flood impacts varies considerably for the GCMs; it could increase or be reduced dramatically. Although the hydropower sector would welcome the increased flow conditions under GFDL, local officials and riverine populations at risk from floods now could be even more at risk in the future, and areas now safe could be added to the roster of flood hazard zones. Under the GFDL scenario, maximum flood flows would increase by 50 percent, rare floods would be twice as common, and flood damages would markedly increase in the many urbanized stretches of the river. Alternatively, the flow reductions predicted by GISS and UKMO might elicit an attempt by managers to reduce the flood control volume in order to maintain energy production, thus maintaining the current flood risk.

Adaptations

Because the current basin development approach discourages integrated planning, it is difficult to take a comprehensive approach to interpreting the significance of potential climate change. The results of modeling identified a range of possible climate impacts and elicited a wide range of potential adjustments. For instance, river basin planners responsible for energy production expressed concern over the prospects of climate change but were confident that the existing and planned hydropower infrastructure and practices, especially for the large reservoirs at Ita, Salto Grande, and Garabi, could be adapted to manage future changes. It is less likely that incremental or modest adjustments could cope with the increased flooding projected in some of the scenarios.

It is interesting to note that because GISS reproduced a very good fit between the $1 \times CO_2$ scenario and the basin's recorded temperature and rainfall patterns, there was a tendency on the part of managers to consider it the most plausible projection of climate change.

The overall conclusion of basin managers attending the adaptations workshop for the Uruguay basin was that climate change must be considered within the evolving context of regional development already unfolding in the river basin and in Latin America. The nature of future regional development is the most critical factor in fashioning effective adjustments. Global warming impacts were viewed as another set of pressures on water resources, probably of less importance than those resulting from population growth, establishment of an international regional economic market, and deterioration of overall environmental conditions in the basin. The potential worsening effect of climate change on water resources should be evaluated on a number of different scales because the benefits of energy and food production, flood control, and environmental quality extend beyond the physical boundaries of the river basin to the national level and across international boundaries.

By linking the potential pressures of climate change with constraints that are already part of the context of development, basin managers tended to focus their recommendations for adaptation on measures that would strengthen the process of water resource planning and management regardless of climate change. The best adaptive climate change strategy was seen, ultimately, as improving the process of integrated planning and management.

THE MEKONG BASIN

The Mekong is Southeast Asia's largest river, as well as the world's twelfth largest, in terms of both discharge (approximately 0.467 BCM[1] annually) and main stem length (about 4,000 km). After flowing in a deeply incised valley for the first 1,850 km of its course, the Mekong enters the lower basin near the Burmese–Laotian border. One-fourth of the entire Mekong watershed is in the upper basin, but this area produces only one-tenth of the runoff. International planning efforts have been restricted to the lower basin, the focus of this study.

Physical and social setting

Mean annual rainfall over the lower Mekong basin ranges from about 1,000 mm in parts of northeast Thailand to as much as 3,000 mm over some of the mountains in northeastern Lao PDR, with a wet season lasting from June to November. Estimates of annual potential evaporation range from 1,500 to 1,800 mm. The climatic pattern of distinct wet and dry seasons is strongly reflected in the runoff pattern. Only upstream of Chiang Saen is Mekong discharge influenced more by snowmelt than by monsoon rainfall.

The lower Mekong basin covers parts of Cambodia, Laos, Thailand, and Vietnam (Figure 5). The region is marked by rich cultural diversity, as well as several different political and economic systems. Despite these differences, the lower basin countries cooperate in water resources planning, having engaged since 1957 in one of the world's unique international river basin development institutions: the United Nations-sponsored Mekong Committee. The committee coordinates basin-wide water resources planning and provides a forum for dialogue among the basin nations. Yet, despite this cooperation, development in the lower Mekong has been constrained by war and funding. Indeed, an overriding feature of the lower basin is its relative lack of development. No dams and few bridges cross the main stem of the Mekong.

The majority of the inhabitants of the lower basin – roughly 80 percent – are engaged in agriculture, mainly paddy rice cultivation. The Mekong Delta in Vietnam is the most agriculturally productive area of the lower basin, and a set of climate-related water resource problems in the delta – salinity intrusion during the dry season, seasonal floods,

[1] BCM = billion cubic meters = 10^9 m^3

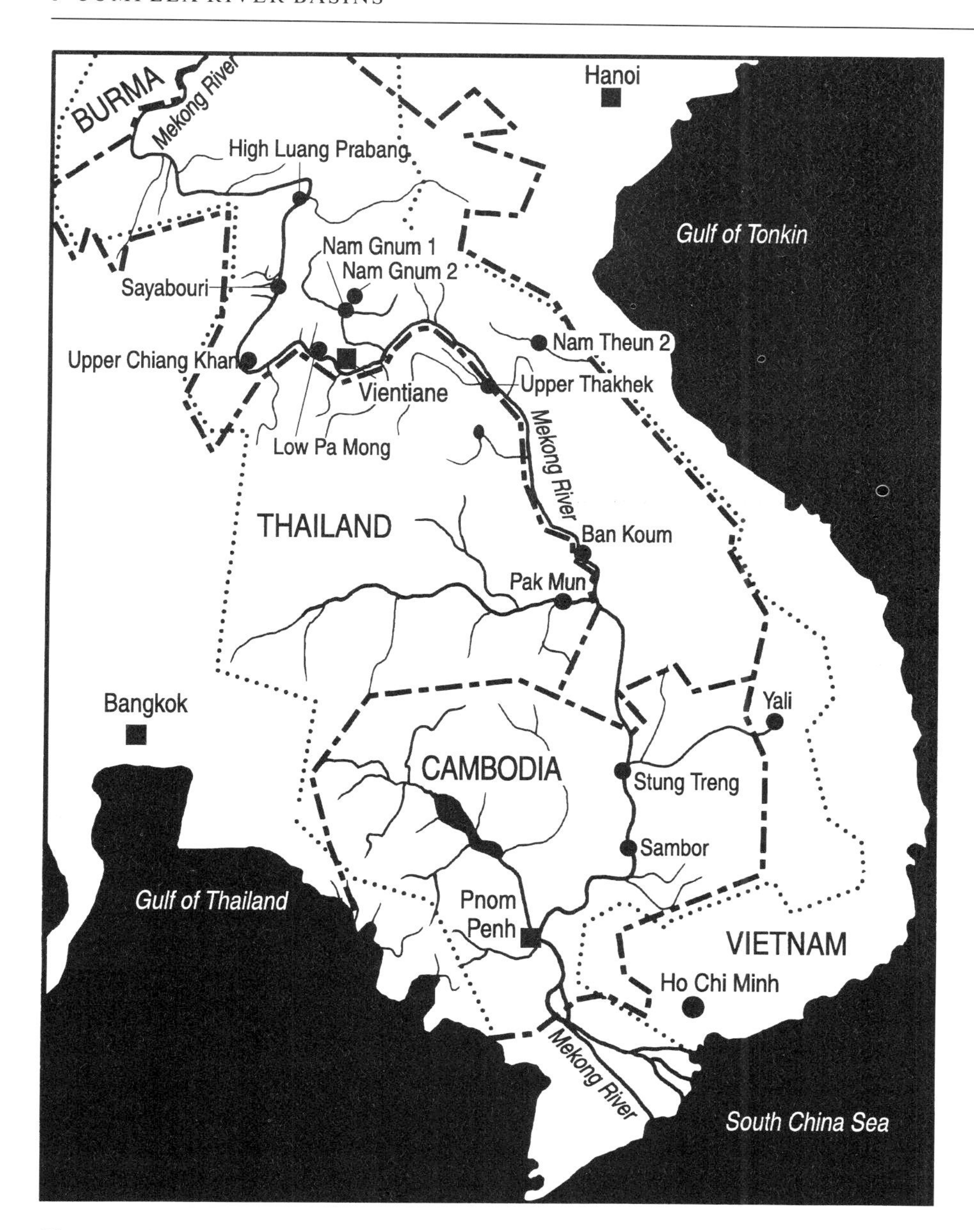

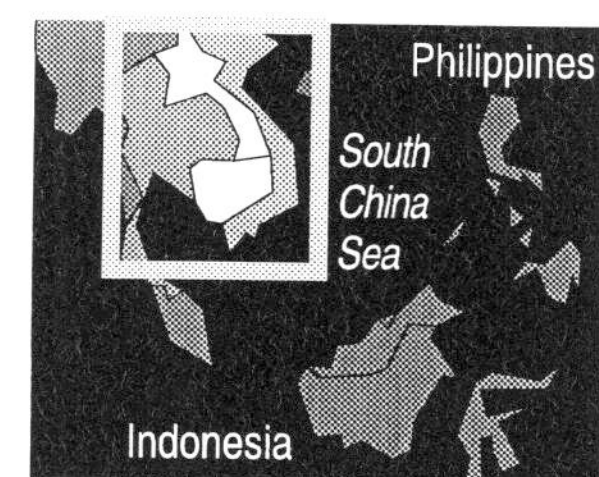

Figure 5 The Lower Mekong Basin showing locations of existing and planned developments.

drainage and waterlogged soils, possible sea-level rise – make it perhaps the area of greatest vulnerability to climate change.

Hydrological impacts

The lower Mekong basin study followed the approach of the other river basin case studies. Output of selected GCMs along with arbitrary increments were used as scenarios of future climate. To translate the GCM output to runoff, the nearest grid cell values for precipitation were assigned to each climate station utilized in operational runoff forecasting by the Mekong Committee. Due to limitations on the number of basin simulation computer runs, only the mid-point values of the GISS transient runs were utilized, and due to the similarity of the runs, only Transient A data are reported here. This can be construed as a 'cooler' equilibrium of doubled greenhouse effect. For the sensitivity analysis, values of precipitation at each station were simply multiplied by 1.2 (for a 20 percent increase) or 0.8 (for a 20 percent decrease).

The climate scenarios were used to drive a rainfall–runoff model, which generated sequences of mean monthly discharge for selected stations along the Mekong. Full details of the modeling are given in Phien and Ti (1990). The discharges were also translated into effects on Mekong Delta salinity and flood characteristics. Finally, the altered streamflow sequences were used as inputs to the MITSIM (Massachusetts Institute of Technology River Basin Simulation model), the management model used to assess the potential effects in such critical water resource areas as reservoir operations, hydropower production, and irrigation water deliveries.

It is important to note here that because previous runoff modeling for the Mekong has focused on short-term flood forecasting, there is little experience with assessing the effects of concurrent changes in temperature at a monthly time frame. Though we conducted a sensitivity analysis of the effects of warmer temperatures on runoff using the greatest GCM-derived $2 \times CO_2$ warming (which results in a roughly 3 percent additional decrease in runoff), for the main analysis we used a rainfall–runoff model that does not include temperature effects.

Hydrographs for Luang Prabang, located near the top of the basin, and Pakse, located in the middle of the basin, are shown in Figure 6 (the base period varies among stations but is generally 1960–1980). Comparisons of scenario-perturbed discharges and respective rainfall sequences show that the discharges reported here reflect very well the simulated changes in rainfall, thus validating the runoff model. This correlation is particularly clear for the UKMO case, which introduced the most pronounced effects on monthly flows. Note also that the peak monthly discharge occurs earlier the further downstream it is measured from Luang Prabang to Pakse. The GISS and UKMO scenarios for Luang Prabang introduce an increase in mean monthly discharge for the rising limb (May, June, July) of the hydrograph, but both scenarios retain the normal peak flow time at Luang Prabang. The peak time would be delayed one month in the GFDL scenario. Transient A gives a decrease in monthly discharges for the months prior to the peak and an increase in monthly discharges after the peak. The peak would be delayed one month according to both transients. For Pakse, the GISS and UKMO scenarios introduce a very significant increase in discharge for January, February, June, July, and August. GISS gives a significant decrease in December, and UKMO gives a significant decrease in April. A slighter increase shows up in November, according to GFDL, which also introduces a significant decrease in January.

The GISS Transient A scenario results in a significant increase for the months of February through June and for October. Although the GFDL and GISS Transient A results would defer the peak time one month, to September, the remaining models would not change the time of occurrence of the highest monthly discharge. GFDL tends to delay flood peaks slightly, but the effect is less pronounced than the seasonality changes associated with UKMO and GISS.

Water management impacts

Effects of the runoff changes on water management were assessed in relation to a planned system of 13 reservoirs (one of which currently exists), which constitutes the keystone component of regional water development in the delta (Ti and Phien, 1990). Two chief indicators were used to assess water management effects of the altered flows: percentage of time that the reservoirs are at full capacity and hydropower production – both compared to planned performance under the base condition (1950–1980). A wide range of reservoir content changes result (Figure 7a), as the system simulation model attempts to meet power production goals by recoordinating reservoir fill and draw-down operations. In general, reservoir operations are disrupted by any significant climate change because operating rules, developed under assumptions of climate stability, cannot be expected to function properly under altered hydrology. This disruption shows up as reduced time that the reservoirs are at full levels, especially in those with reasonably large capacity, such as High Luang Prabang (HLP), Nam Ngum 1 (NN1), Yali (YL), Mun (MUN), and Stung Treng (ST), though some scenarios show large increases at some reservoirs as the altered runoff is apportioned by MITSIM's optimizing routine in an attempt to maximize power production. Only Pak Mun (PM) and Yali exhibit significant time empty (not shown). Yali runs dry 7–10 percent of the time under the various scenarios, compared to a planned 4 percent. A large increase in reservoir contents occurs at Nam Theun 2 (NT2) and Nam Ngum 2 (NN2) under some of the scenarios, but due to adjustments in cascade management these increases do not yield higher power production. Less significant effects occur at Upper Chiang Khan (UCK), Upper Thakhek (TK), Ban Koum (BK), and Sayabouri (SYB).

Despite built-in adjustment capacity, reservoir impacts translate into power production declines in most cases because the system is optimized for the base flow. The effects of climate change can best be seen by plotting total power generated under altered climate scenarios against the base climate (Figure 7b). Several of the reservoirs would experience reduced power production under the different scenarios, especially UCK, PM, TK, and YL. For other power plants, the changes introduced by the GCMs are not significant.

Because most of the climate scenarios result in a reduced operating efficiency of the system overall, its benefit/cost ratio is also lowered, with benefits reduced by as much as U.S. $1 billion (Table 10). For hydroelectric power generation, the total costs remain unchanged at $28.9 billion, but the net benefits vary with the different scenarios. In general, there is significant decrease in the total net benefits associated with all of the GCM runs and with the -20 percent rainfall sensitivity run. GISS and GISS Transient A reduce the benefits of the system by roughly U.S. $1 billion.

Another important feature of the basin is its extensive and productive delta, which is annually flooded by the uncontrolled river. Except under the GISS scenario, typical delta

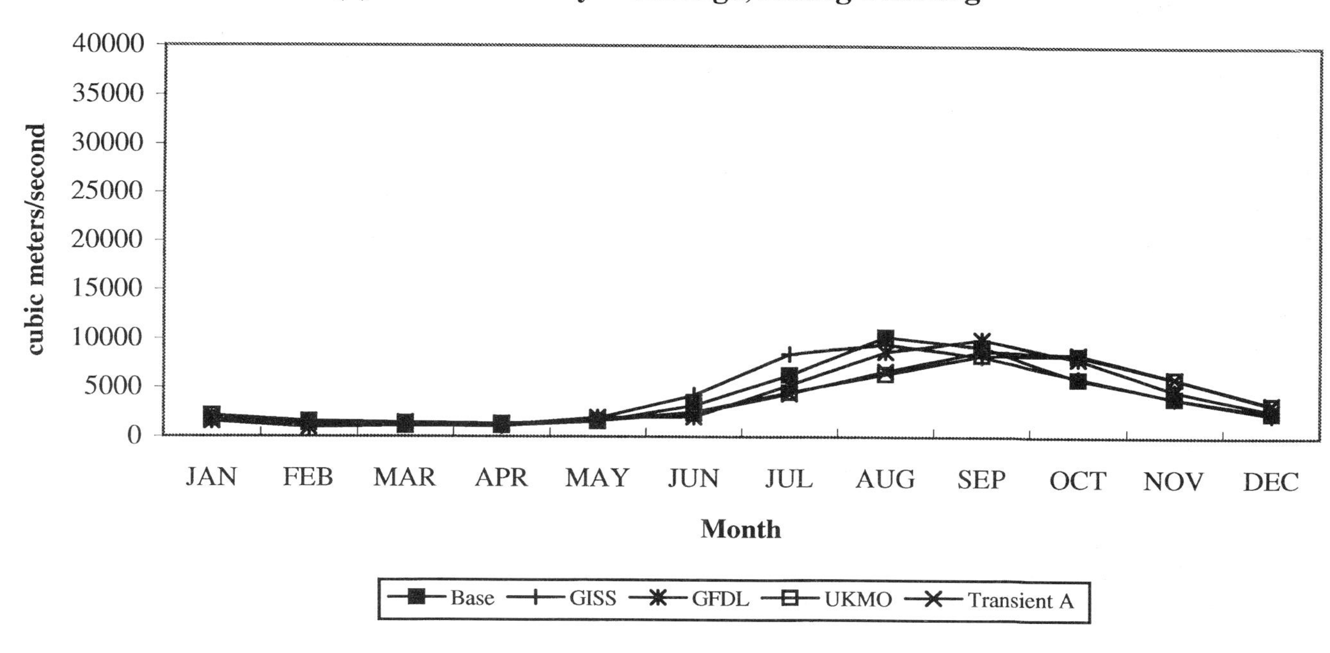

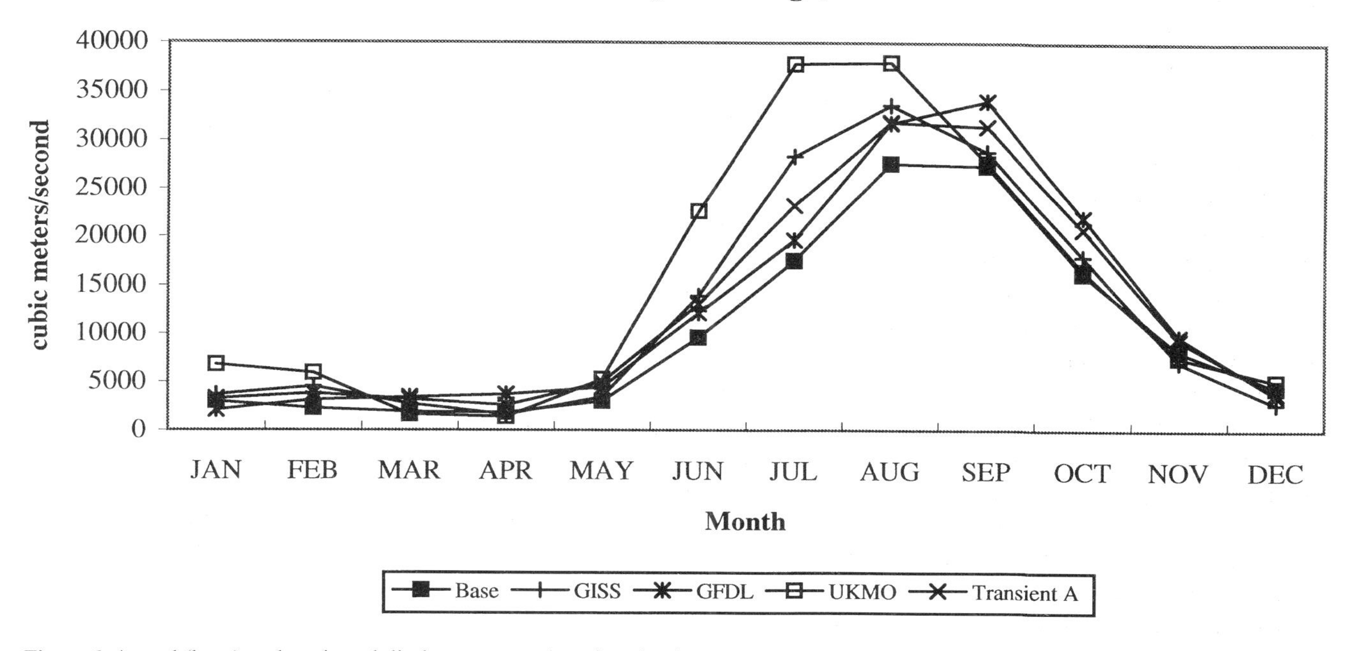

Figure 6 Actual (base) and projected discharge at two locations in the Lower Mekong Basin for four global climate model scenarios. (a) Luang Prabang and (b) Pakse.

floods would be longer lasting and higher than baseline (Figure 8). Although the annual flood is important to rice production and for its ability to repel salt water intrusion, the projected increase of flood duration, especially under UKMO, would cause considerable damage to agriculture and fisheries – an effect that would be exacerbated by sea-level rise.

Adaptations

We assessed the potential for adjusting Mekong basin management to climate change by holding discussions at a workshop of representatives from the riparian countries and international management institutions. The workshop also helped to broaden awareness and interest among senior

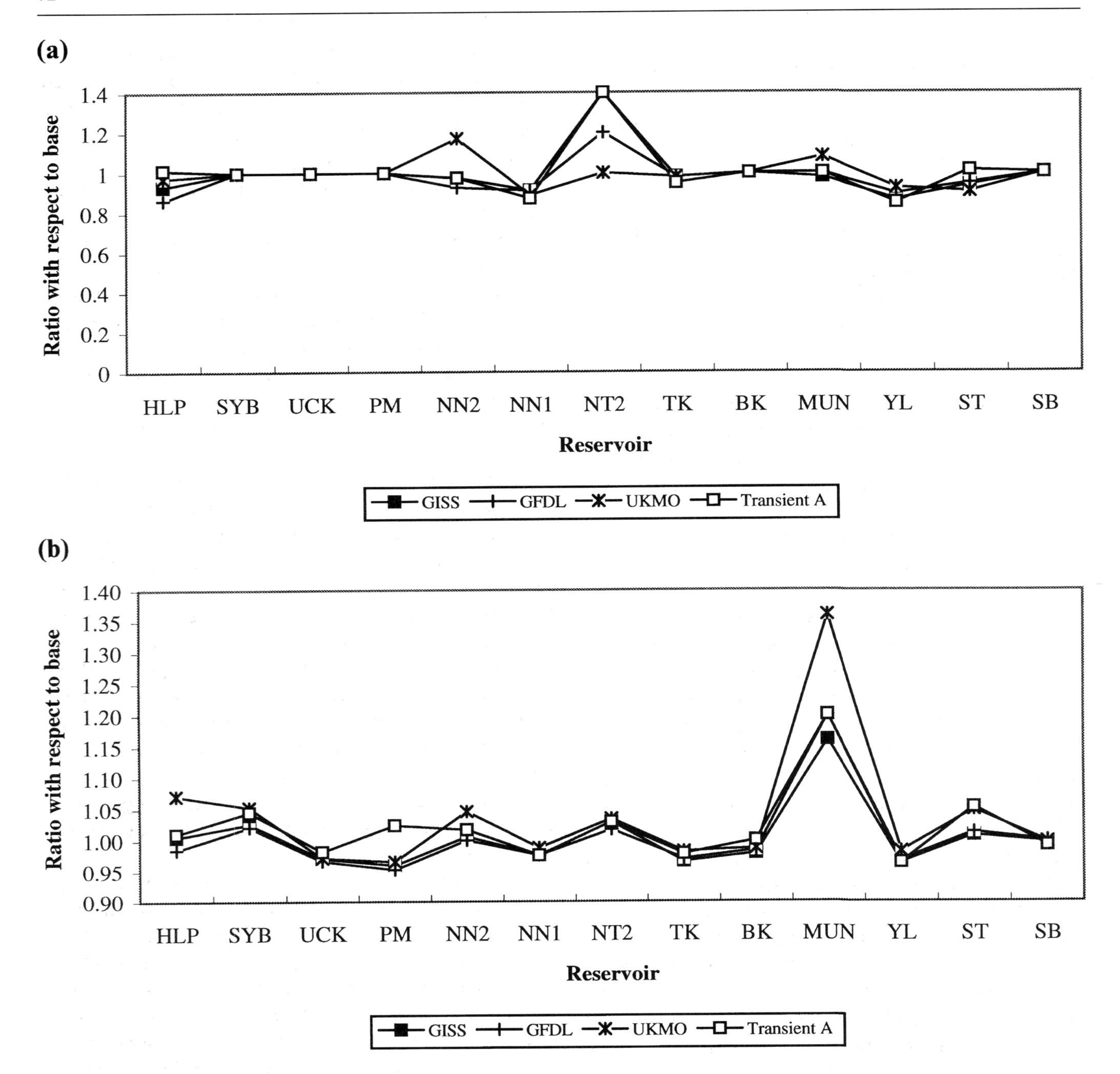

Figure 7 (a) Ratios of months when reservoirs are full under projected runoff compared to actual (base) conditions, and (b) ratios of hydropower production under projected runoff compared to actual (base) for four climate scenarios. Reservoir abbreviations: HLP = High Luang Prabang; SYB = Sayabouri; UCK = Upper Chiang Khan; PM = Pak Mun; NN2 = Nam Ngum 2; NN1 = Nam Ngum 1; NT2 = Nam Theun 2; TK = Upper Thakhek; BK = Ban Koum; MUN = Mun; YL = Yali; ST = Stung Treng; SB = Sambor.

water resources planners in each riparian country and to identify follow-up studies and efforts to incorporate the findings in the basin planning process.

The participants noted the high degree of agreement of the pattern of temperature and precipitation changes projected by the three GCMs used in the study and were particularly concerned with possible shifts in the occurrence of the month of maximum rainfall, discharge, and changes in the length of wet and dry spells. They also expressed concern about the large changes in streamflow generated by some GCM scenarios, which were greater than 20 percent in several months. The participants agreed that this magnitude of change in streamflow warranted further detailed studies of impacts on typical water resources projects in the lower

Table 10. *Benefits of the Mekong cascade under climate change*

	Net benefits (U.S. $1000)	B/C ratio
Base	63,669,296	3.31
GISS	62,713,590	3.18
GFDL	62,979,940	3.19
UKMO	63,548,370	3.21
Transient A	62,598,880	3.16
−20% precip	61,562,430	3.14
+20% precip	72,769,420	3.53

basin. Participants felt that the general magnitude of reduced flow associated with temperature rise (which was not directly modeled in the study, but was estimated in the sensitivity test to be 3–5 percent for the extreme scenarios in the northern area and 1–2 percent for the southern area of the basin) would not greatly affect the conclusions.

Potential increases in floods caused by changes in the timing of flood peaks, as well as magnitudes, also elicited concern. The impacts on low flow were shown to be of little significance when no effects of temperature increase were included, though with temperature increase, low flow may be reduced from 2 percent to 5 percent in the northern region and 1 percent to 3 percent in the southern basin.

Existing reservoir rule curves do not appear to deal well with the different streamflows attending climate change, though some of the cascade projects examined in the study, such as High Luang Prabang and Stung Treng, perform better than others under climate change. Changes in rule curves will be necessary either to maintain or to maximize project benefits under new climatic conditions.

The group expressed concern over the possible effects of greater fluctuations in the annual flow regime on reservoir sedimentation. Participants also voiced concern about changes in water quality resulting from global climate change, as well as attendant impacts on fisheries. As these effects were not assessed in this study, it was suggested they be examined in future research.

Both structural and nonstructural adjustments were envisioned to help water management systems accommodate change:

- 1. Structural: enlarged reservoir storage; sediment control (checkdams, step-reservoirs, bottom spillways, dredging); more smaller dams and hydropower facilities, tributary projects; and pump storage development.
- 2. Nonstructural: reforestation to reduce sediment yield, in order to mitigate the effects of precipitation extremes; rule curve changes to optimize reservoir management under the new conditions; and changes in the mixture of peak and firm power demand and production.

Two typical reservoir projects in the Mekong system (Low Pa Mong and Nam Ngum 1) were selected for a detailed design and management study based on future climate scenarios. The study will consider potential structural (e.g., storage size) and nonstructural (e.g., rule curve) adjustments that would accommodate potential climate and runoff fluctuations.

Regarding the potential increase in the flood problems, it was suggested that methods for flood estimation be standardized and that additional and improved data collection and dissemination are critical. Further suggestions were that more hydrometeorological stations be established especially in Cambodia, that flood warning system(s) be improved and extended, and that studies on flood damage assessment, such as the one proposed by the Mekong Secretariat, be conducted.

An important overall conclusion of the workshop participants was that the global warming issue did not as yet necessitate large-scale, structural adjustments or dramatic changes in day-to-day operations. Their recommendations leaned toward more studies of altered hydrology and the effects of deforestation, better monitoring of critical basin changes, and better understanding of water use and management problems. This is not to suggest a skepticism about climate change on the participants' part. On the contrary, there was a genuine interest in the issue and concern on how to incorporate climatic change into planning. They saw the issue as genuine and felt a need to be better informed about global warming. However, until climatic change begins directly to affect their operations, changes in operations were not seen as necessary.

Finally, although integrated basin development in the Lower Mekong has been slow, it has proceeded by consensus and there have been few large-scale commitments. River basins around the world are replete with problems resulting from development schemes launched with only partial understanding of climate, hydrology, institutional coordination, and environmental and social impacts. River basin development is an exceptionally complex endeavor and can be successful only with a solid base of physical, socioeconomic, and legal knowledge. The gathering and compilation of these data take years, if not decades. A small-scale, long-term, deliberate approach to basin development has a number of advantages and can be viewed as an asset in adjusting to climatic change in the Lower Mekong.

THE INDUS BASIN

A complex irrigation system dominates the main basin of the Indus River in Pakistan, whereas the less developed upper basin is characterized by the high mountain, mostly glaciated

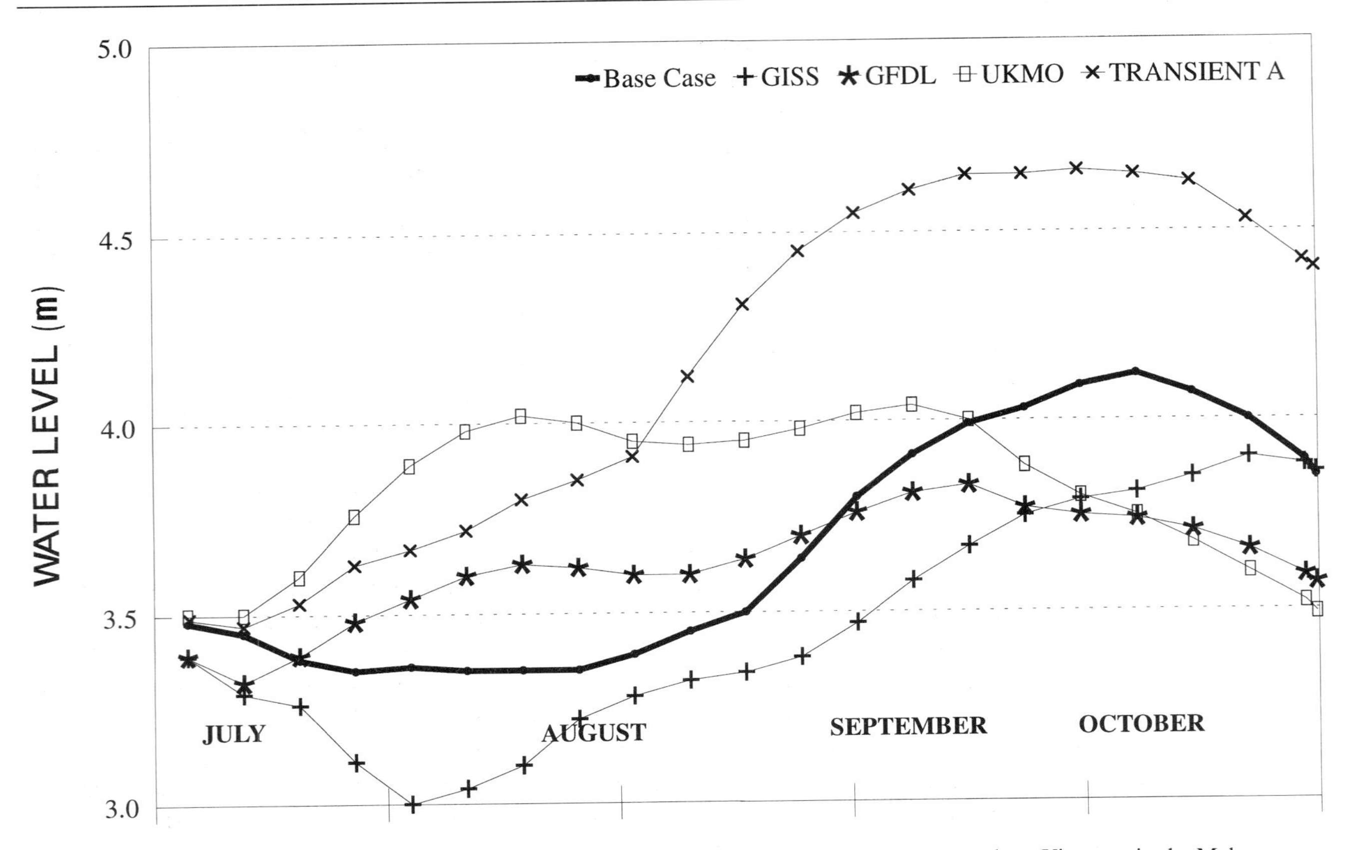

Figure 8 Typical flood stage (Base Case) and flood stage projected for four climate scenarios at Tanchau, Vietnam, in the Mekong Delta. All scenarios except GISS cause an earlier and longer lasting flood. The GFDL scenario causes a dramatic increase in flood height.

watersheds of the Himalaya (Figure 9). Indeed, the main basin is so heavily modified by agricultural development that the Indus study team called it a 'cultural' rather than a natural river (Wescoat and Leichenko, 1992).

Physical and social setting

The Indus River and its tributaries rise in the sparsely populated glaciated mountains of southern Asia. The basin includes portions of Tibet, India, Pakistan, and Afghanistan. It has a total drainage area of 960,000 km^2, some 60 percent of which lies in Pakistan. The mountainous upper basin is influenced by continental climates of central and western Asia, which have a westerlies pattern of circulation, late winter snowfalls, cold winters, and short, warm summers (Rao, 1981). This regional climatic pattern becomes highly complex within the high mountain ranges that make up the headwaters of the Indus watershed (Alam, 1972).

In the upper watershed of the Indus River main stem, pastoralists and traders traverse the mountain pastures and passes between spring and fall. Families in the decentralized village communities farm along the lower elevation glacier-

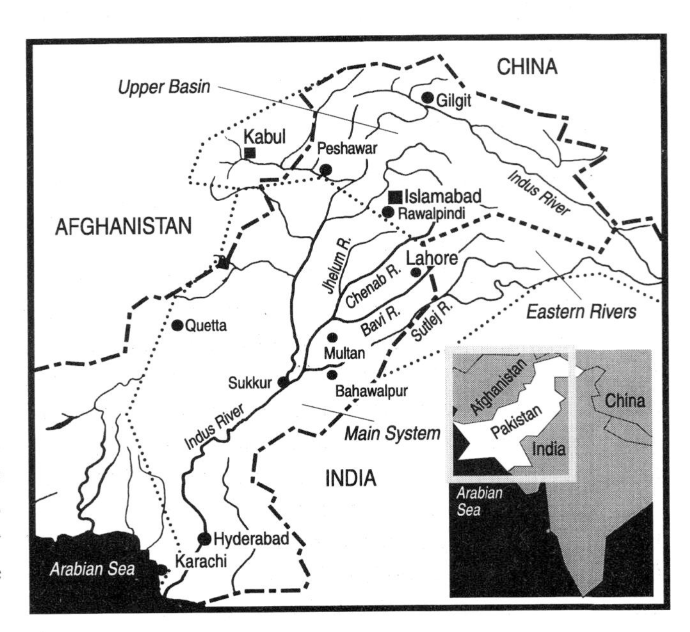

Figure 9 The Indus Basin. Simulated water development effects in this study refer to the 'Main System' area. International boundaries are disputed.

fed tributaries during the brief summer growing season (Allen, 1990; Vander Velde, 1989). There are no important river regulation devices on the Indus main stem until it reaches Tarbela Reservoir, which, with a live storage capacity of 10.94 BCM, is Pakistan's largest storage device.

The upper reaches of the Indus and its tributaries debouch onto the plains at regulatory structures known as 'rim stations' that mark the large, integrated irrigation system. The climates, landscapes, peoples, and water systems of the Indus plains differ dramatically from those of the upper basin. The plains have mild winters and intensely hot summers, relieved only by monsoon winds and precipitation in July and August (Pakistan Meteorological Department, 1991). About 173 BCM of water pass through the rim stations each year. Except for a few areas with sufficient precipitation for rainfed farming, this 173 BCM constitutes the bulk of the water supply for Pakistan, and irrigation on the plains constitutes the largest use of water in the country.

Precipitation generally decreases southward from the rim stations with the coastal cities of Hyderabad and Karachi receiving only 155 and 250 mm/yr, respectively. By the time the Indus River reaches the city of Thatta, the 173 BCM of water that entered the system at the rim stations have been depleted by almost 70 percent. About 123 BCM of water are annually consumed through evapotranspiration from rivers, reservoirs, and roughly 162,000 km^2 of irrigated land. A small fraction of this water is consumed by cities. The residual 40 MAF (million acre-feet) flow into the Indus Delta, an eroding and deteriorating ecosystem that, despite freshwater depletion, supports vital fish, shellfish, dolphin, turtle, and bird habitats.

Hydrological impacts

The regional climate is dominated by the southern Asia monsoon, a climatic feature poorly simulated by GCMs. Because the scenarios inadequately simulate the distinct monsoonal seasonality, the GCMs were treated somewhat differently than in the other basins: monthly temperature changes and annual precipitation ratios were selected from the most representative GCM grid cell in the basin and applied throughout. Warming averaged 2.0 to 4.7°C in the upper basin, and 2.0 to 3.6°C in the main system. Precipitation scenarios ranged from −20 percent to +30 percent.

Five climate change scenarios were examined in the upper basin and the main system. Two scenarios used the GISS and GFDL global climate models (GCMs), and three scenarios assumed an arbitrary +2°C warming and changes in precipitation of +20 percent, −20 percent, and no change (Table 11).

Average annual temperature increases in the five scenarios range from 2.0 to 4.7°C in the upper basin and from 2.0 to 3.6°C in the main system. The highest monthly increase in the upper basin is 6.6°C and in the main system, 6.2°C. Annual precipitation changes range from −20 percent to +20 percent in the upper basin and from −20 percent to +30 percent in the main system. Although it was presumed that climate warming would most likely increase precipitation, the uncertainties associated with the impacts of climate warming on precipitation led to the inclusion of one scenario with decreased precipitation. Runoff changes were modeled for the main upper basin river – the Jhelum – and inferred to the other, less instrumented and less studied upper watersheds.

Table 11. *Climate change scenarios in the Indus basin*

Climate scenario	Upper basin Temp. (°C)	Upper basin Precip. (%)	Main system Temp. (°C)	Main system Precip. (%)
GISS	4.72	+10	3.20	+30
GFDL	4.46	+20	3.57	+20
2°C/+20 % P	2.00	+20	2.00	+20
2°C/NC	2.00	NC	2.00	NC
2°C/−20 % P	2.00	−20	2.00	−20

Hydrological impacts in the Jhelum River basin were quantified with the University of British Columbia (UBC) watershed model (Quick and Pipes, n.d.). The UBC model uses daily temperature and precipitation data from four subbasins of the Jhelum, which are aggregated into ten-day estimates of runoff into Mangla Reservoir from the whole basin. When temperature and precipitation data were altered according to the GCM climate change scenarios, annual runoff from the Jhelum was found to increase by 14 to 16 percent. Jhelum runoff increased (Figure 10), even with reduced precipitation, as the warming heightened the glacial discharge. Total water yield from the upper basin was found to increase 11 to 16 percent with the GCM-derived scenarios and to decline by 19 percent with the warmer/drier sensitivity scenario (Table 12).

Hydrological impacts of climate change in the upper basin depend to a large extent upon snow, ice, and precipitation processes. In basins with seasonal snow cover and some glaciation, such as the Jhelum and Chenab, warming would cause snowmelt to begin earlier in the year and to finish earlier in the year as the snowpack is depleted. Ice melt from extensively glaciated basins, such as the Indus, would continue later into the year, effectively 'mining' the water stored in glaciers.

These results are subject to two important qualifications. First, the Jhelum model was designed to forecast ten daily flows rather than multiyear hydroclimatic processes. The

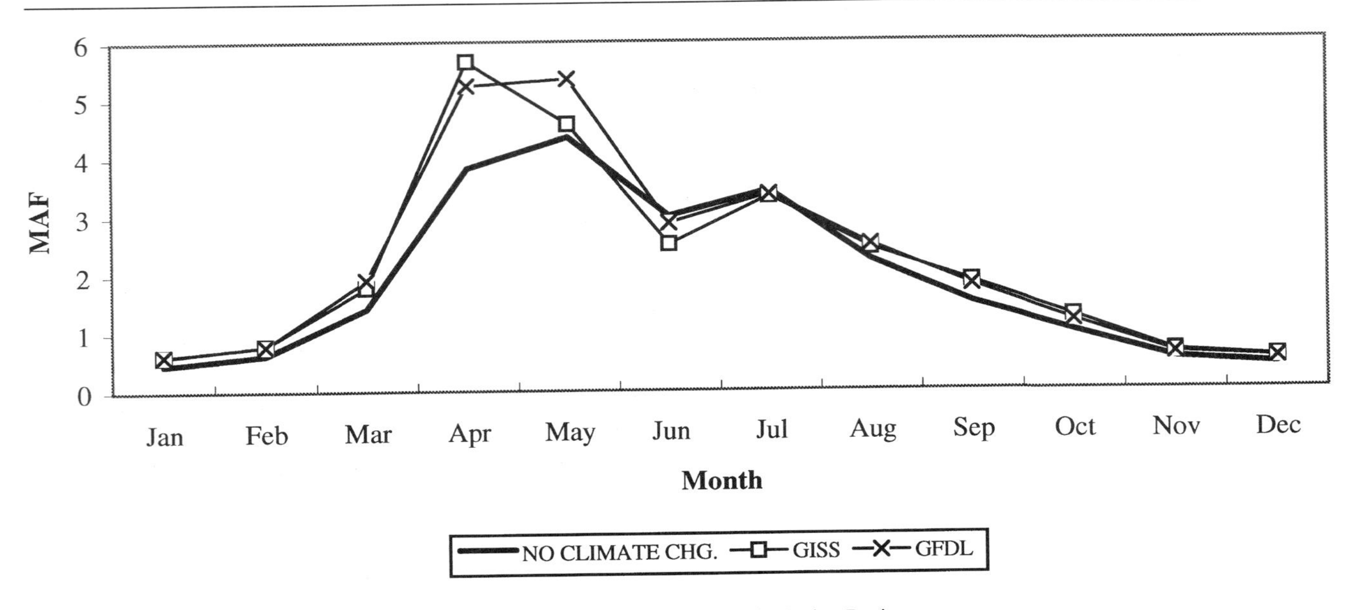

Figure 10 Jhelum River runoff for the two climate GCM scenarios used in the Indus Basin.

model has some carryover of snow cover and soil moisture between years, but not longer-term processes. Second, the model was calibrated with limited data from four hydrometeorological stations.

The upper Indus, Kabul, Swat, and Chenab rivers have not been modeled at all, so subbasins of the Jhelum were used as rough analogs for those basins. Three climate change scenarios, including the two GCM scenarios, indicated that total water yield from the upper basin could *increase* by 11 to 18 percent. Most of this increase would occur between September and April. A simple 2°C warming was found to have little effect on upper basin yield, presumably because snow and ice melt at higher elevations compensate for water losses, at least over the short term. The +2°C/−20 percent precipitation scenario projected a possible 19 percent *decrease* in runoff, which suggests that runoff might be more sensitive to changes in precipitation than in temperature.

The GISS and GFDL models suggest that total annual runoff from the upper basin could increase by 11 to 16 percent. Most increases occur between September and April. Runoff is especially sensitive to changes in precipitation.

If precipitation does not increase, unglaciated valleys would experience water shortages in late summer (*kharif*). Areas of limited glaciation would experience increased flow over the short term, followed by diminished flow. Extensively glaciated basins might yield longer-term water supplies, but the hydrology of those basins requires further investigation.

Management effects

The effects of these scenarios on hydropower were assessed for existing projects (Table 13). Overall basin impacts were also assessed using the World Bank Indus Basin model (IBMR), an investment planning tool, in relation to the 1988 management and use baseline and three year-2000 development scenarios: no new projects but increased demand; minimum planned investment; and maximum planned investment. Different allocations of water among canals (with constant provincial allocations) were also tested. Two key impact indicators were calculated: the return of future investment with and without climate change (Figure 11), and total discharge to the Indus Delta (Table 14), which gives an aggregate indication of water use in the system and represents an important environmental factor (i.e., less water inflows degrade delta environmental quality).

Although increased runoff could be advantageous for water supply and hydropower production, it could aggravate problems of flooding, waterlogging, and salinity in the basin. The increased cost of flood embankments could be as high as U.S. $100 million. Waterlogging and salinity currently affect over 25 percent of the irrigated land in the basin. Even with an overall water surplus, shortages could occur in local areas of the highly productive Punjab rice–wheat zone and in unglaciated valleys of the upper basin. These areas currently lack adequate storage, conveyance, and irrigation management.

Climate change could reduce the benefits of future water development (Figure 11). Under the GCM scenarios, climate change reduces the expected economic benefits of water development by 45 percent under the minimum investment plan and by 42 percent under the maximum investment plan. Changes in water allocation could help water-scarce areas, but they could also increase water consumption and thereby reduce freshwater inflows to the delta by 12 to 75 percent.

Table 12. *Impacts on upper basin runoff and water supply*

Climate scenario	Total runoff (BCM)	Percent change
Existing*	160.1	—
GISS	177.7	+11
GFDL	185.8	+16
2°C/+20% P	188.9	+18
2°C/NC	159.0	−1
2°C/−20% P	129.5	−19

Note:
* Based on Ahmad *et al.*, 1990, and Kalabagh Consultants, 1988; includes Indus, Kabul, Swat, Jhelum, and Chenab Rivers.

Table 13. *Hydropower production and climate change*

	Hydropower	
Climate scenario	(GWh)	Percent change
Without climate change (1985–86 to 1990–91)	14,672.4	—
2°C/+20% P	17,527.8	+19.5
2°C/NC	15,377.3	+4.8
2°C/−20% P	12,210.3	−16.8
GFDL	17,908.6	+22.1
GISS	17,514.6	+19.4

Note: Power generated at Tarbela and Mangla with historic water allocation.

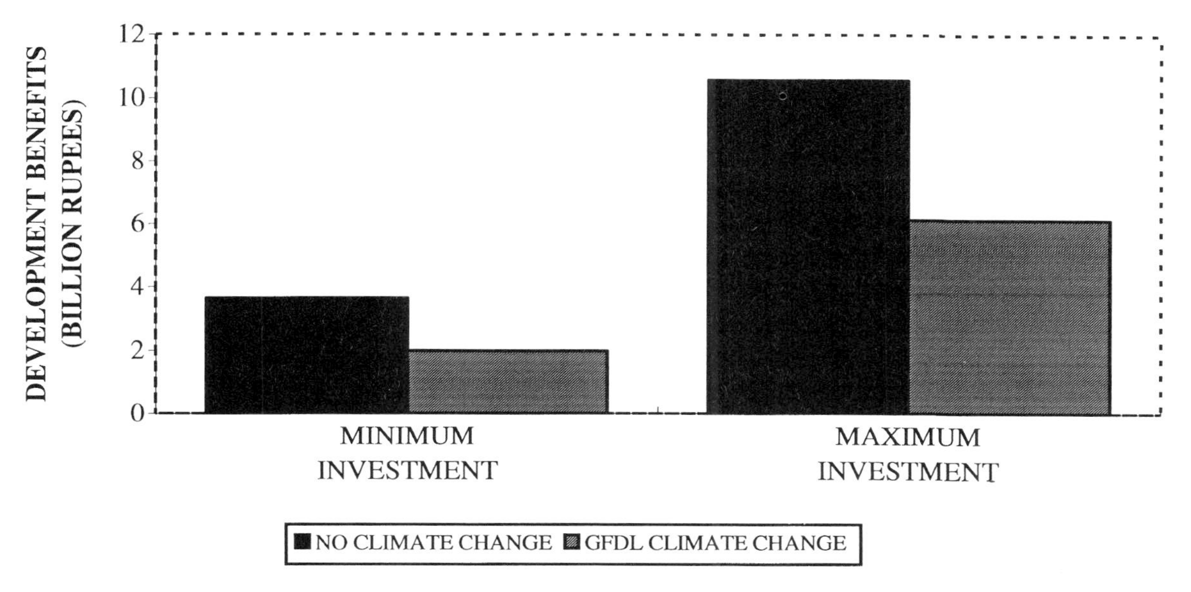

Figure 11 Changes in expected benefits from planned Indus Basin water developments caused by the GFDL scenario. The 'base' is equal to the total value added by irrigation agriculture in the year 2000 if no new projects are built. Because the 'No New Projects' scenario was computationally infeasible, this base was estimated. A 3.25 percent annual economic growth rate was assumed.

Unless adaptive measures are taken, the ecology and economy of the delta could be harmed.

Adaptations

Several individual studies of potential adjustments to climate change were conducted by the Indus study team (Wescoat, 1991). Because most scenarios point to an increase in available water, it appears that the main system could adjust to climate change without major disruption of water supplies, though flood hazards and groundwater problems (e.g., waterlogging and salinity, which already affect 25 percent of the cropped area) could worsen. However, such adjustments are contingent on implementation of at least middle-range development plans.

The first category studied was those adjustments embedded in the output of the IBMR runs. In an optimization model, the model solution includes adjustments to climate change, as well as the impacts after adjustment. We observed, for example, that farmers respond to most scenarios of climate change by increasing their use of groundwater. They respond to increased water supply by producing more sugarcane and to diminished supplies by producing less rice.

The efficacy of these adjustments may be judged in terms of the amount of crops produced, the value added through production, and the land and water resources consumed. It was observed, for example, that for some scenarios increased canal diversion did not necessarily lead to a proportionate increase in crop production or in the value-added through

Table 14. *Flows below Kotri into the Indus Delta in billion cubic meters (BCM)*

	No climate change	+2°C no change P	+2°C +20 % P	+2°C −20 % P	GISS +30 % P	GFDL +20 % P
1988 Water management						
100% allocations	50.247	47.335	56.963	infeas.*	61.421	65.781
80% allocations	39.063	n/a**	n/a	12.988	n/a	n/a
Year 2000 with no new project						
100% allocations	infeas.	infeas.	54.577	infeas.	infeas.	48.466
80% allocations	35.834	33.189	n/a	18.440	44.185	n/a
Year 2000 with minimum investment						
100% allocations	51.090	infeas.	47.788	infeas.	infeas.	65.217
80% allocations	29.398	25.426	n/a	15.603	35.186	n/a
Year 2000 with maximum investment						
100% allocations	37.357	infeas.	38.118	infeas.	infeas.	51.475
80% allocations	19.200	15.270	n/a	11.187	28.847	n/a

Notes:
* Infeas. indicates that the Indus Basin model could not calculate a feasible solution for these more extreme scenarios.
** n/a means a feasible solution was obtained for the 100% allocation, so a reduced allocation of 80% was not necessary.

production. This finding is disturbing because it suggests that increased resource supplies, or investment, might not yield the economic benefits expected.

The second category of modeled adjustments studied was those represented by the alternative water development plans and water allocation rules. These conscious planning and policy decisions were not originally designed as adjustments to climate change but may be interpreted as such post hoc, because if one plan or policy results in greater net benefits than another, it may be regarded as a potential adjustment to climate change.

The 80 percent allocation rule, for example, may be regarded as an effective adjustment to regional water scarcity. It would enable large increases in water use to occur and the total value added to be maintained even under the +2°C/−20%P scenario. But the 80 percent water allocation rule comes at a cost: increased crop production in areas with saline groundwater, reduced flows into the Indus Delta, and possibly inequitable water distribution within canal commands (e.g., between the head and tail reaches of the canals). Thus, 'adjustments' may create new impacts that require further adjustment.

If water development plans are evaluated in terms of value added (total and per unit area), then the year-2000 Minimum Investment plan performs well for a wide range of baseline and climate-change scenarios. The year-2000 No Projects plan cannot maintain historic surface water diversions. In those cases where the model could not reach a feasible solution, we cannot know the full magnitude of the impacts.

A study of dam operations and impacts on hydropower indicated that there was considerable capacity for adjustment to altered inflow regimes. Most scenarios of climate change suggest additional reservoir storage and hydropower production. However, this study did not cover the more difficult problem of adjusting to altered rates and patterns of reservoir sedimentation due to watershed change. Those impacts could not be estimated as part of the study, but they represent a high priority for future research and policy analysis.

A study of crop cultural practices emphasized the need to develop new crop varieties and to initiate adaptive agricultural research on alternative rotations, cropping calendars, and cultivation methods before major losses are incurred. The benefits of such innovations will not likely be realized without more effective coordination between the agricultural and water sectors. This prerequisite for sound irrigation management has often been stated, but the slow progress made in accomplishing this objective reinforces the wisdom of starting the process of adjustment soon.

Research on irrigation management in the upper basin and main system sheds light on the types of adjustments that occur in response to water supply variability and quality problems today – some of which may be applicable under altered climate conditions. In the upper basin, it will be important not to disrupt traditional social adjustment processes. In the main system, it is important to understand the problems that water managers face today if we are to assess where management capacity is increasing and where it is declining. Increasing management capacity may also represent a source of potential adjustment to climate change, whereas decreasing capacity to deal with current problems probably signals increasing vulnerability to climate change. Technical and economic adjustments may be identified as beneficial, but if the institutional apparatus and management capacity do not exist to implement them, those adjustments will not fall within the range of choice for dealing with climate impacts.

Recent irrigation management research suggests that if more surface runoff becomes available, it will be necessary to manage it more effectively if we are to obtain the potential benefits of improved water distribution and irrigation water quality and, at the same time, avoid canal breaches, waterlogging, and secondary salinization problems. If less water becomes available, however, the water quality, delivery, and financial problems faced by users could require a massive improvement in management capacity.

The final task of this research was to ask: in view of the possible impacts that can occur even after all conventional adjustments have been made, what sorts of explicitly climate-related adjustments should be considered? It was suggested earlier in this chapter that society may respond to the combination of climate change and pressing water management issues rather than to climate change itself. The results here confirm the wisdom of that approach. In most cases, major water development decisions had a greater relative environmental and social impact than did climate change – at least over the short term and within the constraints of the available models. Even so, it would be prudent, given the impacts identified, to take the prospect of climate change into account in every major water policy or planning study in the future.

In addition to this 'adaptation' strategy, it will be important for countries that depend upon large river systems, such as Pakistan, to participate actively in international policy initiatives to control the sources of greenhouse warming, as well as to share experience and resources for adjusting to the effects of warming.

Climate change is a relatively slow process, occurring over decades at a rate that allows society time to consider the impacts that might occur and the adjustments that might be effective. In a complex system like the Indus, the time available for evaluation is a precious resource. Some problems, such as the interprovincial water accord, require decades to solve. Others, such as water pricing and salinity control, remain unsolved after decades of work. When climate change is also taken into account, it becomes clear that the time available to deal with climate and related water problems must not be wasted.

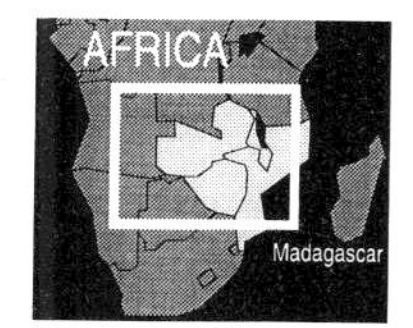

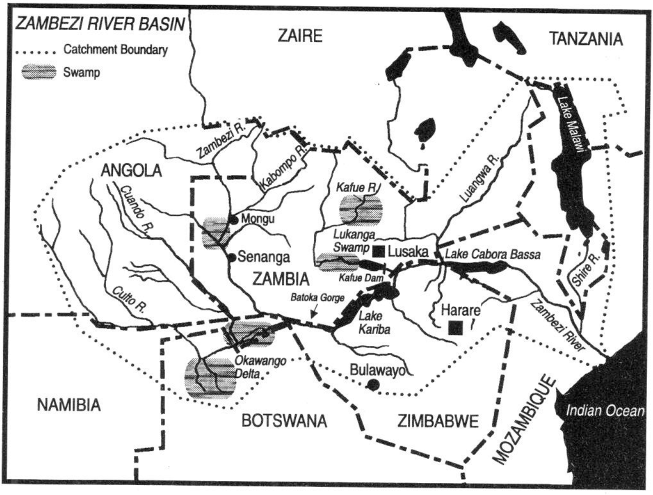

Figure 12 The Zambezi River Basin. Climate change impacts were projected for the basin above Lake Kariba with the existing Kariba dam and with a proposed new dam and reservoir at Batoka Gorge.

THE ZAMBEZI BASIN

The Zambezi basin (Figure 12) is the fourth largest river system in Africa and includes within its boundaries parts of eight countries: Angola, Botswana, Malawi, Mozambique, Namibia, Tanzania, Zambia, and Zimbabwe. The 2,600 km Zambezi River main stem is increasingly the focus of development, but little of its potential has yet been tapped. Two dams, the Kariba and Cabora Bassa, span the main stem, while others, including the Batoka Gorge project, are in final planning stages. Although hydroelectric power has been the primary reason for dam construction, basin states have increasingly been looking to the river to supply the region's growing domestic, industrial, and agricultural needs.

Physical and social setting

The Zambezi basin is situated in a tropical region of southern Africa; precipitation occurs from November to April, due to the southward migration of the Intertropical Convergence Zone. Mean rainfall in the catchment exhibits a decrease from north to south, with the basin above Kariba – the focus of this study – receiving an average of 1,400 mm of precipitation per year (Gandolfi and Salewicz, 1990; UNEP, 1987). Basin temperatures follow the seasonal pattern as well, with

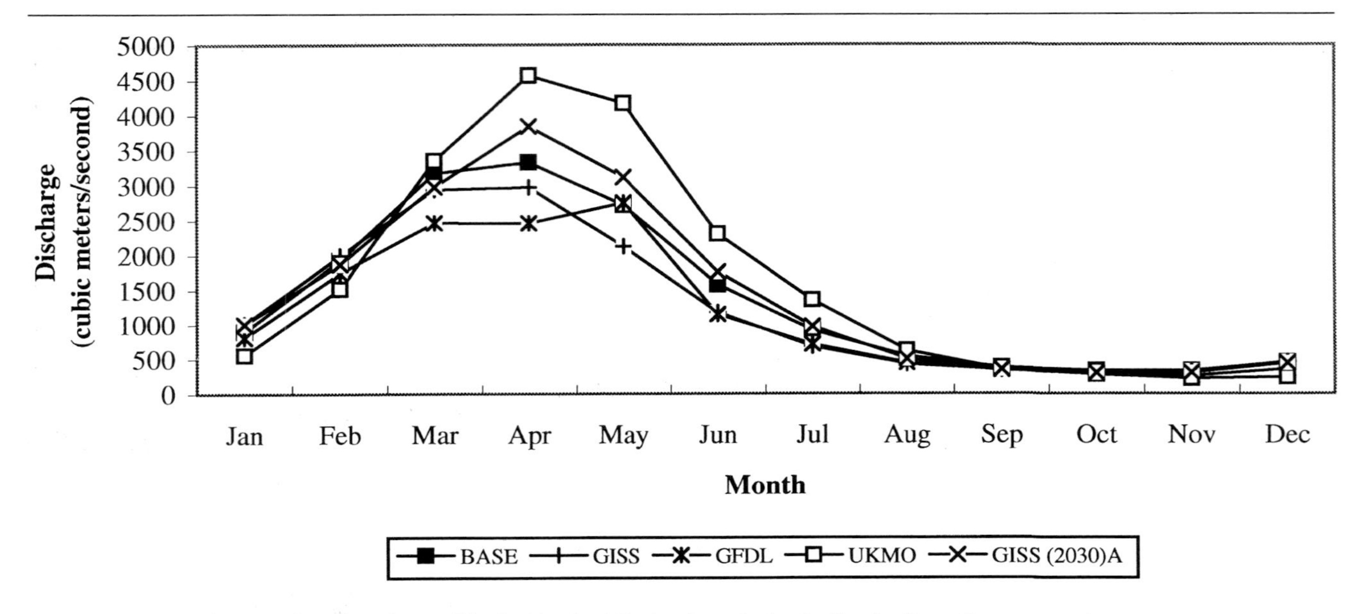

Figure 13 Actual (base) and projected runoff in the Zambezi Basin above Lake Kariba for four climate scenarios.

the highest temperatures occurring in October, just before the rainy season. Mean annual temperatures vary from 18°C to 24°C; extremes may range from 5°C to well over 30°C.

The upper Zambezi basin contributes almost 75 percent of the total annual inflow to Kariba Reservoir. Its hydrology has been described as one of the world's most complex due to intermittent feeder streams and the flood-regulating effects of the Barotse Plain and the Chobe swamps (Balek, 1977). The Barotse Plain, located in Zambia, is a seasonal swamp with a maximum surface area of approximately 7,500 km^2 (Pinay, 1988) and is capable of storing up to 17 BCM of water at the peak of the flood season (Gandolfi and Salewicz, 1990). The Chobe River system of Botswana also operates to regulate the Zambezi's flow. At flood stage, the Zambezi backs up into the Chobe River causing it to reverse flow and fill its upstream swamps. As Zambezi floodwaters subside and flows return to normal, floodwaters are returned to the Zambezi (Pinay, 1988).

These two physical phenomena not only attenuate the peak of the annual flood hydrograph, they also function as immense evaporation pans and sediment traps. Indeed, up to two-thirds of the upper Zambezi precipitation is lost to evaporation before the river passes over Victoria Falls. The water that does make it over the falls carries a very low sediment load due to deposition in the Barotse and Chobe systems (du Toit, 1983).

Hydrological impacts

The GCM and arbitrary climate change scenarios were translated into average monthly discharge of the Zambezi at Victoria Falls, the exclusive indicator of hydrological impacts. Water availability was appraised through the use of a rainfall–runoff model that operates on inputs of monthly values for temperature and precipitation. The model is a deterministic, computer-based simulation that estimates river flow at Victoria Falls based on averaged upper basin rainfall and temperature data. For the GCM scenarios of climate change, these values were adjusted by using precipitation and temperature data for select grid cells overlaying the modeled basin.

The hydrograph in Figure 13 illustrates the potential climate change impacts on river flow measured at Victoria Falls. The seasonal runoff pattern remains relatively unchanged under the GISS scenario; however, GFDL and UKMO project a one-month delay in peak flow due to temporal shifts of the rainy season. Significant differences in river discharges are also noted: GISS and GFDL project decreases in flow, whereas UKMO projects an increase. Finally, although projected monthly precipitation for some of the scenarios (e.g., GISS and GFDL) is well above the historic norm, there is a net deficit in river flows due to higher surface temperatures, which increase the rate of evapotranspiration.

Management impacts

Impacts of climate change on water availability were evaluated using a river basin simulation model that estimates hydropower output and changes in reservoir levels based on river flow input from the rainfall–runoff model. Kariba and the proposed Batoka project were modeled using the Massachusetts Institute of Technology River Basin Simulation model (MITSIM) in order to assess how changes in Zambezi River flow could affect hydropower production and fisheries. Like the rainfall–runoff model, MITSIM uses mathematical

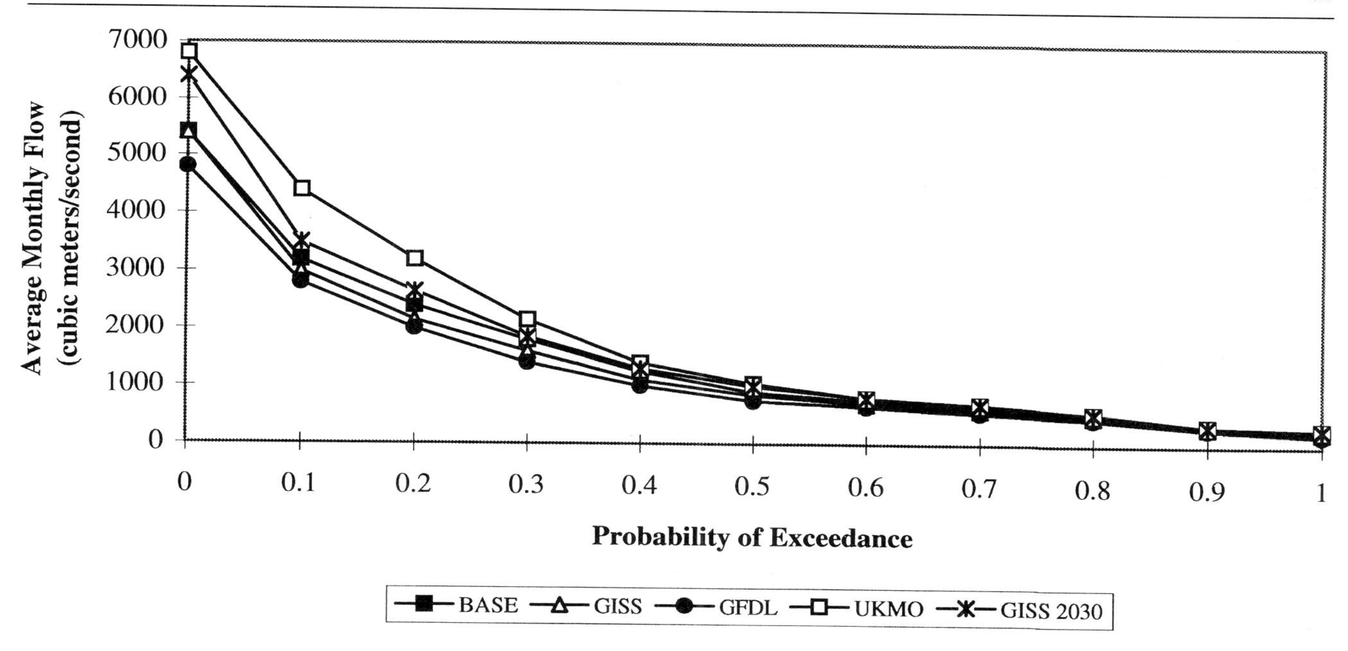

Figure 14 Actual (base) and projected flow-duration curves for the Zambezi under the four climate scenarios.

equations to describe the interrelationships among hydrological features, management structures, and operating rules (e.g., inflow, reservoir storage, and power production) in a planning/management undertaking. Because the Zambezi is used primarily for power production, the analysis utilized only the MITSIM parameters of storage level and power output to assess climate change impacts.

We used flow duration curves (Figure 14) as the chief input to the reservoir simulation model. The curves display the distribution of monthly stream flows for the various scenarios, showing, for example, that under the GISS and GFDL scenarios, the Zambezi's flow never exceeds that of the historic (base) period. The curves reveal a shift in the reliability of given discharges. For example, the probability that a flow of 1,000 m^3/s will be equaled or exceeded under the base scenario is 44 percent, whereas under the GFDL scenario this reliability drops to 40 percent. The UKMO scenario projects that this reliability will increase to 51 percent. These changes in flow reliability would affect reservoir management for hydropower.

IMPACTS ON HYDROPOWER PRODUCTION

Hydropower is expected to supply up to one-quarter of the region's future energy needs (Central African Power Corporation, 1980), and we examined the impact of climate change on hydroelectric power production at Kariba and the planned Batoka reservoirs. Although still in the planning stages, Batoka is important to this analysis because it allows for comparison, not only among scenarios but also between different management regimes.

Power production for Kariba and Batoka was calibrated using a power plant load factor of 65 percent and an efficiency of 95 percent. Power outputs for the various scenarios are described in Table 15 as percentages of target power generation, defined as 7,208 and 9,093 GWh per annum for Kariba and Batoka, respectively. These targets, which are based on the average output of Kariba during the 1980s, do not indicate Kariba's maximum capacity but serve to emulate its historic operating objectives.

Under the base scenario, production at Kariba varies from 87 percent to 97 percent of target power. Kariba's power output decreases by about 11 percent under the GISS and GFDL scenarios. Both the UKMO and the GISS Transient (used as a cooler, 2030 scenario) project small increases in power generation.

Average annual Batoka power output is no better than 78 percent of target power for any GCM scenario. Seasonality of flow, however, has marked effects on production, as outputs fall from about 95 percent in the wet season to about 49 percent in the dry season. This decrease is the result of Batoka's limited storage capacity. The UKMO scenario, which projects an overall increase in total annual flow, gives the largest reduction in dry-season power generation of any of the scenarios. This is a result of an increase in discharges during the high-flow season and a reduction in discharges during the low-flow season.

The scenarios of the impacts of climate change on hydroelectric power production indicate that the installation of the 1,600 MW 'run-of-river' Batoka plant will increase overall system output (i.e., Kariba-Batoka) by approximately 100 to 135 percent over the existing 1,266 MW units at Kariba. Potential impacts of climate change will be most significant

Table 15. *Summary of power generation statistics for GCM scenarios for Kariba and Batoka*

Percent of target power gen. for period:	Base	GISS	GFDL	UKMO	GISS-A (2030)
Avg. of 19-year model run					
Kariba	93	83	81	95	94
Batoka	78	78	74	74	78
Dry Season (Aug–Dec)					
Kariba	87	68	67	96	88
Batoka	53	51	47	45	52
Wet Season (Jan–Jul)					
Kariba	97	94	91	95	98
Batoka	97	95	94	95	97

Notes:

Target Power:	Monthly	Yearly
Kariba (GWh)	600	7,208
Batoka (GWh)	757	9,093

at Batoka during the dry season as a result of inadequate storage to buffer low-flow periods during the year.

IMPACTS ON THE KARIBA FISHERY

Kariba Reservoir is an important fishery for both Zambia and Zimbabwe, providing a source of inexpensive protein to the entire southern African region. Fishing at Kariba consists of large-scale enterprises that harvest the Lake Tanganyika sardine (*kapenta*) in open waters and individual undertakings that exploit fish species in the shallower, inshore areas.

Fluctuations in the level of Kariba Reservoir, which have an annual amplitude of about 3 m and extremes of more than 11 m (Pinay, 1988), are critical to fish production. Under these large fluctuations, the reservoir's shoreline acts as an artificial floodplain and an important zone of transition between the terrestrial and aquatic ecosystems. The wetting and drying of soils in the fluctuation zone stimulate the growth of shoreline grasses. These grasses, in turn, attract herbivores, including elephants, impalas, buffalo, and zebras (McLachlan, 1970), which, in the course of grazing, deposit dung. As reservoir levels rise during the flood season, large quantities of organic material (e.g., dung and decomposed grasses) are released into the reservoir, adding nutrients that are important to the fish (Pinay, 1988). This significant interaction between reservoir and terrestrial systems has been observed by Balon (1978), who found that in shoreline areas visited by game animals, inshore fish production was 35 times greater than in areas where the terrestrial/reservoir interaction was not as great.

Reservoir drawdown in excess of 6 m, however, is detrimental to fish production because it exposes large areas of the reservoir bottom, limiting habitat and destroying important inshore nesting areas. Many areas of Kariba Reservoir are characterized by gently sloping shorelines, and as much as 10 percent (630 km^2) of the total reservoir area can become exposed by a drop in water level from 488 m to 482 m.

Fish production in Kariba Reservoir is also restricted by the amount of nutrient input from the Zambezi. The river, which supplies over 70 percent of the inflow to Kariba, provides important nutrients that support zooplankton and thus provide a critical link in the fishery food chain (Marshall, 1982; Balon and Coche, 1974).

Under conditions of climate change, there could be less water entering Kariba. Figure 15 indicates that water levels at Kariba Reservoir could average as much as 6 m below base levels (note, for example, the difference between base and GFDL levels for June). Fish populations would likely decrease in association with long-term reservoir declines.

A dam at Batoka Gorge would regulate inflow to Kariba, and, certainly, more consistent flows could benefit the fishery. However, the dam would essentially block any nutrients flushed from the upper basin. In addition, Batoka Dam would create its own reservoir and potentially viable fishery. The fishery potential of the reservoir is difficult to assess; however, Hosier (1986) estimated that yields would likely be low due to the reservoir's small size.

Adaptations

A workshop was held in Zimbabwe to give a broad array of basin policy makers and researchers a chance to explore how to respond to climate change. Because most climate scenarios predicted decreased flows and power production, and because of the importance of electricity for basin development, the workshop participants almost unanimously called for more reservoir development in the basin as the key adaptation. The scenarios of joint operation of Batoka and Kariba helped fuel the discussion. Representatives of the Zambezi River Authority (ZRA), along with other participants, noted that with or without climate change the proposed Batoka-Kariba power generating scheme is critically important for optimal energy production. The Batoka plant will produce firm power during periods of high river flow, thus allowing Kariba operators to limit releases and retain sufficient storage for power generation during low, dry-season flows. It was suggested that even more dams could be feasible and that a cascade system regulating the entire middle Zambezi from above Victoria Falls to Cabora Bassa Dam in Mozambique should be considered. In addition to dam construction, pumped storage was suggested as a means of optimizing power generation from the Zambezi system.

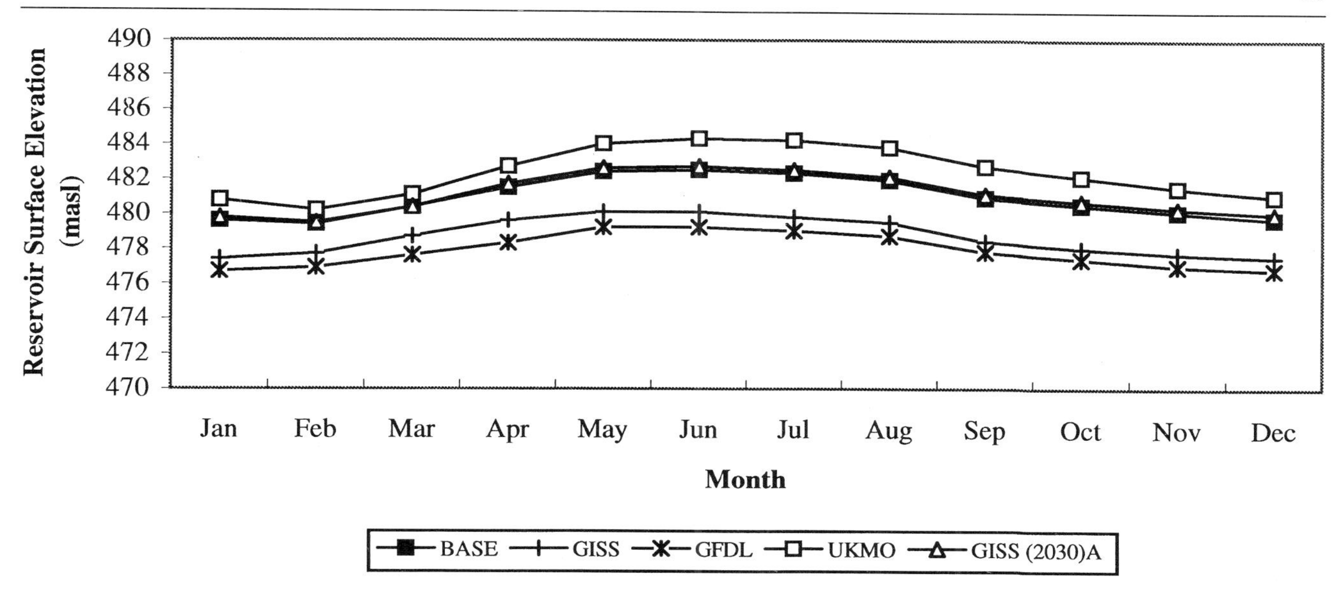

Figure 15 Lake Kariba reservoir levels under actual (base) conditions and the four climate scenarios. The operating rule curve, in this case, indicated both preferred and maximum levels.

Nevertheless, participants agreed that it would be dangerous to rely primarily on hydropower as the region's principal energy source, especially with the uncertainty of climate change. In lieu of constructing large impoundments for energy development, the following possible developments were put forth:

1. An increase in turbine efficiency at Kariba
2. Installation of turbine intakes below present intake levels at Kariba to increase active storage and take advantage of the dead storage
3. Construction of relatively small run-of-river hydropower units
4. Importation of energy produced in the Zaire River basin
5. Exploitation of Zimbabwe's abundant coal reserves and the development of more thermal power plants
6. The development of nuclear energy utilizing the abundant fuel sources in Zambia and Zaire
7. Increased use of alternative fuels (Zimbabwe already produces fuel substitutes) and solar energy – especially at the village level
8. Conservation and more efficient energy uses

It was agreed that a mix of some or all of the above would be needed if the region is to continue to develop.

Participants also acknowledged that the effects of climate change could be dwarfed by the impacts of poorly planned water projects that do not consider the hydrological interrelationship of the basin as a whole. Planned diversions of the Zambezi flow could reduce hydropower and fisheries production at Kariba much more severely than the reductions demonstrated by the modeled scenarios of climate change.

The overriding concern of basin managers was the need for cooperation among basin states in resource development. Participants agreed that in order to cope with the impacts of climate change, basin states would have to be willing to cooperate in the future development of the Zambezi water resource. States would have to begin looking beyond national needs and understand the interrelated needs of the basin as a whole.

There also emerged a sense that responses to climate change in the Zambezi basin will probably not be planned, anticipatory actions. More likely, basin managers will respond in reactive modes influenced by the path that regional development takes over the next few decades and affected by their ability to recognize and evaluate climate impacts.

THE NILE BASIN

The Nile has been the lifeblood for many great civilizations. Biblical stories of famines and the discovery of ancient devices for measuring river flows illustrate how the river dominated the well-being of ancient Egypt and remind us that the Nile has naturally varied between high and low flows, in essence causing Egyptian civilization to experience both short- and long-term climate change. Depending on a host of cultural and physical factors, the Nile civilizations were sometimes able to adapt, but in other cases the riparian societies suffered greatly and declined.

Physical and social setting

The Nile basin covers a surface area of almost 2.9 million km^2, almost one-tenth the area of Africa. The main stem of the river flows north for a distance of 6,500 km, from 4 degrees south latitude to 31 degrees north latitude, and extends from 21 degrees 30 minutes east longitude to 40 degrees 30 minutes east longitude (Figure 16). The Nile is an international river in the strongest sense: the main stems of the White Nile, the Blue Nile, and the Nile, plus their tributaries, cover parts of Tanzania, Uganda, Rwanda, Burundi, Zaire, Kenya, Ethiopia, Sudan, and Egypt (Shahin, 1985).

Although the watershed is large, the portion contributing to streamflow ends at Hassanab at the confluence of the Atbara River and the Nile. The contributing portion has an area of only about 1.6 million km², due to the fact that north of 18 degrees north latitude, rainfall is effectively zero. The 'specific discharge' of the Nile (the long-term average annual flow divided by the area of the watershed) is the lowest for all river basins with areas greater than 1 million km². Even if the noncontributing portion is omitted, the discharge is about the same as for the Missouri basin, which has the second lowest discharge, and the discharges for both of these basins are nearly half the discharge of the basin with the next lowest value. What makes this even more interesting is that the Congo River basin, which shares a significantly long common watershed divide with the Nile, has approximately 10 times the specific discharge of the Nile (Kalinin, 1971).

This great contrast in neighboring river basins can be explained by climate and topography, and the Nile's climate and topography make for a complex hydrology that is very sensitive to climate change. The Nile is marked by extremes: mountainous plateaus and very flat plains. The Equatorial Plateau and its system of lakes have a very delicate water balance, with direct evaporation from the lake surfaces almost equal to the direct precipitation on the lakes. A small shift in either rainfall or evaporation can result in dramatic changes in the supply of water from Lake Victoria, as was observed in the 1960s, when a rapid increase in the flow from Victoria occurred. Piper *et al.* (1986) state that the '1961–1964 rise is not unique and that similar fluctuations have occurred in the past.' Indeed there is some evidence from Paleoclimatic records that in recent times the Victoria basin produced no outflow at all.

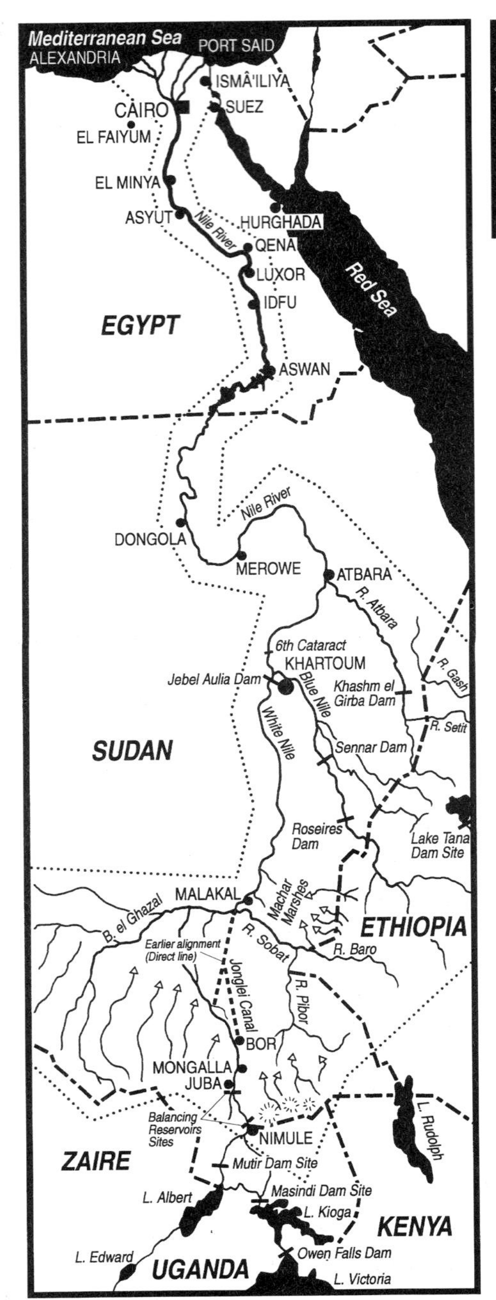

Figure 16 The Nile River Basin.

Like the Zambezi, this dry basin has several swamps that lose large amounts of water to evaporation. The Equatorial Plateau, dominated by Lake Victoria, supplies 29 out of the 84 BCM of water reaching Aswan (Shahin, 1985). The water then plunges from the plateau onto the plains of southern Sudan, and on average 50 percent of the flow spills into the flat swamplands of the Sudd, with the remaining 50 percent reaching the White Nile at Malakal.

The area of the swamps varies from year to year, with a permanent area of approximately 33,600 km². The swamps evapotranspire on average 51 BCM annually (Chan and Eagleson, 1980). The contribution of the Sudd, which makes up 44 percent of the contributing Nile basin, is on average 15 BCM; combined with nonspilled equatorial flow, the average flow of the White Nile at Malakal is 29.44 BCM. This flow continues with small seepage and evaporation losses until it joins with the Blue Nile at Khartoum to form the Nile (Shahin, 1985).

The Blue Nile rises in the Ethiopian Plateau and makes a very steep descent to the White Nile at Khartoum. The Blue Nile basin covers only 16 percent of the contributing basin, but it contributes on average 52 BCM, or 62 percent of the Nile flow at Aswan (Shahin, 1985). Finally, the Atbara

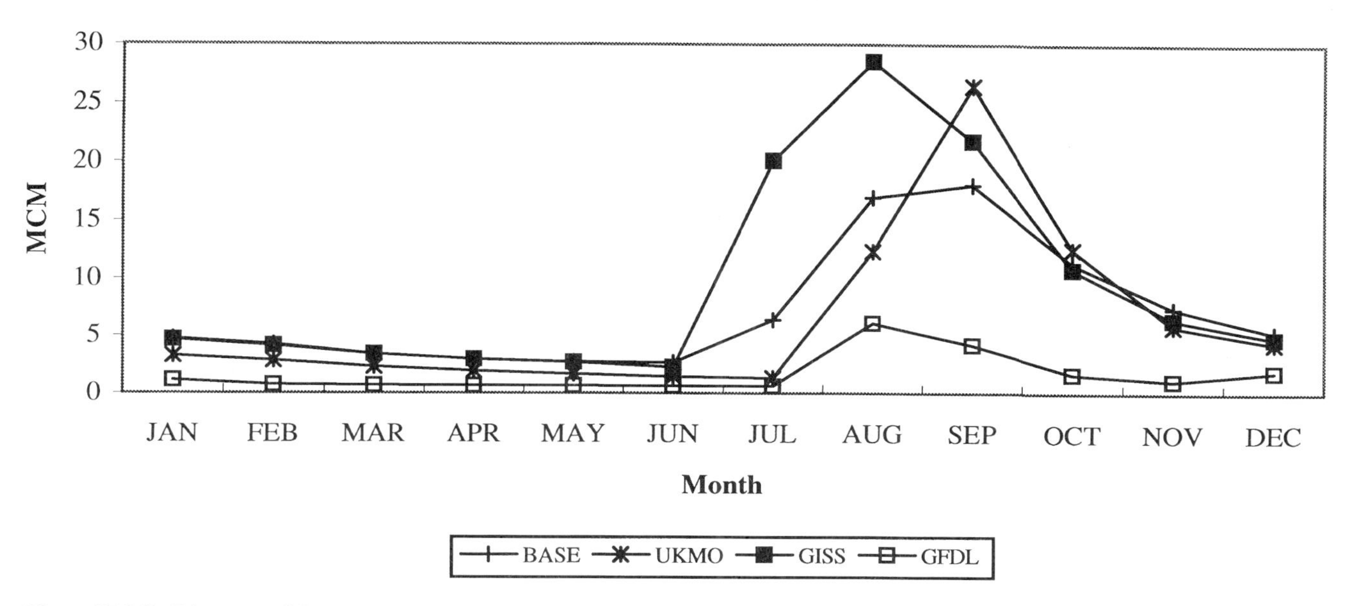

Figure 17 Nile River runoff for actual (base) conditions and three climate scenarios.

River, with a 113,000 km^2 basin, joins the Nile River north of Khartoum and contributes 12 BCM. The Atbara river also rises in the Ethiopian Plateau and has the same climatic and topographic features as the Blue Nile basin. Under natural conditions the Nile obtains no additional contributions of water after the Atbara, with only seepage and evaporation losses occurring over the 2,500 km stretch to the Mediterranean Sea. The Nile at Hassanab, just downstream of the confluence of the Atbara River, has an average flow of 88 BCM. As the Nile flows the remaining 1,200 km to enter Lake Nasser, it loses 4 BCM.

Hydrological impacts

We used a hydroclimatic approach based on a water balance method to model the basin. It was divided into five major subunits: the equatorial lakes basin, the Sudd basin, the Blue Nile basin, the Atbara basin, and the White and main Nile basins. The equatorial lakes basin is divided into four catchment areas and three lakes, and each was modeled separately. In addition, the Sudd basin is a simplified representation of the entire White Nile catchment from Mongalla to Malakal, including the River Sobat. All the swamps are represented as one and are modeled via the water balance approach of Sutcliffe and Parks (1987).

All of the catchments were lumped in a one-dimensional water balance model with spatially weighted model parameters. Average monthly precipitation values were used directly in the model, but temperature was used to derive average monthly PET values according to the Thornthwaite method (Shaw, 1983). Soil moisture and groundwater were also considered.

The equatorial lakes were modeled using a monthly mass-balance approach with catchment inflow, direct lake precipitation, and direct lake evaporation equal to PET. The model used the lakes' elevation–area–storage curves and nonlinear outflow storage curves (World Meteorological Organization, 1977). No calibration was needed. The swamps of the Sudd were modeled using the mass-balance and nonlinear reservoir approach of Sutcliffe and Parks (1987). Inflow to the swamps was the catchment runoff plus the spill from the Bahr el Jabel. Contribution to the White Nile flow from the Sudd at Malakal was the discharge from the swamps.

Finally, the GCM scenarios were put into a GIS system for the basin, and normal and GCM-derived climate surfaces were created and overlain on the basin. The runoff model was then used to generate multiyear time series of monthly streamflow at Aswan, based on the historical series from 1900 to 1972 (Figure 17). Both GISS and UKMO predict increases in high flows, though UKMO shows significant reductions in flow for the remainder of the year. The most striking result comes from the GFDL scenario, which shows flows reduced by over 75 percent due to the model's high temperature and low precipitation projections for the basin.

Management impacts

Altered Nile flows were then assessed using a Nile River simulation model, with the simulation focused on hydropower and irrigation releases at Aswan. System performance

Table 16. *Nile River management results in percent of base scenario*

Scenario	Nile yield (%)	Sudan yield (%)	Egypt yield (%)	Evap. (%)	Power (%)	Energy (%)
Base	100	100	100	100	100	100
UKMO	88	65	88	164	79	79
GISS	130	162	118	159	128	127
GFDL-A	23	103	0	0	0	0
GFDL-B	23	0	22	78	6	6
GFDL-C	23	38	17	44	4	4

was recorded for each month of the modified series. Annual average system performance is reported in Table 16 as a proportion of base performance.

Two important issues must be taken into account when interpreting these predicted performance statistics. First, an agreement exists between Egypt and Sudan on the allocation of Nile River flows. This allocation is based on the current annual mean flow at Aswan of 84 BCM, taking into account losses, evaporation, and seepage caused by storing the water in Lake Nasser. Formal procedures also exist to deal with any enhancements of the Nile flow due to the planned upper Nile development projects; they call for an equal sharing between Sudan and Egypt of any flow above the current 84 BCM. In the performance analysis reported in Table 16, it was assumed that the allocation of discharge change would be: GFDL-A, 100 percent to Sudan; GFDL-B, 50/50; and GFDL-C, 100 percent to Egypt. In terms of the GISS scenario, which projects a 30 percent increase in Nile flow, Sudan and Egypt would each receive 11.5 BCM in additional allowable annual withdrawals after accounting for the increased evaporation due to the warming over the reservoir. Allocations in cases of decreased flow at Aswan, as projected by UKMO and GFDL scenarios, are less easily assessed. We assumed a response similar to the case of surplus: The loss (including losses from the reservoir) would be shared equally.

The good news is that the High Aswan Dam will allow both Egypt and Sudan to fully capture the benefits of the 30 percent increase in flow of the Nile projected by the GISS scenario. These benefits are a boon in both irrigation and hydropower. In the UKMO scenario, irrigation supply to Egypt is reduced by only 12 percent. Given current irrigation practices, a 12 percent improvement in irrigation efficiency seems easily accomplished through management practices and with little capital investment. Under the GFDL-A management scenario, where Egypt forgoes its right to Nile water, the Sudan would have 103 percent of current water.

The bad news is that the temperature increases over Lake Nasser greatly increase potential evaporation from the reservoir surface, which would amplify the effects of reduced Nile flows. The 12 percent UKMO flow reduction translates into a 35 percent reduction in irrigation supply for the Sudan and a 12 percent reduction in Egyptian irrigation supply. Egyptian hydropower production would be reduced by 21 percent.

Finally, the GFDL scenario, which projects a 77 percent reduction in flow, would be a disaster for water management in Egypt and for Sudan under any of the allocation alternatives.

Adaptations

A wide range of potential adjustments were discussed at a workshop in Cairo attended by Egyptian government officials, water managers, academics, scientists, and international experts. It included discussions of changed allocation, management in Egypt and Sudan, and infrastructural changes in the upper and lower basin. Generally, the managers felt that Egypt could adapt quite easily to a 10 to 15 percent reduction in Nile flows simply by conserving water, altering release schedules, and making other nonstructural adjustments. Flow reductions of 20 percent or greater were deemed much more destructive and would result in major social and economic impacts.

We discuss two types of adjustments here: changes in water allocation, and structural adjustments in the upper and lower basin.

ALLOCATION ADJUSTMENTS

As discussed above, any calculation of climate change impact in the Nile is complicated by assumptions about intricate water allocation and institutional arrangements, chiefly between Sudan and Egypt. The greatest adjustments would be needed under the GFDL scenario, which shows a 77 percent decrease in Aswan inflows. This scenario is obviously so critical that institutions, and some infrastructure, would simply fail. Yet, some adaptive response is inevitable. One option is to give all the water to Sudan, thus wasting less in Lake Nasser evaporation and allowing at least one country to 'win.' This approach, however, is unlikely to appeal to Egypt. Alternatively, Sudan might take no water, so that Egypt would get 22 percent of its current Aswan yield, and Aswan Dam outlet works could be modified (lowered) to make use of the reduced lake level. The third case might be a 50/50 split of the reduction, in which Sudan would get 38 percent of current water and Egypt would receive 17 percent. This response might be the likeliest one.

Despite these options, such a major decline in flow would probably mean that 'all bets are off' in terms of past arrangements; formal and informal agreements would presumably be changed, and tensions might result. One potential overarching adjustment, then, is a renegotiation of international agreements, probably aimed toward reduced development by Sudan such that it would not take its full share of the current supply. Indeed, Sudan would require substantial investment to take all of its current allocation – probably more substantial than is available and feasibly applied in this region.

POTENTIAL STRUCTURAL ADJUSTMENTS

Modifications to the engineering system are likely to occur in response to climate change. For example, the GFDL decline is so great that the current infrastructure would fail as reservoir levels fall below the current outlet works threshold. Thus we examined new engineering systems. We even discussed tunnels to be dug at the base of the High Aswan Dam to allow for releases, but such a change would take the Lake Nasser Power Plant out of operation. However, the power plant at the smaller Aswan Dam would remain in operation. More logically, however, less dramatic changes would bring about further developments in the upper and lower basin and perhaps add impetus to complete projects now on hold.

Upper basin adaptations (Sudan and Ethiopia)

There is potential for infrastructural change in the upper basin that could mitigate some of the effects of the negative climate change scenarios. Most obvious is completion of the Jonglei Canal project (which is now over half completed but is in limbo). The logic of the canal depends on the fact that half of the discharge of the equatorial lakes now spills into the Sudd swamps and is evaporated. The goal of the canal is to circumvent the Sudd, thus shunting water around this large evaporator and, in the process, allowing an additional irrigation area. The canal might help little under the two White Nile dry scenarios (UKMO and GFDL), because, with these scenarios, the equatorial lakes become a closed basin and little or no flow spills into the Sudd, thus negating the original engineering logic of Jonglei. However, in a case less drastic than GFDL, the canal might reduce the overall loss of runoff due to precipitation and temperature changes. Note that GFDL presents such a bad picture because it shows a decrease in Blue Nile flow and it projects that the White Nile will dry up completely. The other scenarios show increased precipitation and runoff in the Blue Nile, but there is little engineering option to take advantage of increased flow from the Blue – especially in Ethiopia – in any way that would increase inflows to Aswan.

The proposed Sudd 'collector' canal system may provide some additional water under the dry scenarios. Its overall goal is to keep water out of the Sudd as much as possible, but it requires significant engineering works, which have been difficult to plan and implement in the region.

Lower basin adaptations (Egypt)

Changes in the management of Lake Nasser and the associated irrigation system are possible in response to climate change. With the GISS and UKMO scenarios, there would probably be little change – the Lake Nasser operation could adjust to reduce these impacts (the adjusted UKMO loss, for example, is only 13 percent). Of course, such adjustments then cascade into the irrigation system. However, with UKMO, some adjustment might also be needed to the outlet works so that they can accept low flows.

GFDL would be such a disaster as to not offer feasible, readily imagined adjustments. It would require physically moving the outlet works and drastically changing the seasonal cropping pattern in the irrigation system, perhaps to allow releases to be concentrated to better fight evaporation loss. UKMO and GFDL also cause losses in hydropower, most likely eliciting the construction (already discussed in water planning) of run-of-river, low-head downstream hydropower barrages. There would be more pressure for thermal power generation.

Among the other adjustments suggested at the Cairo workshop were:

1. Increased irrigation efficiency through canal linings and better management
2. More reuse of drainage water
3. Better use of the Nile valley aquifer
4. Changes in crop types
5. Development of western desert groundwater resources
6. Desalinization

This study has shown that, despite potential adjustments, Nile River flows throughout the basin are extremely sensitive to temperature and precipitation changes. The GCM scenarios provide widely diverging pictures of possible future river flows, from a 30 percent increase to a 78 percent decrease. Faced with such diverse scenarios, it is difficult for current water managers in the basins to adopt any response policy other than 'wait and see.'

Due to pressures of increased population and economic growth, capital investment projects to increase water supply from the Sudd are under serious consideration. The results from this study might suggest that capital investment be made in decreasing water demand via more efficient irrigation management, which is one response to climate change, rather than investing in projects that may become ineffective as a result of climate change.

ADAPTING TO CLIMATE CHANGE IMPACTS

Discussions of potential adaptations in each of the case studies above suggest that basin managers are indeed ready and able to envision environmental futures quite different from the present, but that they also tend to respond in traditional ways (increase dam size or reenergize projects that are now on hold). Variously defined in the climate impacts literature, adaptation can be taken, in the broadest sense, to mean any action that seeks to reduce the negative effects, or to capitalize on the positive effects, of climate change. Smith *et al.* (see Chapter 8) differentiated between two types of adjustment:

1. *anticipatory*:
   adjustments in operational or physical aspects of systems made in advance of climate change or impacts, presumably because planners are convinced that change will threaten the future integrity of the system. A common anticipatory adjustment, not related to climate, is oversizing in anticipation of future demand increase. Anticipatory adjustments might be made at different time-scales and phased-in as evidence for climate change accumulates.
2. *reactive*
   (or 'on call'): once impacts, shortages, or failures occur, decision-makers assess the situation and may adjust accordingly. Some reactive adjustments, such as lowering reservoirs during periods of increased flooding, are relatively quick and easy to effect. Others, such as building new infrastructure, take considerable time.

It is also useful to discriminate between *incremental* adjustments, minor changes that can be made in anticipation or response and that involve little investment or system change (e.g., changing the elevation of water intakes), and *major* adjustments requiring more time and investment (e.g., enlarging dams). All of these types of adjustments were suggested and discussed at the adaptation workshops convened in the case study basins.

Differential sensitivity and adaptability

The ability to adapt to climate change without untoward losses depends on the nature of the change, on regional sensitivity, and on the ability to marshal the resources needed to anticipate, measure, and react to impacts. Water resources are managed at the regional and local scale, and their sensitivity and adaptability, either in anticipation or in reaction to climate change, vary greatly among places due to environmental and social characteristics. In light of these variations, it is important not to overgeneralize about how global climate change might reduce the capacity of water systems to provide critical services in the study basins and elsewhere.

Theories of social response to global environmental change are poorly developed, and concepts such as sensitivity, adaptation, and resilience are applied in an ill-defined and inconsistent manner. Although most analysts differentiate between adaptation (social accommodation of climate change) and mitigation (efforts to reduce climate change – see Committee on Science, Engineering, and Public Policy, 1992), less attention has been paid to the conceptual integrity of adaptation ideas and to the potential for adjusting climate-sensitive resource systems as the climate changes. Ecologists and systems analysts differentiate between *resilience*, the ability of a system to return to predisturbance status without lasting, fundamental change, and *robustness*, the ability of a system to continue to function in a wide range of conditions. Resilience implies that resource systems may fail, but also that they may recover quickly to some 'normal' range of operation. Robustness implies maintenance of system properties and outputs, even under unusual stress, by virtue of strength and control rather than flexibility. System robustness is increased by 'hardening' with increased investment, structural strength, and operational control. There is, generally, a trade-off between the two characteristics: hardened, overbuilt systems operate normally under different conditions but require large investment, and their failures, though rarer, imply longer recovery times.

A commonly prescribed adaptation to climate change in water resources management is simply to maintain and enhance resource system characteristics that offer flexibility (see Chapter 8). Maintaining flexibility might include everything from simply delaying some plans until the threat is clarified to changing standards for design buffers or safety margins in dams and other systems. Such an approach would seem illogical to a planner seeking maximum benefit/cost ratios or cost-efficiency, and decision-makers have often responded by dismissing such ideas as impractical in the 'real' world of water resources development or by arguing that standard design practices already make most systems sufficiently robust to absorb projected climate change (Hanchey *et al.*, 1988). There is, thus, a need to link such strategic adjustments to more tactical adjustments (e.g., altered reservoir levels) such as those described below.

Adaptability in the case study basins

Differences in regional water development mean that some areas could adjust to climate change more readily than others. Yet, it is difficult to predict *a priori* which areas have the greatest adaptive capacity. On the one hand, the more heavily developed rivers, like the Uruguay and Indus, have established infrastructures that offer greater control over water and the ability to block negative effects (reactive adjustment). On the other hand, systems in early stages of development, like the Mekong, encompass greater potential for future adjustment; with increasing understanding of the potential for climate change, managers can incorporate

Table 17. *Critical factors of water resources development and management in the river basins*

Basin	Sea-level rise	Salt water intrusion	Hydro-electric power	Fisheries	Navigation transport	Irrigation	Municipal industry	Domestic water supply	Riverine flooding	Env. quality
Uruguay			⊗	x		x	x	x	⊗	
Mekong	x	⊗		x	⊗	⊗		x	⊗	x
Indus	x	x	x	x		⊗		x	⊗	x
Zambezi	x		x	⊗		x			x	
Nile	x	x	⊗		⊗	⊗	x			

Notes:
⊗ = Most important
x = Important

adaptive designs into current plans (anticipatory adjustment).

Adjustments to the impacts of climate change projected in the case studies were explored in workshops attended by water resource managers in each basin. Most basin planners argued for further study and monitoring of global warming effects rather than for making major adjustments now, unless those adjustments served pressing contemporary water management problems as well. In cases where climate impacts could be accommodated by modest infrastructural and management changes, basin managers adopted a 'study-and-wait-and-see' attitude.

From initial surveys of the basins and discussion at the adjustment workshops, we developed a roster of the most critical factors of water resources development and management in each case study basin (Table 17). It is difficult to rank such sensitivities, although each basin does have one or two key water management goals sensitive to climate change (e.g., hydropower production in the Zambezi and Mekong, and irrigation in the Indus). A sense of the hydrological sensitivity of the basins is provided by examining initial results from climate–runoff models based on alternative scenarios of future climate (Table 3). Clearly, the drier basins are more sensitive hydrologically. Yet, it happens that one of the dry basins, the Indus, has a very highly developed canal system that could adjust to even quite marked climate change. Moreover, although the more arid basins tend to be more sensitive to climate change, a significant amount of water comes from glacier melt in the Indus, and climate warming there actually increases runoff.

All five case studies illustrate the critical role that hydrological characteristics and development levels play in vulnerability to climate change. We rank their joint hydrological sensitivity and management adaptability in Table 18. In the Mekong and Indus, for example, impacts associated with GCM scenarios appear relatively small in comparison with the potential impacts, positive or negative, of water development. If the Mekong system of 13 reservoirs is built, including a cascade of eight reservoirs on the main stem, water managers could probably accommodate the hydrological impacts addressed in this study.

Table 18. *Overall basin sensitivity and adaptability*

Basin	Hydrological sensitivity	Structural robustness	Structural resiliency	Adaptive capacity
Uruguay	moderate	high	high	moderate
Mekong	low	low	high	high
Indus	moderate	high	moderate	high
Zambezi	high	low	low	low
Nile	high	high	low	low

The Indus may be vulnerable to climate change because most of the available water is already heavily managed and used. Yet, even in this setting, new water projects could, if implemented, do more to change the patterns and problems of water resource use than would climate change. New reservoir and other structural improvements are constrained, however, by financing limitations, institutional capacity, and social conflict. Moreover, the Indus, while also ostensibly quite 'resilient' (because of system complexity, redundancy, and diversity), is losing some robustness due to infrastructural degradation.

Two of the basins – the Nile and Zambezi – are not only hydrologically quite sensitive, but they also rely on single, large reservoirs whose performances are sensitive to climate fluctuation. Many of the scenarios described herein suggest that water management in the Nile and Zambezi would have to be changed substantially to accommodate global warming. The worst-case scenario for the Nile (the 77 percent reduction in river flow projected by GFDL) portends economic disaster.

The Zambezi especially illustrates the vulnerability (low robustness and low resiliency) of a one-element management system – in this case a reservoir facing operational problems and high demand. Simulations based on use of an additional reservoir (Batoka) show the value of operating at least two management systems in tandem (thereby increasing robustness), but the net value of the additional investment is, as in the Indus, degraded by climate change. The Nile shows large robustness by virtue of a huge kingpin project (Aswan), but it may not be very resilient or adaptable.

Although the Uruguay appears, overall, relatively less sensitive and more adaptable than the other basins, its areas of urban and industrial development are especially vulnerable to floods and droughts likely to accompany climate change. Flood and drought (or any extreme event) analysis is constrained by the fact that atmospheric scientists cannot determine how global warming would affect climate variability and extremes. Yet, an attempt to recalculate flood frequencies for the Uruguay (Table 9), based on the GCM scenarios, shows that dramatic increases in flood heights could occur. The Uruguay basin also illustrates resilience associated with multiple – some relatively small – management structures and employment of several relatively small irrigation projects rather than large, integrated projects. Each small project, however, is individually relatively less robust.

Finally, in some respects the Mekong can be considered one of the most adaptable of the basins studied simply because development there has proceeded so slowly, and future options are relatively more open. Its current uses are relatively less robust but quite resilient (e.g., recessional rice production), and the basin has strong national and regional institutions (e.g., the Mekong Secretariat and the Thai Irrigation Ministry) that will facilitate development and, if needed, adaptation to climate change.

RIVER BASIN PLANNING AND POLICY IMPLICATIONS

Three questions related to basin planning adjustment in the face of global warming remain largely unanswered by the case studies. First, what are the institutional and organizational capacities for adjustment to climate change in each basin? Second, what new problems might be caused by adjustment to climate change? Adjustments that radically increase agricultural water use efficiency, for example, could reduce flows necessary to ecosystems maintenance. Third, what is the 'range of choice' for adjusting to climate change, and how is it affected by other factors (e.g., global economics and finance, environmental concerns, international relations)? Impact studies should include alternatives that 'expand' the range of choice for dealing jointly with current water problems, climate change, and other forces affecting water regional development.

Three recommendations stand out. First, future water planning should concentrate on the physical and social *dynamics* of evolving river basin systems, which implies a shift from approaches based on climate averages and simple trends to approaches that take into account short-term fluctuations, medium-term shocks, and long-term trends. Inherent in this recommendation is the fact that basin development must be seen as a dynamic and adaptive process.

Second, future water planning should draw more heavily on the *lessons of experience* in various basins. There are now decades of project experience with water allocation, pricing, irrigation management, institutional restructuring, structural investment, education, and training in developing regions around the world. Some developments have had unanticipated benefits; others, unanticipated costs. As climate changes, will water managers anticipate the impacts and adjust by drawing upon the lessons of experience? The answer to this question requires more detailed study of how managers have responded to problems in the past.

Third, climate change could lead to *cooperation or conflict* in the world's international river basins. Cooperation may allow for renewed efforts to integrate water and land resources management, the consideration of two-way interactions among regional and global factors in development, and the definition of sustainability at the regional and global scale – all critical to successful, sustainable development in the face of a changing climate. Conflict will almost inevitably worsen the impacts of environmental changes by reducing regional resiliency and adaptability. Fortunately, there are hopeful signs in the world's great, international river basins. Joint projects are being pursued in most of the case studies described here, and integrated basin planning still stands as the highest aspiration of regional development, a goal made more critical by the potential for dramatic social and environmental change in the world's rapidly developing areas.

REFERENCES

Ahmad, M., A. Brooke, and G. Kutcher. 1990. *Guide to the Indus Basin Model Revised.* Washington, DC: The World Bank.

Alam, F. C. 1972. Distribution of Precipitation in Mountainous Areas of West Pakistan. In *Distribution of Precipitation in Mountainous Areas*, vol. 2, 290–306. Geneva: World Meteorological Organization.

Allen, N. 1990. Household Food Supply in Hunza Valley, Pakistan. *Geographical Review* 80:399–415.

Balek, J. 1977. *Hydrology and Water Resources in Tropical Africa.* Amsterdam: Elsevier.

Balon, Eugene K. 1978. Kariba: The Dubious Benefits of Large Dams. *Ambio* 7(20):57–64.

Balon, E. K., and A. G. Coche. 1974. *Lake Kariba: A Man-Made Tropical Ecosystem in Central Africa.* The Hague, Netherlands: Dr. W. Junk b.v., Publishers.

Central African Power Corporation (CAPC). 1980. *Zimbabwe Power Development Plan.* Report by Merz and McLellan and Sir Alexander Gibb and Partners. Salisbury, Zimbabwe: CAPC.

Chan, S., and P. S. Eagleson. 1980. Water Balance Studies of the Bahr el Ghazel Swamo. Report No. 261. Cambridge, Mass.: Massachusetts Institute of Technology, Department of Civil Engineering.

Committee on Science, Engineering, and Public Policy. 1992. *Policy Implications of Greenhouse Warming.* Washington, DC: National Academy Press.

du Toit, R. F. 1983. Hydrological Changes in the Middle-Zambezi System. *The Zimbabwe Science News* 17(7/8):121–6.

Gandolfi, C. and K. A. Salewicz. 1990. *Multiobjective Operation of Zambezi River Reservoirs.* IIASA Working Paper WP-90-31. Laxenburg, Austria: IIASA.

Hanchey, J. R., K. E. Schilling, and E. Z. Stakhir. 1988. Water Resources Planning Under Climate Uncertainty. In *Preparing for Climate Change*, ed. Climate Institute. Washington, DC: Government Institutes, Inc.

Hosier, Richard H. 1986. Energy Planning in Zimbabwe: An Integrated Approach. *Ambio* 15(2):90–6.

Kalabagh Consultants. 1988. Kalabagh Dam Project. Technical Memoranda: *Recorded Flood Flows in the River Indus at Kalabagh Dam Site.* Lahore, Pakistan: Kalabagh Consultants.

Kalinin, G. P. 1971. *Global Hydrology* (translated from Russian, Israel Program for Scientific Translation Ltd.). Springfield: U.S. Dept of Commerce, NTIS. Cited in M. Shahin. 1985. *Hydrology of the Nile Basin.* New York: Elsevier Science Publishing Company, Inc.

Kates, R. W., J. Ausubel, and M. Berberian, eds. 1985. *Climate Impacts Assessment.* New York: John Wiley and Sons.

Marshall, B. E. 1982. The Influence of River Flow on Pelagic Sardine Catches in Lake Kariba. *Journal of Fish Biology* 20:465–9.

McLachlan, A. J. 1970. Some Effects of Annual Fluctuations in Water Level on the Larval Chironomid Communities of Lake Kariba. *The Journal of African Ecology* 39(1):79–90.

Michel, A. 1967. *The Indus Rivers.* New Haven, Conn.: Yale University Press.

Motta, J. C., and C. E. M. Tucci. 1984. Simulation of the urbanization effect in flow. *Journal of Hydrological Sciences* 29:131–47.

Pakistan Meteorological Department. 1991. *Climatic Normals, 1961–1988.* Lahore, Pakistan: Pakistan Meteorological Department.

Perritt, R. 1989. African River Basin development: achievements, the role of institutions and strategies for the future. *Natural Resources Forum.* August:202-8.

Phien, H. N., and L. H. Ti. 1990. *Establishment of Hydrologic Scenarios and Input Data for Simulation.* Technical report prepared for the Mekong Secretariat. Bangkok.

Pinay, G. 1988. *Hydrobiological Assessment of the Zambezi River System: A Review.* IIASA Working Paper WP-88-089. Laxenburg, Austria: IIASA.

Piper, B. S., D.T. Plinston, and J. V. Sutcliffe. 1986. The Water Balance of Lake Victoria. *Hydrologic Science Journal* 31:25–37.

Quick, M., and A. Pipes. n.d. Manual: *UBC Watershed Model.* Vancouver, Canada: Department of Civil Engineering, University of British Columbia.

Rao, Y. P. 1981. The Climate of the Indian Subcontinent. In *Climates of Southern and Western Asia*, vol. 9, 67–123. World Survey of Climatology. New York: Elsevier.

Schaake, J. 1990. From Climate to Flow. In *Climate Change and U.S. Water Resources*, ed. P. E. Waggoner, 177–206. New York: John Wiley and Sons.

Shahin, M. 1985. *Hydrology of the Nile Basin.* New York: Elsevier Science Publishing Company, Inc.

Shaw, E. M. 1983. *Hydrology in Practice.* Wokingham, England: Van Nostrand Reinhold Co. Ltd.

Smith, J. B., and D. A. Tirpak, eds. 1989. *The Potential Effects of Global Climate Change on the United States*, Appendix A: Water Resources. Washington, DC: U.S. Environmental Protection Agency.

Sutcliffe, J. V., and Y. P. Parks. 1987. Hydrological modeling of the Sudd and Jonglei Canal. *Hydrologic Science Journal* 32:143–59.

Tegart, W. J. M., G. W. Sheldon, and D. C. Griffiths. 1990. *Climate Change: The Intergovernmental Panel on Climate Change Impacts Assessment.* Canberra: Australian Government Publishing Service.

Ti, L. H., and H. N. Phien. 1990. *Study of Impacts of Climate Change on Water Resources in the Lower Mekong Basin.* Bangkok: Mekong Secretariat.

United Nations. 1950. *Proceedings of the United Nations Scientific Conference on the Conservation and Utilization of Resources*, vol. 1 *(Plenary Meetings)* and vol. 4 *(Water Resources)*. New York: United Nations, Department of Economic Affairs.

United Nations Economic and Social Council. 1970. *Integrated River Basin Development.* Doc. No. E/3066. New York: United Nations.

United Nations Environment Program (UNEP). 1987. *Diagnostic Study on the Present State of Ecology and the Environmental Management of the Common Zambezi River System.* UNEP/IG.78/Background Paper 1. Nairobi, Kenya: UNEP.

Vander Velde, E. J., Jr. 1989. *Irrigation Management in Pakistan Mountain Environments.* Country Paper – Pakistan, no. 3. Lahore, Pakistan: International Irrigation Management Institute.

Waggoner, P. 1990. *Climate Change and U.S. Water Resources.* New York: John Wiley and Sons.

Wescoat, J. L., Jr. 1991. Managing the Indus River Basin in Light of Global Climate Change: Four Conceptual Approaches. *Global Environmental Change: Human and Policy Dimensions* 1: 381–95.

Wescoat, J. L., Jr. 1992. Beyond the River Basin: The Changing Geography of International Water Problems and International Watercourse Law. *Colorado Journal of International Environmental Law* 3:301–30.

Wescoat, J. L., Jr. and R. M. Leichenko. 1992. *Complex River Basin Management in a Changing Global Climate: The Indus River Basin in Pakistan: A National Assessment.* Collaborative Paper Series No. C-5, Center for Advanced Decision Support for Water and Environmental Systems and the Institute of Behavioral Science. Boulder, Colo.: University of Colorado.

White, G. F. 1957. A Perspective of River Basin Development. *Journal of Contemporary Problems* 22:157–87.

White, G. F. 1964. Contributions of Geographical Analysis to River Basin Development. *Geographical Journal* 129:412–36.

World Meteorological Organization (WMO). 1977. *Hydrologic Model of the Upper Nile Basin*, vol. 1-B, Appendices. Geneva: WMO.

4 Global Sea-level Rise

ROBERT J. NICHOLLS

School of Geography and Environmental Management, Middlesex Univ., UK

Laboratory for Coastal Research, University of Maryland, USA

STEPHEN P. LEATHERMAN

Laboratory for Coastal Research

University of Maryland

SUMMARY

Accelerated sea-level rise is one of the more certain consequences of greenhouse-induced climate change and would have many adverse impacts on the world's coastline. The impacts and cost of a rise in sea level have been assessed for many industrial nations, but much less is known about the potential effect on developing nations. To address this problem we initiated a number of vulnerability assessments in developing countries in collaboration with in-country experts. The assessments included (1) national overviews and selected case studies in Brazil, Egypt, Bangladesh, India, Malaysia, and China; and (2) detailed national assessments using aerial videotape-assisted vulnerability analysis (AVVA) in Senegal, Argentina, Nigeria, Uruguay, and Venezuela. Scenarios of sea-level rises of 0.2, 0.5, and 1.0 m by the year 2100 were considered, using a 1 m rise as a standard scenario. Existing patterns and levels of coastal development were assumed. Coastal areas that are vulnerable to loss of use including the resident population, under a given sea-level rise scenario, were termed at risk.

Our results show that a number of developing countries are particularly vulnerable to the effects of sea-level rise. The following conclusions are noteworthy:

- Most of the land at risk is in deltas and would be subject to inundation. However, high-value land along many sandy coasts is also at risk of erosion.
- On a global scale, significant reduction of coastal wetlands appears likely and may result in negative impacts on other coastal resources, such as fisheries.
- Countries with large agricultural populations in deltas, such as Egypt, Bangladesh, and China, are particularly vulnerable. For a 1 m sea-level rise:
    - *In Egypt*: 12 to 15 percent of the agricultural land and over 6 million people will be at risk, including half the population of Alexandria.
    - *In Bangladesh*: Land that provides 16 percent of the national rice production and is inhabited by 13 million people will be at risk. The Sunderbans, the second largest mangrove swamp in the world and one of the last refuges of the Bengal tiger, could be lost.
    - *In China*: 72 million people, major cities such as Shanghai and Tianjin, and tens of thousands of square kilometers of agricultural land will be at risk, primarily in the four major coastal plains.
- Should it be selected as a protection option, sand nourishment of tourist beaches could constitute a significant component of costs.
- Uruguay and Senegal face a larger burden of protection costs (relative to gross investment in 1990) than Nigeria, Venezuela, and Argentina.
- Building setbacks could greatly reduce vulnerability to sea-level rise.

These conclusions are based upon the present pattern and distribution of coastal development. However, future major coastal development in the countries studied is almost certain. All such development would be best regulated within integrated coastal zone management plans with a primary aim of minimizing vulnerability to future sea-level rise. Protection is technically feasible and likely in developed areas, although increasingly large populations would be living beneath sea level and could face catastrophic consequences in the event of failure of coastal defenses. The standard of design for any protection should reflect this risk. In deltaic settings, novel and relatively untried approaches, such as sediment management, may be useful. It is important to note that there are limits to human adaptation. In particular, it is difficult to counter the projected loss of coastal wetlands, even in developed countries.

Much more detailed information on the coastal zones of the world is required to assess vulnerability and formulate responses to sea-level rise with enough detail to be useful to planners. In particular, further work is needed on:

- the direct impacts of sea-level rise that are not addressed in this study, including salinization of freshwater supplies.
- the interaction of sea-level rise with other manifestations of climate change, such as changing storm intensity and frequency.
- the dynamic analysis of optimum timing of different responses to the impacts of sea-level rise (e.g., the economic practicality of different response options, including beach nourishment, land raising, and seawall emplacement, on decadal time frames; Yohe, 1990).

The uncertainty associated with future sea-level rise demands flexible coastal zone policies that can be adapted to changing conditions. Global sea level is already rising, most sandy shorelines are retreating, coastal wetlands are being lost, and coastal populations continue to grow rapidly. Therefore, many of the policies that would be helpful in responding to accelerated sea-level rise are already necessary. Some of these potential responses, such as land use planning, will require long lead times to become fully effective. Therefore, it is essential that we, in the near future, evaluate the range of possible responses to sea-level rise and act accordingly; otherwise we may reduce the options of future generations in countering the predicted accelerated sea-level rise.

INTRODUCTION

Ever since the dawn of civilization, the coastal zones of the world have been important to humankind. Today, a large portion of the world's rapidly growing population lives in the coastal zone, and 13 of the world's 20 largest cities are located on the coast. Population increase and development impart major stresses on the resources of the coastal zone. The additional impact of potential climate change on these resources is therefore of great concern (Figure 1).

Unlike other manifestations of climate change, sea-level rise is already a problem. Best estimates are that over the last century sea level has risen globally by 1.8 ± 0.1 mm/yr (Douglas, 1991). This rise in sea level can be correlated with a global rise in temperature (Gornitz *et al.*, 1982) and is due to the thermal expansion of seawater and the melting of land-based ice, predominantly small glaciers (Warrick and Oerlemans, 1990). Even small increases in sea level have global impacts over long time scales (Figure 2). For instance, best estimates are that over 70 percent of the world's sandy coasts are presently eroding (Bird, 1985). Whereas many factors contribute to this global problem, the long-term trend of sea-level rise is considered by many as an important underlying factor in worldwide shoreline recession (e.g., Vellinga and Leatherman, 1989).

Global warming could accelerate sea-level rise by causing thermal expansion of ocean water, melting mountain glaciers, and melting ice sheets or causing ice sheets to discharge into the ocean. A number of quantitative estimates of future sea-level rise have been developed, with estimates ranging from as low as 0.1 m to as high as 3.7 m by the year 2100 (Warrick and Oerlemans, 1990). The Intergovernmental Panel on Climate Change (IPCC) estimated in 1990 that sea level would rise 0.3 to 1.1 m by the year 2100 (Figure 3); the estimate assumed that Antarctica would be a net water sink because increased snowfall would more than offset any increase in melting (Warrick and Oerlemans, 1990). The IPCC 1992 update implies a reduction in these numbers by 20 to 30 percent (Wigley and Raper, 1992). Further adjustments are to be expected as our understanding of future climate and its effects on sea level improves.

Rising sea level can directly inundate (or submerge) low-lying wetland and dryland areas, erode shores by increasing offshore and longshore loss of sediment, increase the salinity of estuaries and aquifers, raise coastal water tables, and exacerbate coastal flooding and storm damage (Table 1). As our understanding has increased, there has been a growing realization of the serious effects and costs associated with even small accelerations in sea-level rise. Assessments of this impact on some of the most vulnerable industrial nations estimate that the total undiscounted cost of a 1 m rise in sea level would be $12 billion for the Netherlands, at least $74 billion for Japan (IPCC, 1992), and as much as $200 to $475 billion for the United States (Titus *et al.*, 1991). Much less work has been done in estimating the impact of sea-level rise on developing nations. Many of the nations that seem particularly vulnerable have large and expanding coastal populations with much future development likely for their coasts. Yet they lack the resources to determine vulnerability, much less to implement appropriate responses (Carter, 1987; Holdgate, 1989; anonymous, 1990; IPCC, 1992).

To better quantify this risk, we initiated studies with in-country experts on the potential impacts of a range of scenarios of sea-level rise. Particular emphasis was given to a 1 m rise because this is a standard scenario for other studies (IPCC, 1990; 1992). The principal collaborating scientists are listed in Table 2. Studies were conducted in 11 developing countries (on three continents) having a range of coastal characteristics and levels of development (Figure 4). In the countries considered, national populations ranged from 3 million in Uruguay to 1.1 billion in China, while coastline lengths ranged from about 650 km in Bangladesh to about 18,000 km in China. Predictions of the impacts of sea-level rise assumed existing patterns and levels of coastal development. All of the studies emphasized the impacts and possible responses to the problem of coastal land loss. Bangladesh,

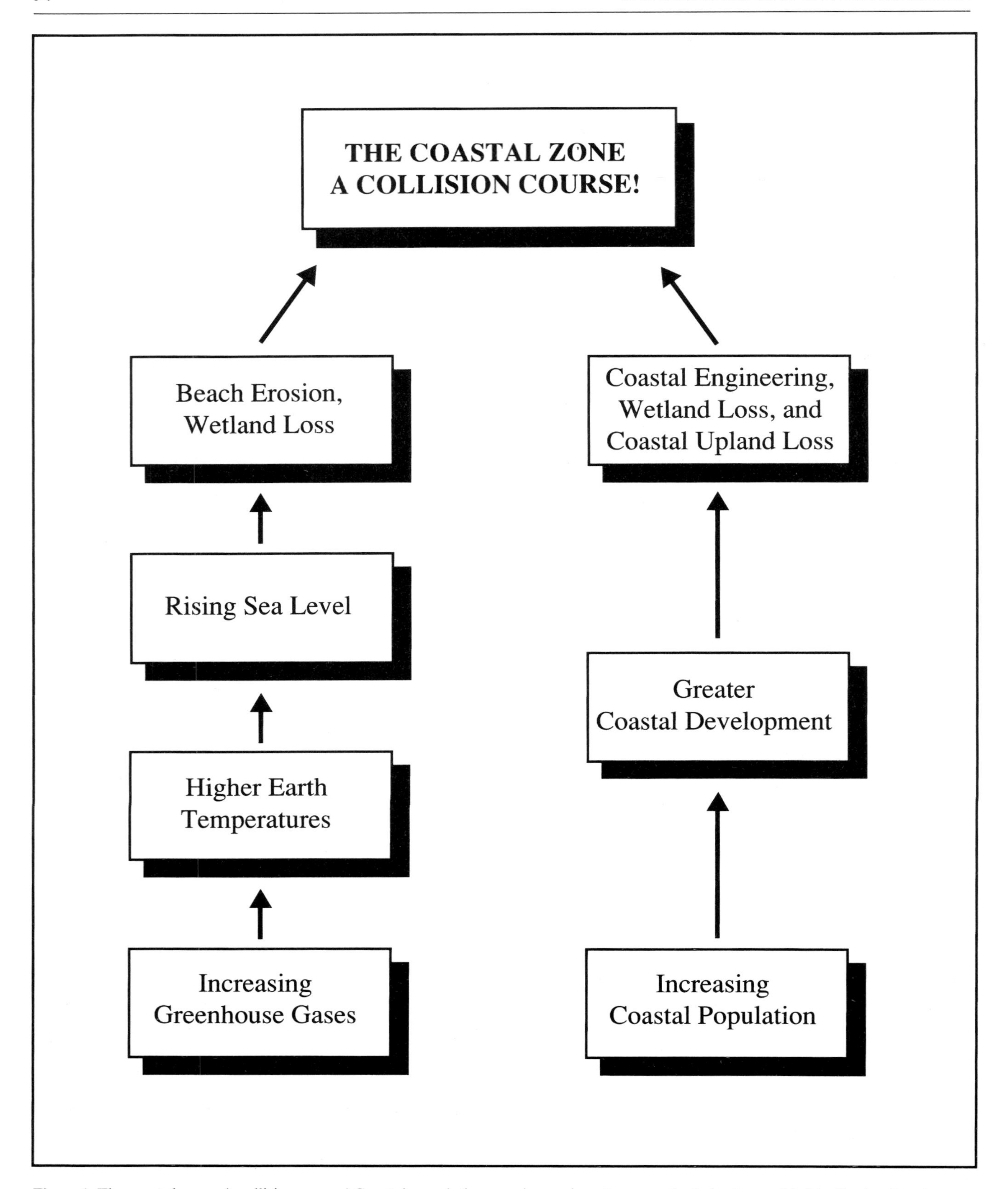

Figure 1 The coastal zone: A collision course! Coastal populations are increasing at unprecedented rates worldwide. Sea level is also rising. Although the rate of the rise in sea level is slow today, it is expected to increase due to greenhouse-induced warming. Together, these two changes will place unprecedented stress on the coastal zones of the world.

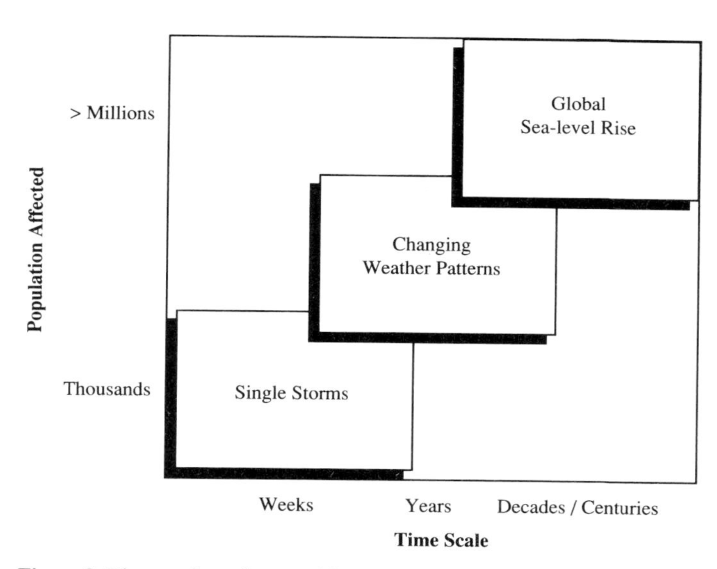

Figure 2 Time scale and coastal hazard (adapted from Nicholls *et al.*, 1993). Although sea-level rise is a slow process, a steady rising trend will, over time scales of decades to centuries, have a direct impact on many millions of people and an indirect impact on hundreds of millions of people living in the coastal zone. The faster the rate of rise, the larger the number of people affected.

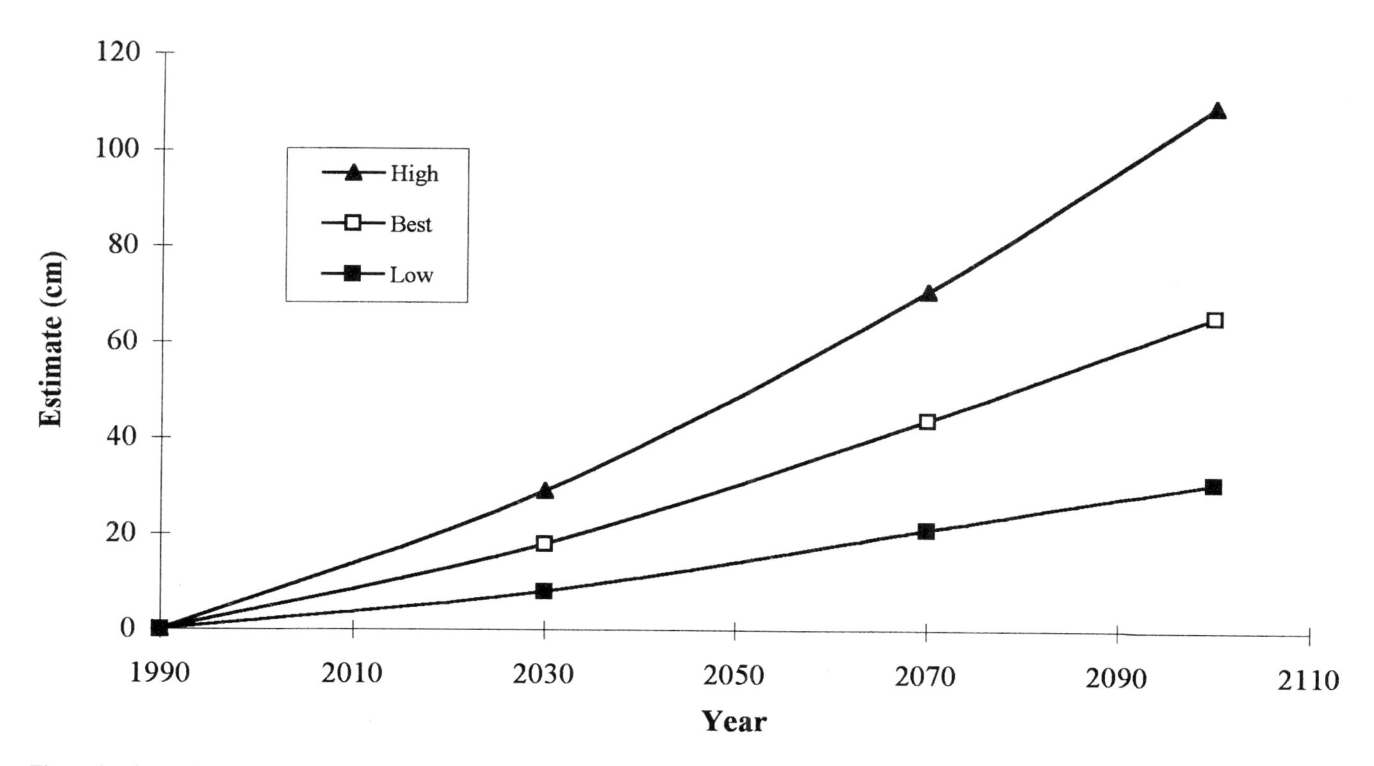

Figure 3 The IPCC 1990 sea-level-rise scenarios (after Warrick and Oerlemans, 1990). They assume that greenhouse gas emissions grow under business-as-usual assumptions.

Table 1. *Physical effects of sea-level rise*

- Inundation of low-lying areas ← Land loss
- Erosion of beaches and bluffs ← Land loss
- Salt intrusion into aquifers and surface waters
- Higher water tables
- Increased flooding and storm damage

Table 2. *Principal collaborating scientists*

Name	Institute	Country
José Arismendi	Instituto de Ingenieria	Venezuela
Virendra Asthana	Jawaharalal Nehru University	India
Larry F. Awosika	Nigerian Institute for Oceanography and Marine Research	Nigeria
Karen C. Dennis	University of Maryland	U.S.A.
Mohammed El-Raey	University of Alexandria	Egypt
Gregory T. French	University of Maryland	U.S.A.
Mukang Han	Peking University	China
Saleemul Huq	Bangladesh Centre for Advanced Studies	Bangladesh
Say-Chong Lee	Drainage and Irrigation Department	Malaysia
Zamali Midun	Drainage and Irrigation Department	Malaysia
Dieter Muehe	Universidade Federal do Rio de Janeiro	Brazil
Claudio F. Neves	Universidade Federal do Rio de Janeiro	Brazil
Isabelle Niang-Diop	Université Cheikh Anta Diop	Senegal
Enrique Schnack	Laboratorio de Oceanografia Costera	Argentina
Claudio R. Volonté	Canelones, Uruguay	Uruguay
Boacan Wang	East China Normal University	China
Wyss Yim	University of Hong Kong	Hong Kong

Brazil, China, Egypt, India, and Malaysia were subjected to national overviews, the results of which depended largely upon existing data. These were followed by more detailed and site-specific case studies of Recife, Alexandria Governorate (Egypt), the North China Coastal Plain, the Shanghai area, and Hong Kong.

In common with other researchers around the world (IPCC, 1990), we found that there is a lack of basic data on the coastal zones of most developing countries. Therefore, we developed a method for estimating the impacts of rising sea level: aerial videotape-assisted vulnerability analysis (AVVA) (Nicholls *et al.*, 1993; Leatherman *et al.*, 1995). The coastline is videotaped from a small plane, and coastal characteristics are interpreted from the video record of coastal zones between occasional ground-truth stations. The video record and any available map data are used interactively with models of land loss due to sea-level rise and response options to quantify vulnerability to sea-level rise. Detailed assessments of Uruguay, Argentina, Venezuela, Nigeria, and Senegal were prepared using this technique, involving aerial videotaping of over 6,000 km of shoreline. These studies provide some of the most detailed analyses to date concerning the vulnerability of developing countries to accelerated sea-level rise.

This chapter presents the results from all 11 countries. More detailed methodology, results, and discussion are found in Nicholls and Leatherman (1995*a*). An integrated national study of the impacts of climate change, including sea-level rise, on Egypt is presented by Strzepek *et al.* (see Chapter 7).

LITERATURE REVIEW

A number of studies have addressed, to varying degrees, the impacts of and responses to sea-level rise. Within the United States, case studies such as those of Charleston, SC. (Kana *et al.*, 1984), Galveston Bay, Tex. (Leatherman, 1984), and Ocean City, Md. (Titus *et al.*, 1985) paved the way for a national analysis of the potential impacts of sea-level rise (Smith and Tirpak, 1989; Titus *et al.*, 1991). The latter study estimated that a 1 m rise in sea level would have an undiscounted present cost for the United States of $200 to $475 billion, assuming that measures would be taken to protect

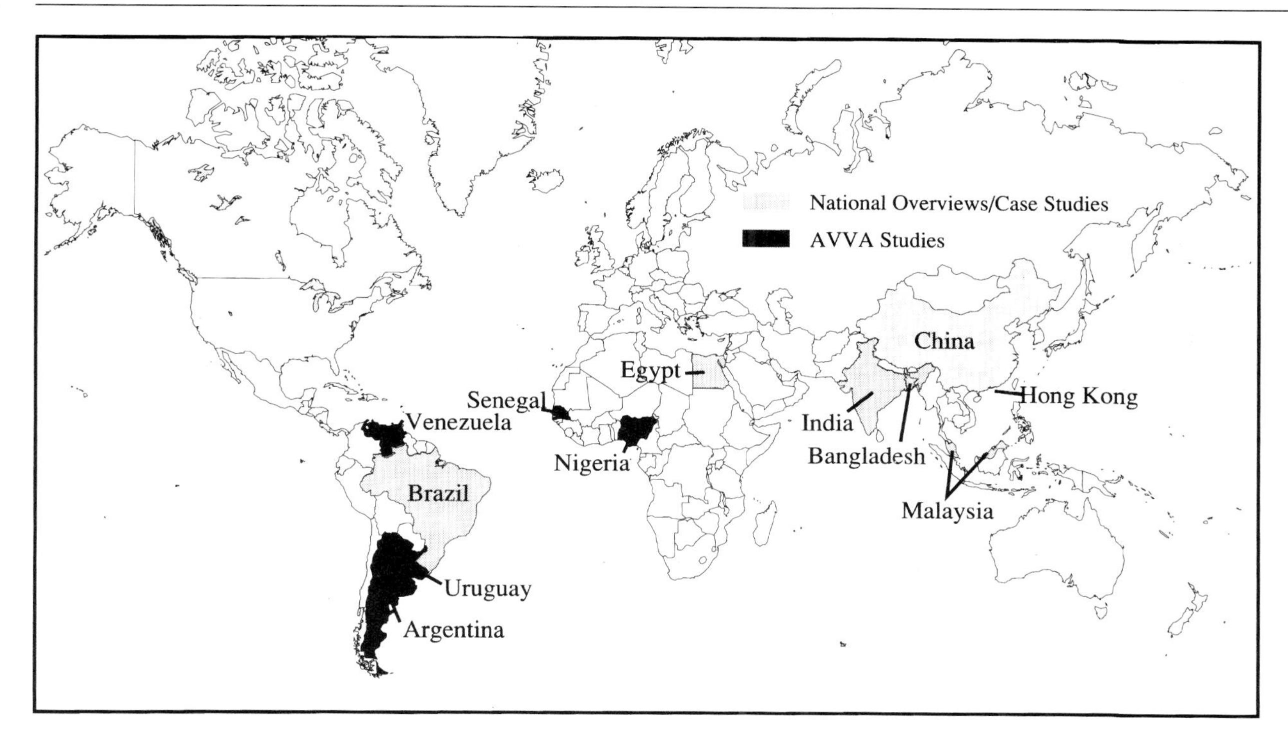

Figure 4 The study countries.

presently developed areas. Although this cost is probably affordable for a wealthy country such as the United States, significant areas of coastal wetlands (29 to 69 percent of the existing area) would be lost, even allowing for inland migration. Similar studies have been undertaken in other developed countries, most particularly the Netherlands (Wind, 1987; Peerbolte *et al.*, 1991).

Broadly based studies have also been initiated in developing countries to identify the vulnerability of certain areas to sea-level rise, particularly heavily populated low-lying deltas such as the Nile (Egypt) and Ganges (Bangladesh) (Broadus *et al.*, 1986; Milliman *et al.*, 1989). These areas have enhanced vulnerability to sea-level rise due to (1) high rates of coastal land subsidence; and (2) actual or potential sediment starvation caused by upstream dams such as the Aswan High Dam on the Nile. Although these earlier desktop studies provided estimates of the population at risk, the more detailed results reported herein indicate an overall underestimation in those studies of this significant problem.

A number of small, low-lying island states, such as the Maldives, are threatened with extinction by 1 to 2 m rises in sea level (Pernetta and Sestini, 1989; Pernetta, 1991). Recognition of this potentiality led to the formation in February 1991 of the Alliance of Small Island States (AOSIS), an active and a powerful political voice for these countries.

The ongoing work of the Coastal Zone Management Sub-Group of the Intergovernmental Panel on Climate Change (IPCC) has provided a focus for the studies of impacts and possible responses to sea-level rise (IPCC, 1990; 1992; Titus, 1990). They provided the first worldwide estimate of the cost of basic coastal protection measures in response to a 1 m rise in sea level – \$371 to \$606 billion (IPCC, 1990) – as well as guidelines for vulnerability analysis (the IPCC Common Methodology). A global vulnerability analysis was also instituted (IPCC, 1992). The results of these studies reinforce concern about the vulnerability of the developing world to sea-level rise.

There is a clear need for more comprehensive studies to better understand the impacts, costs, and possible responses to sea-level rise. Our studies aim to fulfill such a role and identify coastal resources and infrastructure vulnerable to sea-level rise in 11 developing countries. This is an important first stage in formulating plans to more efficiently manage the coastal zone in a sustainable manner, including minimizing vulnerability to future sea-level rise.

METHODOLOGY

The sea-level-rise project, relying heavily on in-country collaborators (listed in Table 2), included three phases:

1. National overviews of sea-level rise in Bangladesh, Brazil, China, Egypt, Malaysia, Nigeria, India, and Senegal.
2. In-depth case studies of the North China Coastal Plain, the Shanghai area, Hong Kong, Alexandria Governorate (Egypt), and Recife (Brazil).
3. National assessments using AVVA of Argentina, Nigeria, Senegal, Uruguay, and Venezuela.

Models and approaches used

Estimates of the impacts of sea-level rise over the next century contain inherent physical and socioeconomic uncertainties even when comprehensive coastal data sets are available. This is illustrated by studies of the Netherlands (Peerbolte *et al.*, 1991). The impacts of sea-level rise are fully manifest only over time scales of decades to centuries. Impacts and possible responses to sea-level rise must be considered over the same time scale. These time periods are longer than conventional planning horizons, and for some potential response options they are longer than the period of available experience. Even the long-term costs of simple engineering responses such as seawalls and beach nourishment are subject to large uncertainties. Therefore, the models and approaches utilized have to be both general and applicable where limited data are available. In addition, considerable judgment is required.

The national overviews reported herein largely relied on coarse topographic maps for the rough estimation of inundation combined with expert judgment regarding the facilities that are likely to be vulnerable. The case studies involved the detailed study of areas that the national overviews identified as vulnerable to sea-level rise. Physical and economic data were collected in all cases. The AVVA technique, described previously, provided new physical data and hence allowed improved estimates of the vulnerability to rising sea level.

Vulnerability analysis

We examined the potential impacts of different scenarios of sea-level rise. This examination constituted a vulnerability analysis: an 'estimate of the degree of loss or damage that would result from the occurrence of a natural phenomenon of given severity' (Organization of American States, 1991).

The Common Methodology reported by IPCC (1992) developed the concepts of vulnerability analysis for accelerated sea-level rise and other impacts of climate change. Our procedure for vulnerability analysis is similar to the Common Methodology (Nicholls *et al.*, 1993). For each specific future sea-level-rise scenario, it takes the present pattern and level of coastal development and a response option and applies them at the national level to estimate the vulnerability of the coastal land, its use, and its population (providing economic and social indicators of the magnitude of vulnerability). It does not consider future coastal development.

The concepts of *loss* and of *at risk* and *displaced* populations require definition. We can estimate land vulnerable to loss for any sea-level-rise scenario. A response option may protect none, some, or all of this vulnerable area. For a specified response option, we consider any unprotected land and its uses to be *lost* and the resident population *displaced*. If no response option is specified, then the vulnerable land, its uses, and the resident population are considered *at risk* (synonymous with vulnerable), except coastal wetlands, which are always considered to be *lost* (cf. Vellinga and Leatherman, 1989; IPCC, 1992). This terminology is followed throughout this chapter.

Sea-level-rise scenarios

We utilized three eustatic scenarios (that is, scenarios of worldwide sea-level rise): 0.2 m, or no acceleration in sea-level rise (after Douglas [1991]); 0.5 m and 1.0 m by the year 2100. (Additional results of a sensitivity analysis using a 2 m scenario are presented in Nicholls and Leatherman [1995a]. These results are approximately double those we obtained for the 1 m scenario.) Where possible, these eustatic scenarios were converted into relative (or local) sea-level-rise scenarios, which includes any local uplift or subsidence of the land. Within the limits of the available (generally limited) data, the difference between the relative and eustatic sea-level rise was usually negligible, and the eustatic scenarios were applied directly. Important exceptions do occur, such as for Shanghai, China, where up to 1 m of land subsidence is possible by the year 2100 (Wang *et al.*, 1995).

Aerial videotape-assisted vulnerability analysis (AVVA)

One of the major problems with studying the impacts of sea-level rise in developing countries is the lack of detailed information on coastal characteristics, the existing pattern and scale of development, and, most fundamentally, coastal elevation. In these studies we found that the best topographic maps available generally had 10 m or larger contour intervals. Conventional procedures to improve these data, such as aerial photography, were too costly and time-consuming for our purposes.

To overcome these constraints, we developed a reconnaissance technique for vulnerability analysis which we call AVVA. For the purposes of this discussion, the AVVA technique comprises the entire data collection and analytical package, including land loss estimates, their impacts, and the various response options.

The AVVA studies primarily examined the following consequences of sea-level rise: land at risk to inundation and erosion (Table 1), value of land and buildings at risk, and cost of various response options. To express the uncertainty of some results, such as beach erosion and response costs, ranges of values are given.

DATA COLLECTION

We characterized and inventoried the coastal development and geomorphology (including rough estimates of coastal elevations) of the study areas by videotaping the coastline obliquely from a small airplane flown at low elevation above the sea (50 to 100 m). This approach builds on earlier videography studies in the United States, particularly work on the rapidly changing coastal zone of Louisiana (see, e.g., Debusschere *et al.*, 1991). In addition, we have found oblique aerial videotape techniques to be of great value in our work in the mid-Atlantic region (e.g., Downs, 1993). (For more complete details concerning the AVVA technique, see Appendix E of IPCC [1992], Nicholls *et al.* [1993], and Leatherman *et al.* [1995].)

The AVVA technique does *not* utilize photogrammetric procedures. Rather the video record provides a contemporary view of the shoreline. It is analyzed in conjunction with existing and limited current ground-truth data to provide the following information about the coastline: (1) an index of terrain and relief changes, including a *subjective* estimate of the relative topography; (2) description of coastal environments; (3) data on land use practices; (4) description of infrastructure; and (5) estimates of population density. The contemporary pattern and distribution of coastal development are often difficult and expensive to acquire from other sources.

Subsequently, various data concerning coastal geomorphology, in-country protection costs, and socio-economic information such as land use and value were collected on the ground. The number of ground stations for coastal surveys varied between countries. Logistical constraints sometimes restricted the number to single figures. All available maps were also collected, particularly for low-lying areas, where the video record is of most limited value.

It is difficult to define percent error because the accuracy of the data derived from the video record is variable (see Nicholls and Leatherman, 1995a). The information on the horizontal extent of land use, such as the area of developed regions, is generally accurate to better than 10 percent. Importantly, tests we conducted in the Chesapeake Bay indicated that the positional/elevation error is unbiased; hence, there is no systematic departure from the true value when data are aggregated across a country.

LAND LOSS ESTIMATES

Our reconnaissance estimates of land loss only consider the direct effects of sea-level use. In practice, other processes may offset or augment the predicted losses, this effect varying from place to place.

On steep rocky coasts, sea-level rise has no significant land loss impact. On wave-exposed sandy coasts, sea-level rise promotes offshore transport of sand and hence shoreline recession. On sheltered, low-lying coasts, direct submergence (or inundation) is the primary land-loss mechanism. Using the video record, we determined the appropriate land-loss mechanism and applied the quantitative procedures described in the sections that follow.

Erosion

Shoreline and cliff-top recession were estimated using the Bruun Rule concept (Hands, 1983). This method requires an estimate of the active profile width (L) and the active profile height (H) (Figure 5). These values were obtained from field data and available charts. The active profile width presents more difficulties than the profile height because it may extend a kilometer or more offshore and because the estimates of shoreline recession (and nourishment cost) are sensitive to the value used (Nicholls *et al.*, 1995). For our calculations we used a low and high estimate of this width, derived from wave data. These values yield a low and high estimate of shoreline recession. Although wave data were available for each country, they were of limited quantity in terms of both duration and number of sites. Therefore, profile width often had to be extrapolated over large distances using expert judgment. Such extrapolations could lead to errors in the estimates of shoreline recession (and the cost of beach nourishment as explained below).

Inundation

In the case of coastal lowland, we applied a direct inundation, or 'drowning,' concept. For extremely low-lying areas such as deltas, the video record is insufficient to define an estimate of elevation, and information from the video record must be integrated with data from available maps and in-country expert judgment. The 2 m inundation contour, and where possible the 1 m inundation contour (above high water), were estimated from the available data, whereas the smaller inundation contours were interpolated. Thus, land loss due to inundation could be determined.

Coastal wetlands can accrete vertically due to biomass/sediment input and keep pace with slow rates of sea-level rise (Stevenson *et al.*, 1986). Inundation and land loss begin to occur only above some threshold rate of sea-level rise, which is site-specific. Knowledge about local wetland dynamics is generally too fragmentary and incomplete to provide this information. In such cases, a useful assumption is that the threshold value is the baseline scenario (a 0.2 m rise over the next century) (Nicholls *et al.*, 1995). Wetland migration was not considered because of lack of data.

VALUE AT RISK

The coastal areas that are at risk for each scenario can be

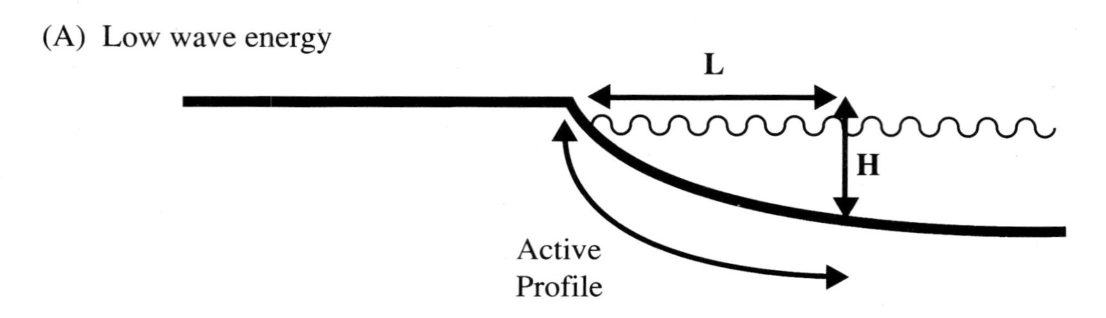

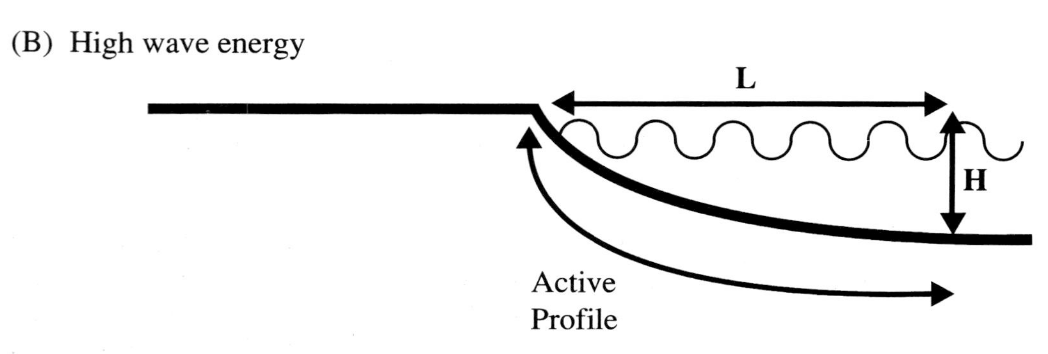

Figure 5 The active profile may extend more than a kilometer offshore. The higher the wave energy reaching the coast, the wider the active profile. Wider active profiles increase shoreline recession and raise the costs of beach nourishment with sea-level rise. Reproduced with permission from Nicholls *et al.*, 1995.

delineated from the video record and available maps. In particular, the number and type of buildings can be directly assessed using the video, and an estimate of the population at risk can be made. The land area and buildings at risk are valued using the best available in-country data. Port and harbor loss or disruption and loss of roads and other utilities was not considered as we did not have the data necessary to make such an assessment. In terms of response, it is highly likely that ports and harbors will be upgraded due to their strategic and vital role within national and international economies. Land values were sometimes difficult to define. In Senegal, land value was not readily available, so value at risk for that country represents only buildings (Dennis *et al.*, 1995a). Wetlands are not valued. Thus, the estimates of value at risk presented in this chapter are probably always minima.

RESPONSE OPTIONS

For areas which might be affected by sea-level rise, three response (or protection) options were considered (as shown for Senegal in Table 3):

1. *No Protection (NP)*
   Existing protection is ignored.
2. *Important Areas Protection (IAP)*
   Areas that currently have medium to high levels of development are protected (e.g., cities, tourist beaches, factories). For tourist areas, beach nourishment is assumed in order to maintain beaches and protect the adjacent coastal buildings and other infrastructure. Elsewhere, seawalls are utilized. The cost of harbor upgrade is taken from IPCC (1990).
3. *Total Protection (TP)*
   All coastal areas with a current population greater than 10 people/km^2 are protected (cf. IPCC, 1990). All additional protection above that listed for IAP is through seawalls except in Nigeria, where relocation appears to be more suitable in many locations (French *et al.*, 1995).

Land loss, value loss, and protection costs were estimated for each response option. By definition, the cost of protection in the NP category is zero, but the loss of land and buildings is maximized. Costs of cleaning up damaged and destroyed infrastructure were not considered. Coastal wetlands are not protected or restored under any of these options.

SEAWALL AND NOURISHMENT DESIGNS

The protection options were developed using in-country costs and are summarized in Table 4. The total cost for seawalls and groins included the cost of purchasing, transporting, and placing rock, plus design costs (10 percent) and maintenance costs to the year 2100 (20 percent over the life of the structure). The cost of sand was typically about \$5 to \$6/

Table 3. *Senegal – length of protection by response option (after Dennis* et al.*, 1995a)*

Development type	Length (km) NP	IAP	TP	Response
Harbors	0	4	4	Harbor upgrade
Existing seawalls	0	7	7	Seawall upgrade
Urban	0	34	34	Seawall
Industrial	0	3	3	Seawall
Tourist areas	0	22	22	Beach nourishment
Fishing villages*	0	0	3	Seawall
Other coastline	0	0	1990	Seawall
Total	0	70	2063	

Notes:
NP – No Protection
IAP – Important Areas Protection
TP – Total Protection
* Under Important Areas Protection, it is assumed that the fishing villages, which are composed of temporary structures, will be relocated progressively inland at essentially no cost.

Table 4. *Unit cost (including estimated maintenance) of the seawall designs and beach nourishment (based on a national average) for the 1-meter scenario (in millions of dollars per kilometer)*

Country	SC	LCWC	HCWC	CL	Nourishment
Argentina	0.33	0.44	1.1/1.7	3.1/4.3	3.6/12.9
Nigeria	0.52	0.69	1.9	n.a.	4.5/16.3
Senegal	0.33	0.44	1.7/2.3	n.a.	8.8/33.2
Uruguay	0.6	0.67	1.7/3.9	4.8/8.1	33.9
Venezuela	0.4	0.53	0.7/1.5	0.5/1.5	5.2/12.3

Notes:
n.a. – not applicable
SC – sheltered coast seawall
LCWC – low-cost wave-exposed coast seawall
HCWC – high-cost wave-exposed coast seawall
CL – cliff coast seawall

m^3 where local supplies of suitable sand were readily available.

Seawalls

We developed three simple seawall designs: a low- and high-cost seawall for wave-exposed coasts (LCWC and HCWC, respectively) and a seawall for sheltered coasts (SC) (Figure 6). The LCWC design assumes no beach erosion, whereas the HCWC design assumes total beach loss. Actual behavior will lie somewhere between these two extremes. We utilized a modified HCWC design for cliff-toe protection, incorporating a 20 m wide berm. The SC design has a lower cost than the LCWC design because, for the former, additional wave run-up is not a major problem.

Beach nourishment

The pumping of sand from offshore to increase the size of the beach and hence counteract erosion is commonly called beach nourishment, although other terms such as beach replenishment are also used (Davison *et al.*, 1992). This is a necessary option to maintain beaches for tourism and protect the adjacent tourist infrastructure. Sand must be added to the entire active profile (Figure 5) to counter the erosion. The volumes of sand required were determined using the low and high active profile widths already defined for the erosion analysis; these volumes were then used to determine a low- and high-cost design. An additional cost of beach nourishment on long open beaches is longshore loss of sand out of the nourished area (Dean, 1983). To avoid this sand loss, we assumed the construction of large (and expensive) terminal groins on open beaches. This feature raised the cost of nourishment up to threefold but resulted in a cost that was more realistic than that determined when only the cost of a one-time pumping operation was considered (Titus *et al.*, 1991; Davison *et al.*, 1992).

Seawall/nourishment cost comparison

The estimated cost of beach nourishment greatly exceeded the estimated cost of seawalls in this study, particularly in Uruguay where the shoreline is subject to high wave energy (Table 4).

The cost of beach nourishment, given long time-scales (decades to centuries) and accelerated sea-level rise, will likely be higher than current costs (Stive *et al.*, 1991; Davison *et al.*, 1992), but the amount of increased cost is uncertain. Three factors may raise seawall costs above those presented here. First, the capital costs are for ideal single-phase construction. In reality, multiphased construction is more likely and would result in higher total construction costs. Second, the maintenance costs for seawalls may have been underestimated. Storm damage, including repeated undermining of seawalls, is the normal experience in Germany, the Netherlands, the United Kingdom, and the United States and necessitates frequent and often significant expenditure on maintenance work. Third, the selection of the appropriate seawall design was based on environmental characteristics only; coastal land use was considered only in the context of the protection options already defined. However, IPCC (1990) assumed that seawalls protecting cities would be significantly more expensive than seawalls protecting other land uses, an assumption that seems reasonable.

A sensitivity analysis that takes into account higher seawall costs is presented below in the discussion of national

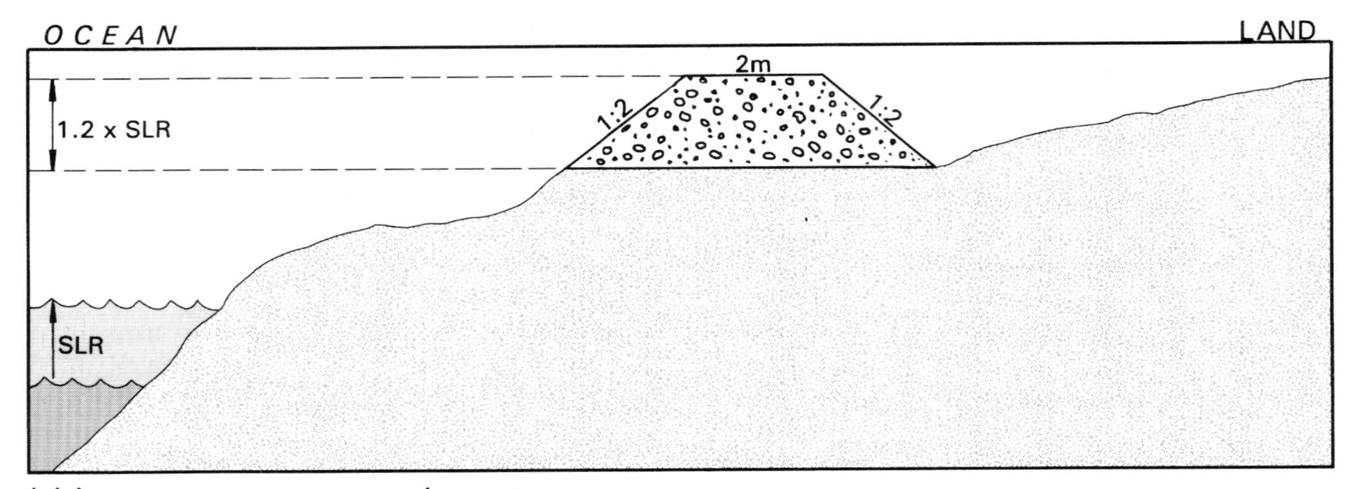

(a) Low cost wave-exposed coast

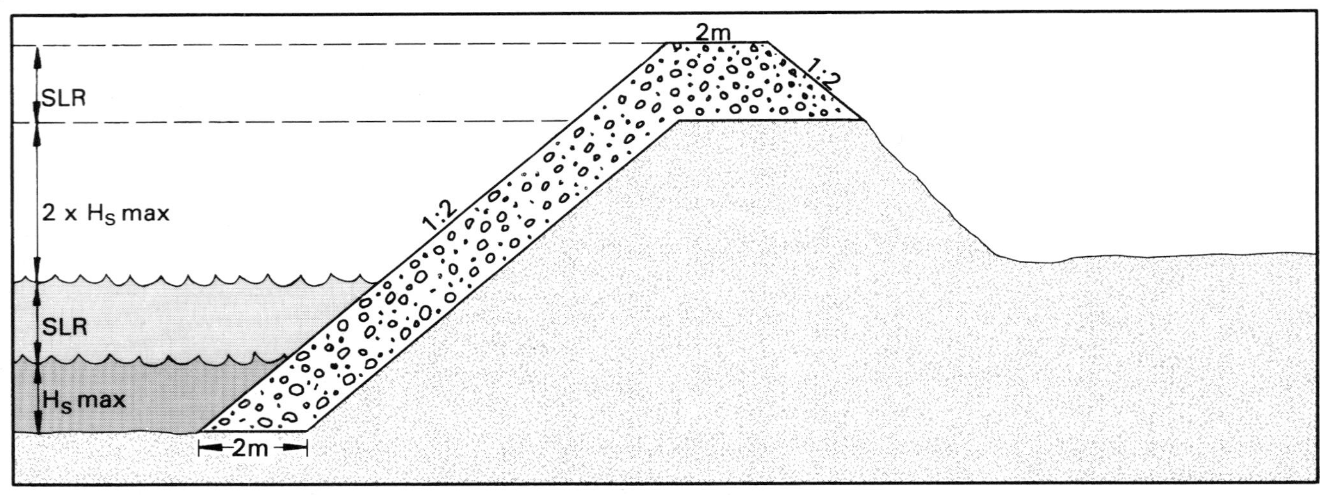

(b) High cost wave-exposed coast

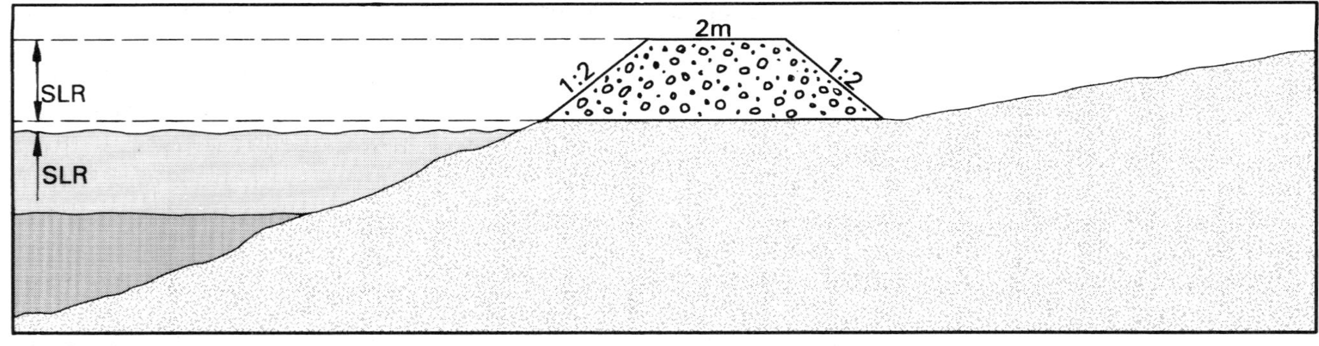

(c) Sheltered coast

H_S max – –Maximum Significant Wave Height
SLR – – – –Sea Level Rise Scenario

Figure 6 Three of the seawall designs: low-cost wave-exposed coast seawall (LCWC); high-cost wave-exposed coast seawall (HCWC); and sheltered coast seawall (SC). Reproduced with permission from Nicholls *et al.*, 1995.

response costs in order to quantify the importance of relative seawall/nourishment costs.

TIMING

These studies lack an explicit time dimension, although the approach used is similar to that of most other vulnerability assessments, including the cost estimates of IPCC (1990). Where an explicit time dimension was required, as a first approximation, we assumed that the costs/impacts would be distributed linearly in time from 2051 to 2100 (50 years). (Note that these costs are not discounted.)

STUDY LIMITATIONS

Some of the more important limitations of this study are as follows.

Other coastal processes

Shoreline position is an integrated response to a number of processes, of which sea-level rise is an important contribution (National Research Council, 1990). Our estimates of land loss due to beach erosion do not consider other processes, such as longshore transport, because the appropriate information was not available. Such factors could increase or decrease the predicted land loss. Future studies should consider more holistic models of shoreline evolution, if possible.

Subsidence in some deltas

Subsidence was not considered in some of the deltas studied, including the Orinoco Delta in Venezuela, the Ganges–Brahmaputra Delta in Bangladesh, and the Niger in Nigeria, due to lack of data. Therefore, the impacts described refer to relative sea-level rise, which could – if subsidence were taken into account – be realized by smaller global rises in sea level (e.g., Milliman *et al.*, 1989).

Wetland migration

This possibility is not considered because of insufficient data. Wetland migration would to some extent compensate for wetland loss, but experience suggests that it cannot compensate for the total area of wetlands lost (e.g., Titus *et al.*, 1991). The area of land available for migration is usually significantly smaller than the existing area of coastal wetlands.

Profile width

Both land loss (due to erosion) and, more particularly, the costs of beach nourishment are sensitive to active profile width (Figure 5). This width becomes uncertain when time scales of a century are considered (Stive *et al.*, 1991). Because of these uncertainties, ranges are given for the magnitudes and costs for each sea-level-rise scenario.

Seawall costs

Seawall costs are uncertain due to several factors: (1) construction timing; (2) maintenance requirements; and (3) appropriate design standards. In addition to these uncertainties, the capital and operating costs of improved drainage are not included in our estimates (cf. Vellinga and Leatherman, 1989).

Beach nourishment design

The simple design approach for beach nourishment utilizing terminal groins can be criticized as impractical, and in practice, regular renourishment will probably be preferred (see Davison *et al.*, 1992). The utilization of structures in conjunction with nourishment is the more likely configuration for higher sea-level-rise scenarios (Weggel, 1986). The high uncertainty in the range of costs results from the fact that we do not know exactly how much sand will be needed to maintain these beaches or the size of the terminal groins necessary to keep the sand where it is required (see Profile width, above).

Level of protection

Our methodology aims only to maintain the existing level of protection (cf. IPCC, 1990), allowing us to consider the costs of sea-level rise as opposed to other existing hazards in the coastal zone. Additional protection above that proposed here would be prudent in many locations to deal with existing coastal hazards such as flooding or storms.

Material availability

It is assumed that the materials for coastal protection options are unlimited. Although this assumption is probably reasonable for rock or hard structures, it is less certain for sand. This is particularly the case in countries such as Uruguay where it is likely that large quantities of beach nourishment will be required. The cost of importing sand could significantly raise the costs above those estimated (e.g., Titus *et al.*, 1985, 1991).

Impacts on adjacent areas

Coastal engineering projects, including our response options, often export the problems of erosion to adjacent unprotected coastal areas (Leatherman, 1991). These impacts were not considered.

RESULTS

The results of this work are reported in detail elsewhere (Nicholls and Leatherman, 1995a). Not surprisingly, all the countries/areas studied were found to be vulnerable to sea-level rise, and the nature of this vulnerability varied between countries. The highlights of these results are discussed below

Table 5. *Land and population at risk from the 1-meter sea-level-rise scenario*

	Land at risk		Population at risk	
Country	in sq. km	%	in millions	%
Argentina	>3,430 to 3,492	>0.1	n.a.	n.a.
Bangladesh	25,000	17.5	13	11
China	125,000*	1.3	72.0*	6.5
Egypt	4,200 to 5,250	12 to 15**	6.0	10.7
Malaysia	7,000	2.1	n.a.	n.a.
Nigeria	18,398 to 18,803	2.0	3.2	3.6
Senegal	6,042 to 6,073	3.1	0.1 to 0.2	1.4 to 2.3
Uruguay	94	<0.1	0.01	0.4
Venezuela	5,686 to 5,730	0.6	0.06	0.3
Total	194,852 to 196,498		94.4 to 94.5	

Notes:
n.a. – not available
* In China, land and population at risk includes increased flooding due to sea-level rise (Han *et al.*, 1995a).
** This is the percent of arable land.

by topic with a particular emphasis on the AVVA studies. Unless otherwise stated, all impacts and responses refer to a 1 m rise in sea level.

Land at risk

Table 5 gives estimates of land and population at risk for ten countries for the 1 m scenario. Land at risk totals over 190,000 km², with most of the land at risk in deltas and to a lesser extent around estuaries (Figures 7 and 8). Loss would occur mainly by inundation. In general, erosion is the major factor for less than 5 percent of the land at risk.

In absolute terms, China, Bangladesh, and Nigeria appear to be most vulnerable to land loss from a 1 m rise in sea level (Table 5). Assuming no protection, each of these countries could lose more than 10,000 km² of land. India, with four major deltas in addition to West Bengal, also appears vulnerable (Asthana, 1989). In addition, more than 2 percent of the existing land area in Malaysia and Senegal appears to be at risk. When land use is considered, Bangladesh, Egypt, China, and India appear particularly vulnerable as much of the potential loss is prime agricultural land and urban areas. In Nigeria, urban areas and oil production facilities are at risk. The land at risk also comprises a large proportion of the existing coastal wetlands in these countries (Table 6). Uruguay would lose only 94 km², mainly through erosion. As discussed below, some of that land is of high value.

The AVVA studies provide estimates of the land loss for each response option. Important Areas Protection and Total Protection do not reduce land loss significantly, for these response options do not prevent wetland loss (Figure 9).

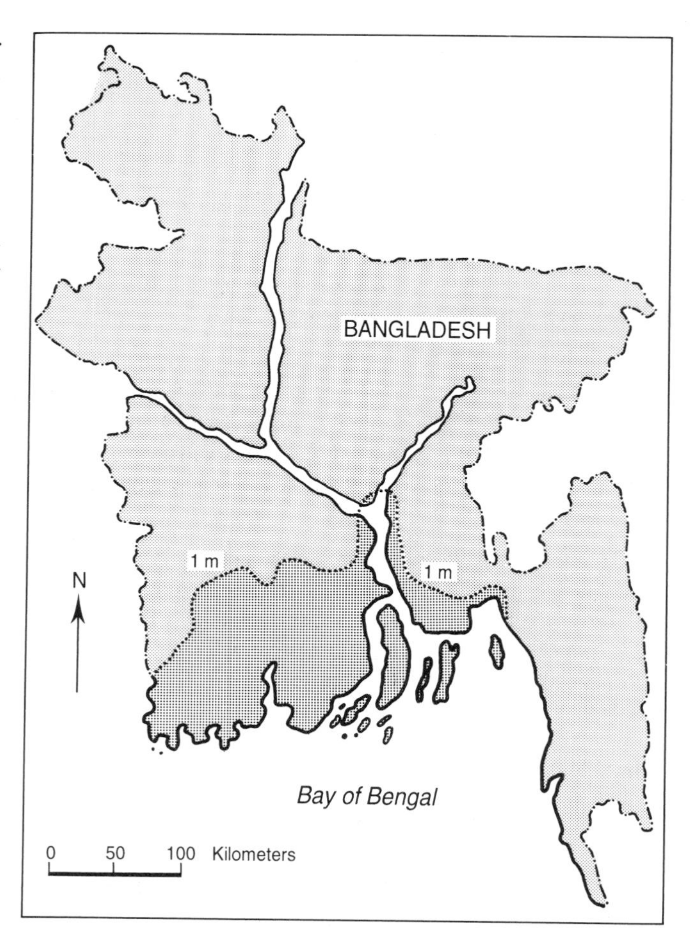

Figure 7 Land at risk in a delta: potential losses in Bangladesh due to a 1 m rise in sea level (after Huq *et al.*, 1995).

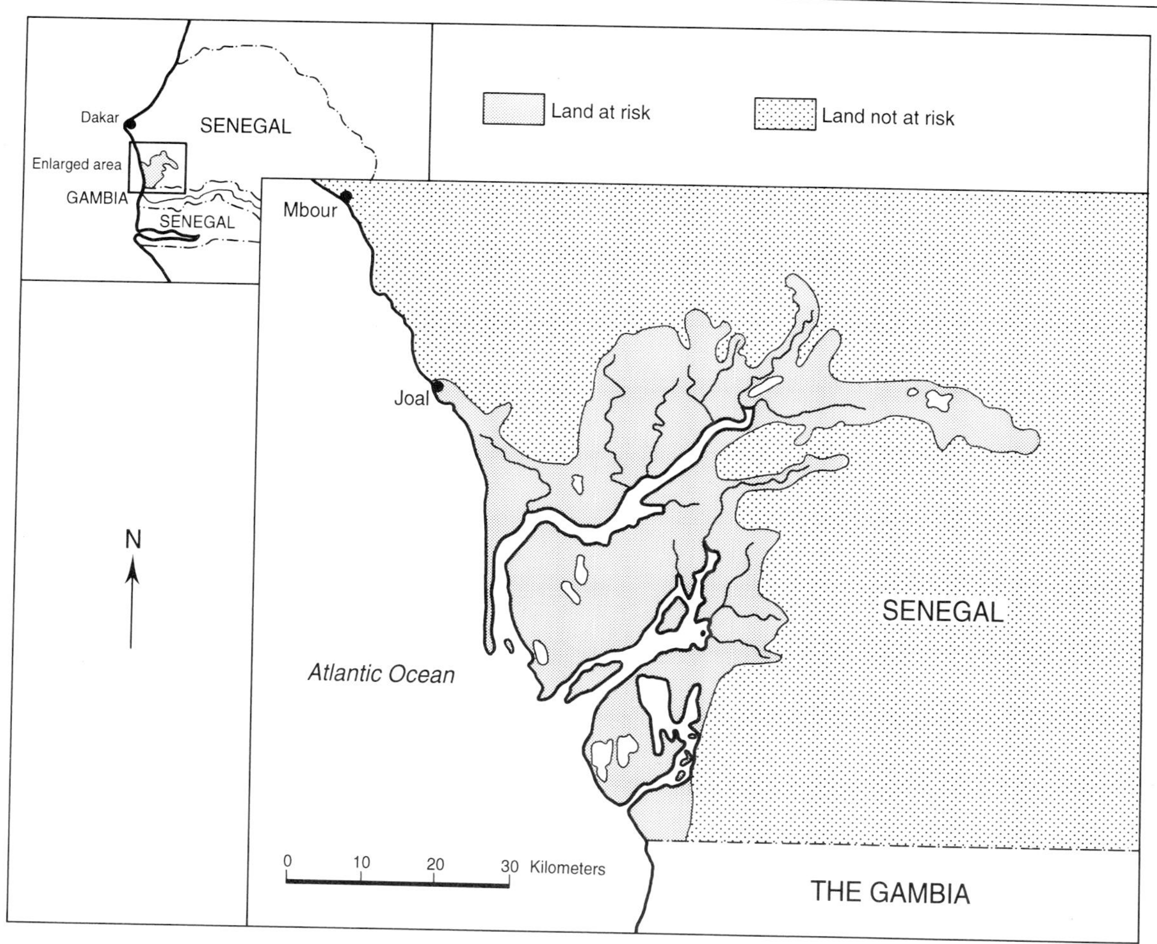

Figure 8 Land at risk around an estuary: Potential losses around the Saloum Estuary, Senegal, for a 1 m rise in sea level. Reproduced with permission from Dennis *et al.*, 1995a.

Even in Uruguay, where erosion is most significant, land loss is reduced by only about one-half under Total Protection. With the same response option, land loss is reduced significantly in Argentina due to protection of the Parana Delta from inundation.

In the heavily populated deltaic areas such as in China and Bangladesh protection would be desirable in the event of accelerated sea-level rise. However, more data are needed to assess the economic feasibility of protecting these deltaic areas and what may be the best solution to prevent land loss.

Wetland loss

Loss of coastal wetlands is occurring today in most of the countries studied due to a range of natural and human processes including present rates of sea-level rise, reduction of sediment input, saline intrusion, dredging and filling for development, direct cutting of mangroves, and, particularly in Asia, direct reclamation. In China, most coastal wetlands have already been reclaimed for agricultural production and, more recently, for aquaculture: These areas are mainly protected by earthen dikes. In Bangladesh and India, the world's largest mangrove forest, the Sunderbans, survives as a national reserve and forms an important refuge for wildlife, including the endangered Bengal tiger. In Peninsular Malaysia, mangrove reclamation for uses such as rice production and aquaculture threatens the estimated 6,000 km^2 of mangroves (in 1990), with significant loss anticipated by the year 2000. In Senegal and Nigeria, salinization is killing large areas of mangroves. In the Niger Delta, lack of sediment deposition due to upstream dam construction is exacerbating problems of wetland loss. Only in South America, in mainly rural locations, do coastal wetlands appear under limited threat. Thus, without any acceleration in sea-level rise, coastal wetlands appear to be under significant stress, which is often associated with human activity.

Coastal wetlands constitute a significant proportion of the land at risk, occupying over 43,000 km^2, in seven countries given a 1 m rise in sea level (Table 6). This land loss would include the Sunderbans. Given the active human destruction of wetlands already discussed, many of these wetlands may be destroyed before accelerated sea-level rise is likely to

Table 6. *Estimated loss of coastal wetlands for the 1-meter scenario, rounded to two significant figures**

	Wetland loss		
Country	sq. km	% of existing wetland	% of land at risk
Argentina	1,100	51	33
Bangladesh	5,800	100	23
Malaysia	6,000	100	86
Nigeria	18,000	77	96 to 98
Senegal	6,000	100	99
Uruguay	23	100	15 to 25
Venezuela	5,600	50	98 to 99
Total	43,000		

Note:
* These results do not include wetland migration. Therefore, some new wetland will develop, except where wetland migration is prevented, either naturally due to steep slopes or artificially due to coastal defenses (see text).

become a major problem. Our results do not include wetland migration or some potential sediment/biomass inputs or sediment reworking that may help existing wetlands to build vertically. Although these processes have the capacity to reduce the total losses described, it is very unlikely that they will compensate for the areas of wetlands that could be lost. Thus, the prognosis for coastal wetlands, already poor given present human impacts, is significantly worsened in the face of accelerated sea-level rise.

Coastal wetlands that have been reclaimed for other uses would still be subjected to risk from accelerated sea-level rise. Large but unquantified areas in China, Bangladesh, and Malaysia fall into this category.

Impacts on deltaic areas

Low-lying deltaic areas, such as occur in Egypt, Bangladesh, China, India, Nigeria, and Venezuela, are particularly vulnerable to sea-level rise. The following impacts all refer to a 1 m scenario. In the Nile Delta in Egypt, 12 to 15 percent of the existing agricultural land is at risk, and over 6 million people at current population levels could be displaced, including half the population of Alexandria. In the Bangladesh portion of the Ganges–Brahmaputra Delta, land that is presently used for 16 percent of the national rice production could be inundated and 13 million people could be displaced. In China, 72 million people and tens of thousands of square kilometers of agricultural land are at risk, primarily in the four major coastal deltaic plains (Figure 10). This estimate includes land and population at risk to increased flooding (Han *et al.*, 1995a). In India, the five deltaic plains have an area of over 21,000 km^2 and a population of over 21 million; however, detailed impacts of a 1 m sea-level rise are uncertain. The Niger Delta in Nigeria contains extensive mangrove swamps and petroleum production facilities. Sea-level rise could inundate or erode over 15,000 km^2 of this land and displace nearly one-half million people. The Orinoco Delta in Venezuela could experience a significant loss of over 5,000 km^2 of predominantly wetlands; at present, however, this area has low population and little development, so human impacts would be minor.

All of these potential impacts would be exacerbated by natural or human-induced land subsidence or by sediment starvation caused by upstream construction of dams and flood-prevention levees (National Research Council, 1987; Milliman *et al.*, 1989). With continued development and resource exploitation in developing countries, these contributory impacts are expected to increase substantially in the coming decades.

Population at risk

The potential for the displacement of millions of people from their homes by sea-level rise, hence creating a major environmental refugee problem, was a major factor in raising consciousness about the potentially adverse consequences of climate and global change (Jacobson, 1988; UNFPA, 1991). We avoid the term 'environmental refugee' and use instead 'population at risk' because protection is both feasible and likely in many cases. This study demonstrates that more than 94 million people are at risk in four countries alone: China, Bangladesh, Egypt, and Nigeria (Table 5). These people primarily reside in the deltaic areas already discussed. In China, some of the population is at risk due to higher flood levels rather than direct land loss (Han *et al.*, 1995a).

These results are for the present population, and significant growth of the populations at risk can be expected before the adverse effects of sea-level rise become manifest. The potential social consequences of sea-level rise, including forced migration, are serious concerns and should be further investigated.

Impacts on selected cities

The protection of cities is expected to be a major cost of accelerated sea-level rise (e.g., Turner *et al.*, 1990), and, given their high value, it appears to be one of the most certain responses. In this investigation we conducted detailed case studies of four major coastal cities that were vulnerable to sea-level rise: Shanghai (Wang *et al.*, 1995), Tianjin (Han *et al.*, 1995b), Hong Kong (Yim, 1995) (Figure 10), and Alexandria (El-Raey *et al.*, 1995) (Table 7). These were ranked,

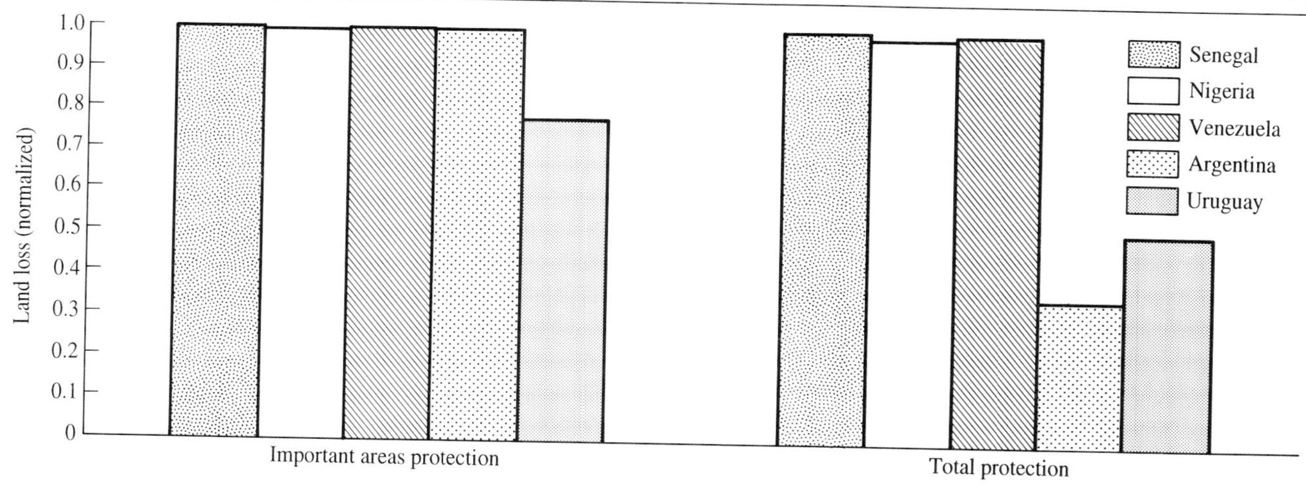

Figure 9 Land loss (normalized using No Protection) versus response option for Argentina, Uruguay, Venezuela, Senegal, and Nigeria and for a 1 m rise in sea level. Adapted from Nicholls and Leatherman (1995b).

respectively, the 5th, 15th, 31st, and 64th largest cities in the world in 1990. We also studied Dakar, Senegal; Lagos, Nigeria; Recife, Brazil (Neves and Muehe, 1995); and Buenos Aires, Argentina. Most of these cities are already vulnerable to coastal flooding. Shanghai and Tianjin were most vulnerable, with significant areas of both cities lying beneath existing high tides and having extensive systems of dikes for protection from flooding. In contrast, Dakar, Lagos, and Recife have the most limited existing protection.

Human-induced land subsidence due to groundwater withdrawal is a major problem in many coastal cities. Present rates of subsidence sometimes exceed the maximum projected rate for worldwide sea-level rise by the year 2100 (Turner *et al.*, 1990). In Shanghai, ground elevations have been lowered by as much as 2.8 m since the 1920s, although this human-induced subsidence now appears to be under control. In contrast, around Tianjin subsidence is up to 5 cm/yr at present! Land subsidence increases the vulnerability of any coastal area to future sea-level rise. Thus, managing subsurface fluid withdrawal to avoid subsidence should be given a high priority in all urbanized areas.

SHANGHAI AND TIANJIN, CHINA

In their study of the effect of sea-level rise on Shanghai Municipality, Wang *et al.* (1995) estimated that a 1 m increase in sea level could inundate 2,145 to 5,945 km^2 of land, including the entire city and up to 96 percent of the Municipality (the large uncertainty concerning area at risk reflects uncertainty about future rates of subsidence). Thus, the city's entire population is at risk. The cost of upgrading the dikes and flood walls to prevent this eventuality is estimated at $281 to $703 million, excluding the costs of drainage. Han *et al.* (1995b) estimated that a 1 m rise in *relative* sea level combined with a 100-year storm surge would submerge the entire city of Tianjin.

ALEXANDRIA

El-Raey *et al.* (1995) estimated that about 2 million people in Alexandria are at risk to a 1 m rise in sea level. In addition, losses of nearly $45 billion at present value could occur, including destruction of the city's beaches, which are a major resort for Egyptians. Maintaining the beaches through artificial nourishment would be an important element in the protection of Alexandria.

HONG KONG

In contrast to Shanghai and Tianjin, Hong Kong has a rugged coastal terrain (Yim, 1995). Development of the city over the last 100 years has included extensive coastal reclamations involving the formation of low-lying, highly developed land protected by seawalls. These areas are now vulnerable to sea-level rise. Significant subsidence of older reclamations beyond that predicted at construction exacerbates this problem. A rise in sea level will increase the risk of flooding, particularly during typhoons, which have historically wreaked havoc on the resident population and caused serious loss of life. (The unnamed typhoon of 1937, for example, killed over 11,000 people.) For a 1 m rise in sea level, the capital cost of raising existing seawalls to maintain the existing level of protection is estimated to be $75 million, excluding the costs of improved drainage.

Value at risk

The present value of buildings and land at risk (from the AVVA studies) is given in Table 8. It should be recognized that these are not fully comprehensive inventories and our estimates are probably low. In absolute terms, substantial assets are at risk, most particularly in Nigeria and Argentina (over $18 billion and $5 billion, respectively, for the 1 m scenario).

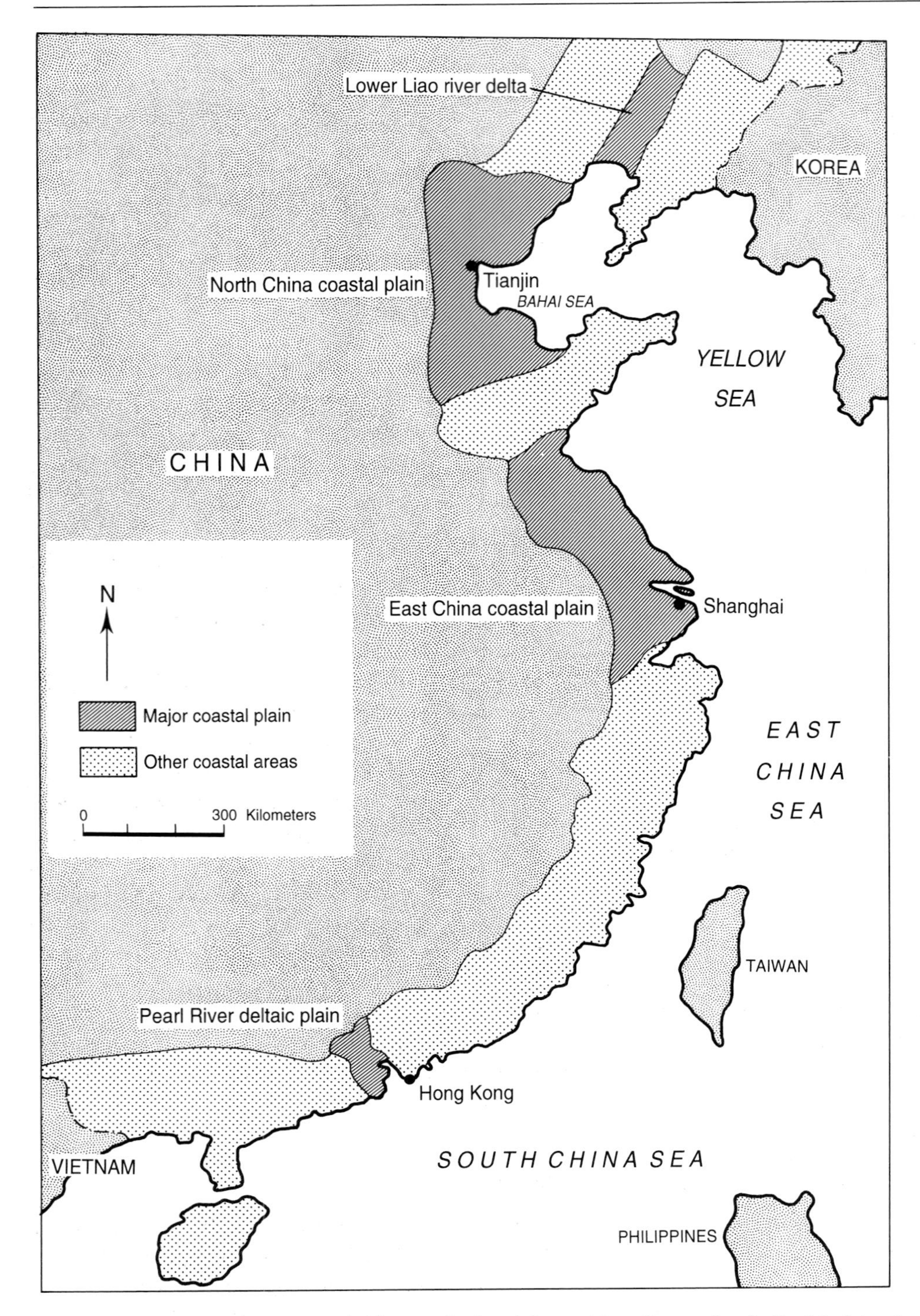

Figure 10 The Chinese coast contains four major low-lying and heavily populated alluvial plains that are vulnerable to sea-level rise: the Lower Liao River Delta; the North China Coastal Plain (including Tianjin); the East China Coastal Plain (including Shanghai); and the Pearl River Deltaic Plain (after Han *et al.*, 1995a).

Table 7. *Population at risk and protection costs for selected cities for the 1-meter scenario*

City	Total population (millions)	Population at risk (millions)	Protection costs* (millions of dollars)
Shanghai	8.1	8.1	281 to 703
Tianjin	4.8	4.8	90
Hong Kong	5.7	n.a.	>75
Alexandria	3.5	2.0	>42

Notes:
* Costs do not include drainage.
n.a. – not available

Table 8. *Value of buildings/land at risk (in millions of U.S. dollars)*

Country	Sea-level rise 0.2 m	0.5 m	1.0 m
Argentina	1,251/1,340	2,621/2,846	5,151/5,585
Nigeria*	3,552	9,003	18,134
Senegal	142/228	345/464	494/707
Uruguay	21	183	1,818
Venezuela	111	224	349
Total	5,077/5,252	12,376/12,720	25,946/26,593

Note:
* In Nigeria, value at risk includes oil and related infrastructure.

The relationship of value at risk to sea-level-rise scenario is shown for Uruguay, Senegal, and Venezuela in Figure 11. The data are normalized using the average value at risk for the 1 m scenario for each country. (Argentina and Nigeria are excluded because the results for these countries were developed using linear distribution assumptions.) Uruguay has a significantly different pattern of risk than that seen for Senegal and Venezuela. The normalized value at risk for the 0.2 m and 0.5 m scenarios is much lower for Uruguay than for Senegal or Venezuela.

This disparity is due to different coastal development patterns in these countries. In Uruguay there is a buffer zone with few buildings along the coast (Figure 12, Plate 5). This zone exists for two reasons: (1) traditional patterns of coastal development; and (2) legislation that created a 250-meter-wide strip for public use in the 1970s (Volonté and Nicholls, 1995). Although this policy was not developed for erosion mitigation purposes, the undeveloped area reduces the value at risk for all scenarios of sea-level rise, particularly scenarios below 1 m (Table 8). It should be emphasized that the undeveloped area often contains a coast parallel road. The implications of the loss of these roads were not evaluated.

In Venezuela, a different pattern of development is common: linear rows of houses built parallel and close to the shoreline. Existing erosion problems are already causing the abandonment of properties (Figure 13, Plate 6). In some areas it is difficult to have a setback because of the steep slopes adjacent to the shoreline. Thus, a small amount of sea-level rise results in disproportionately large values at risk. The Venezuelan government is presently formulating a coastal zone management plan that will begin to deal with some of these problems. Senegal shows a similar pattern of vulnerability to that in Venezuela but with a larger uncertainty. Its large value at risk is partly due to the fact that much of its tourist infrastructure is located close to the ocean in otherwise undeveloped locations (Figure 14, Plate 7).

When the various response options are considered, protection is seen to significantly reduce the value loss to less than 8 percent of the value loss with No Protection (Figure 15).

Table 9. *Costs of Important Areas Protection (IAP) and Total Protection (TP) (in millions of U.S. dollars)*

Country	Rise (m)	IAP	TP
Argentina	0.5	337–883	829–2,150
	1.0	580–1,298	1,829–3,328
Nigeria	0.5	223–319	609–888
	1.0	558–688	1,424–1,766
Senegal	0.5	146–575	407–1,422
	1.0	255–845	973–2,156
Uruguay	0.5	2,070–2,142	2,155–2,729
	1.0	2,905–2,995	3,126–3,793
Venezuela	0.5	556–960	820–1,613
	1.0	999–1,517	1,717–2,634

Protection costs

Costs determined from the AVVA studies for Important Areas Protection and Total Protection are given in Table 9. The most striking result is the high cost of protection in Uruguay, which is up to $2.9 to $3.0 billion even for Important Areas Protection (these costs are only based on the large profile width, and hence represent the upper bound of likely response costs). This is due to the combination of the high potential demand for beach nourishment due to the large existing tourist industry along the coast and the high unit cost for beach nourishment.

The constituent costs for each country under Important Areas Protection are divided into three categories: (1) beach nourishment, (2) seawalls, and (3) harbor upgrade (Figure

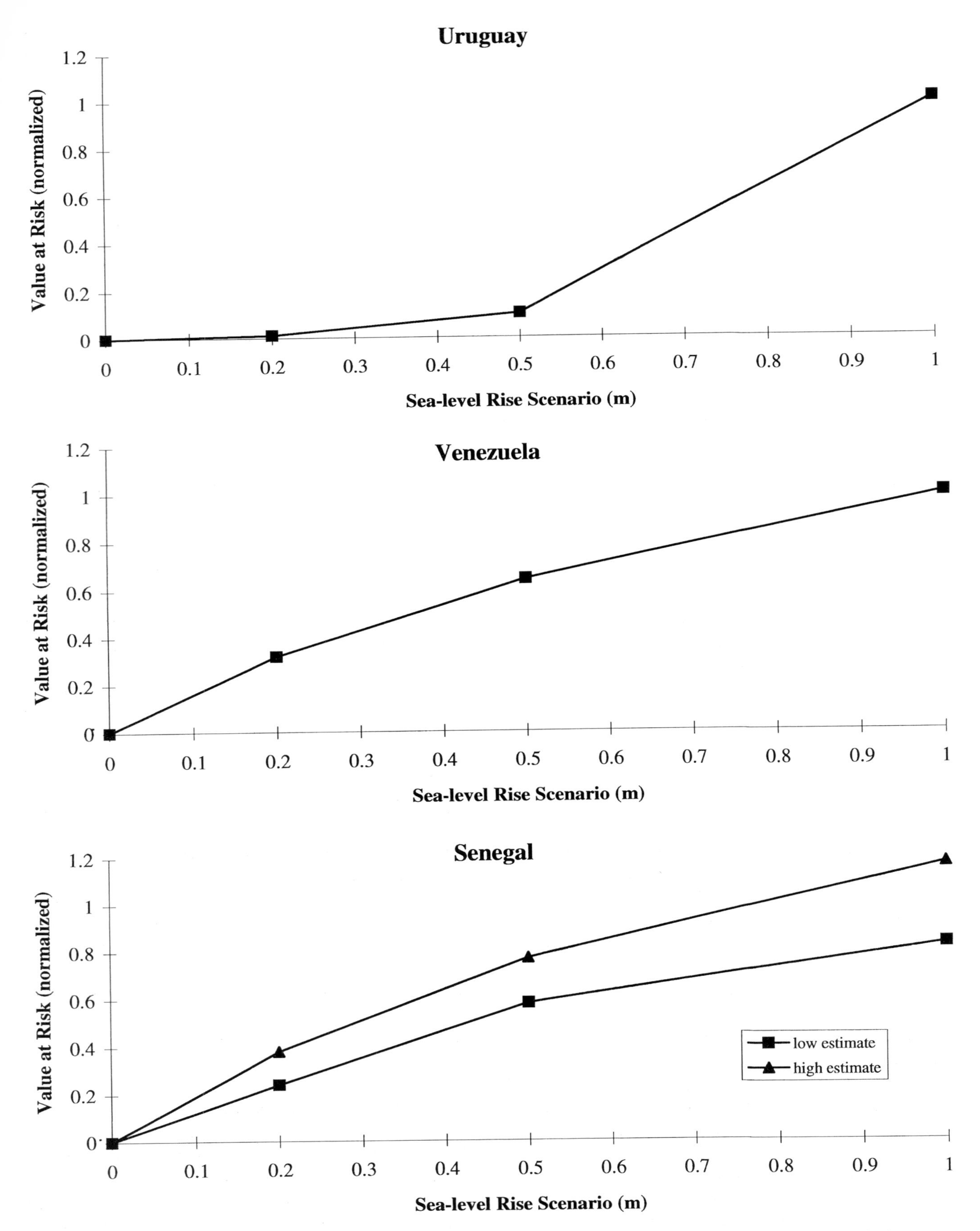

Figure 11 Value at risk (normalized for a 1 m sea-level rise) versus sea-level rise for Uruguay (single estimate), Venezuela (single estimate), and Senegal (high and low estimates). Adapted from Nicholls and Leatherman, 1995b.

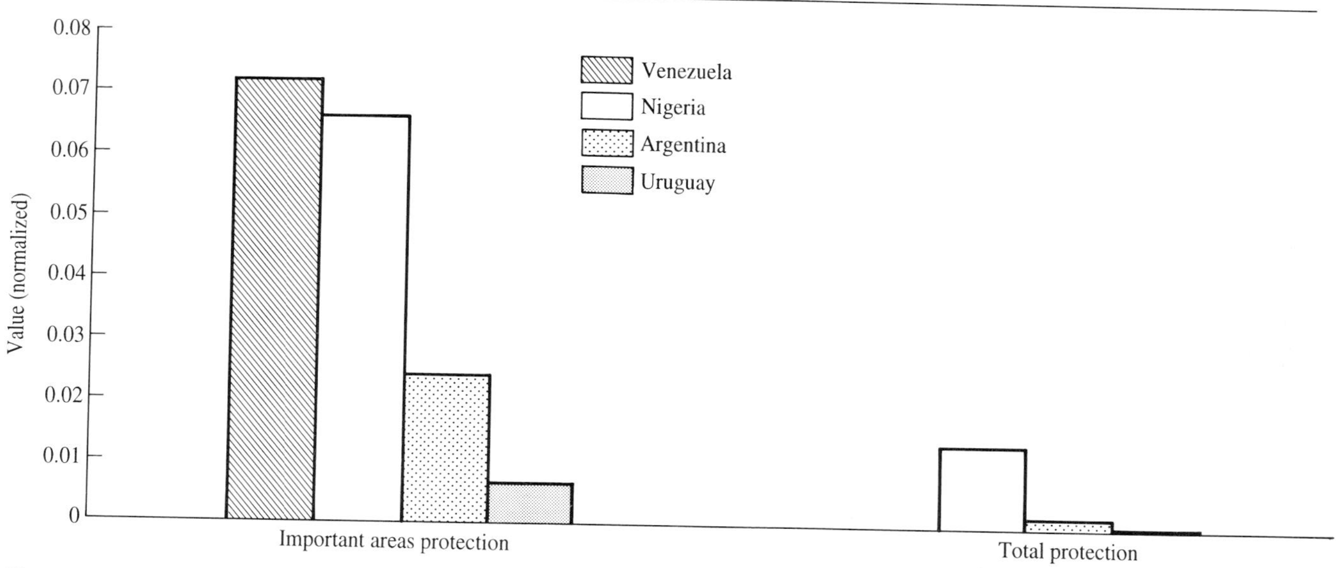

Figure 15 Value loss (normalized using No Protection) versus response option for a 1 m rise in sea level. For Senegal, value loss is zero in both cases. In Nigeria, reduction in value loss includes relocation under Total Protection. Reproduced with permission from Nicholls and Leatherman, 1995b.

16). The magnitude of each cost reflects the relative importance of different activities in the coastal zone in each country. Beach nourishment contributes more than 35 percent to the total costs in each country except Nigeria, where only a 6 kilometer frontage was assumed to be nourished. In Uruguay, beach nourishment contributes over 98 percent of the total costs. Harbor upgrade contributes more than 20 percent of the total cost in Argentina, Nigeria, and Venezuela. Seawall costs are relatively minor except in Nigeria, where they contribute about 50 percent of the national response costs.

The relative increase in cost from that for Important Areas Protection to that for Total Protection is given in Figure 17. The increase represents the cost for seawalls in all cases except in Nigeria. There, 46 to 57 percent of the additional costs are for relocation of the population at risk, primarily in the Niger Delta (French *et al.*, 1995). The increase in costs varies significantly, with values ranging from 155 to 282 percent for Senegal to a modest 8 to 10 percent for Uruguay.

The sensitivity of the protection cost estimates to a doubling of seawall cost is shown in Figure 18. For Important Areas Protection, the cost estimates are relatively insensitive in this regard except in Nigeria, where there is a 50 percent overall rise in cost. Averaged across the five countries, protection costs rise only 5 to 8 percent (or about $400 to 600 million). For Total Protection, seawalls constitute a larger proportion of the response costs. Hence, the cost estimates are more sensitive to a doubling of seawall cost, rising more than 40 percent in all the countries except Uruguay, where costs rise up to 24 percent. Thus, within a factor of two, the cost of Important Areas Protection is relatively insensitive to seawall cost. The cost of Total Protection is much more sensitive to seawall cost.

We cannot know for certain which response options will be chosen in the future. However, if we assume the existing level and pattern of coastal development, it is likely that such plans will more closely resemble the Important Areas Protection option than the Total Protection option.

Vulnerability profiles

Figure 19 shows a two-part vulnerability profile for the AVVA studies based on protection costs and land loss for Important Areas Protection in response to a 1 m rise in sea level. The protection costs are assumed to be distributed uniformly over 50 years (2051 to 2100). This annual cost is compared to the national gross investment in 1990 (the money in the economy available for investment) (World Bank, 1992). As already noted, Important Areas Protection protects little of the land at risk, so the results for land loss are similar to those presented in Table 5. Land loss is given as a percentage of the total national land area.

Protection costs

Uruguay and Senegal face much larger relative protection costs than Nigeria, Venezuela, and Argentina. For Uruguay, they are about 6 percent of the gross investment. These costs do not represent all the costs of sea-level rise, even with present patterns of development. However, economic growth will probably increase future gross investment significantly. Therefore, we cannot assess the affordability of these response strategies, but we can conclude that the

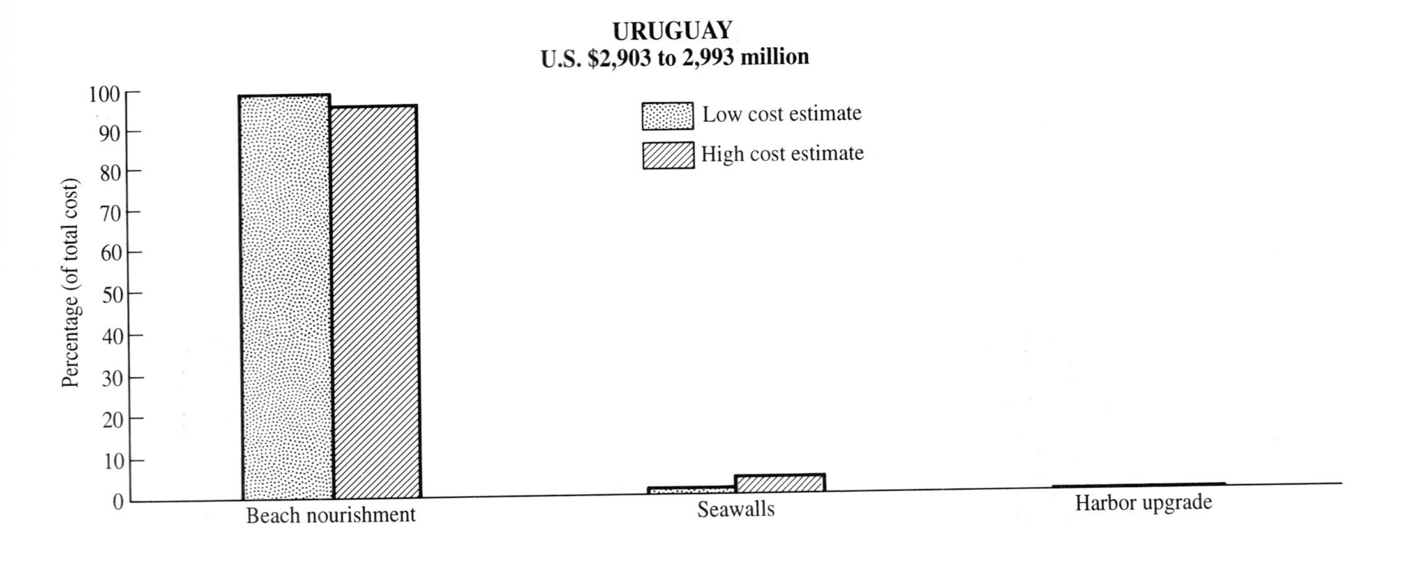

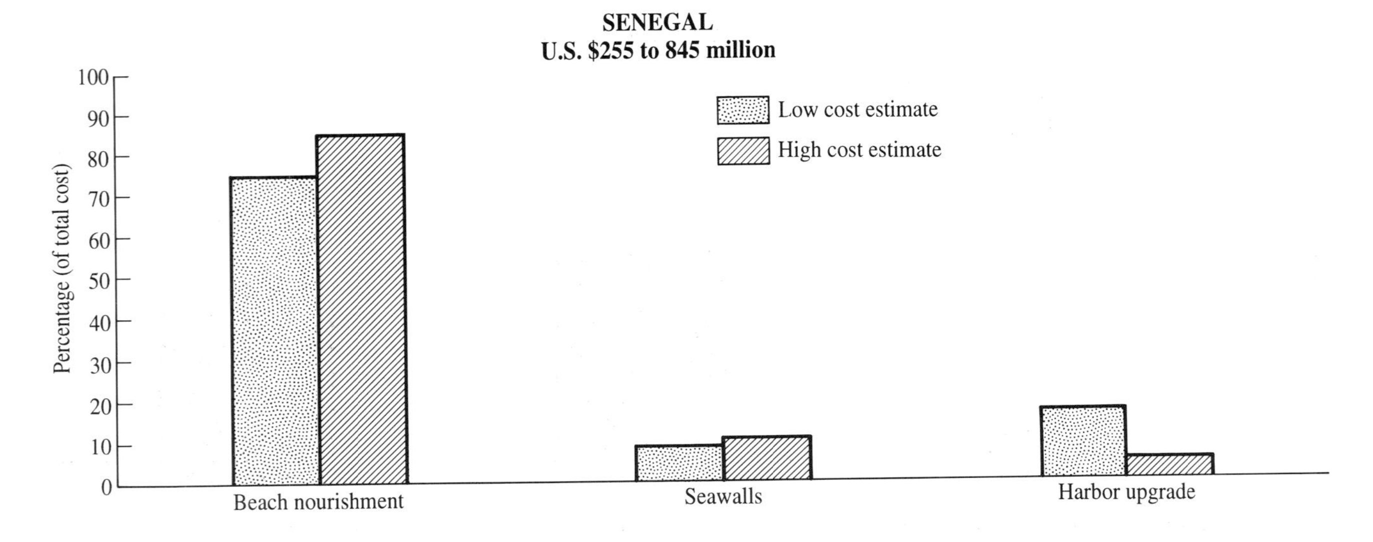

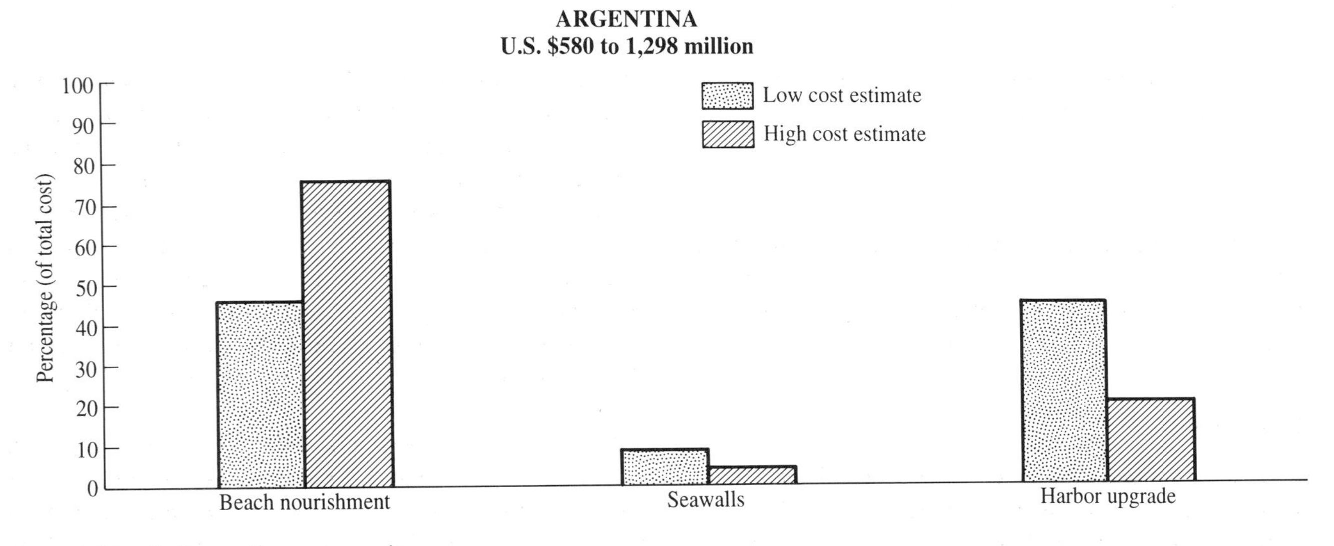

Figure 16 (part) For caption see opposite.

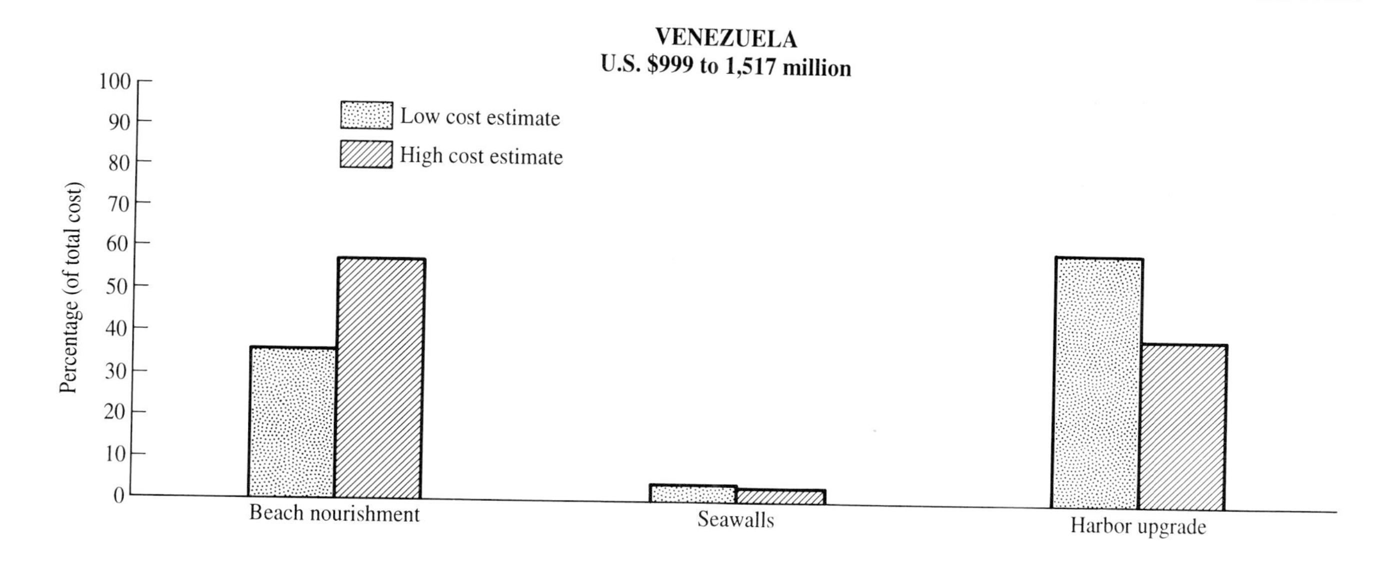

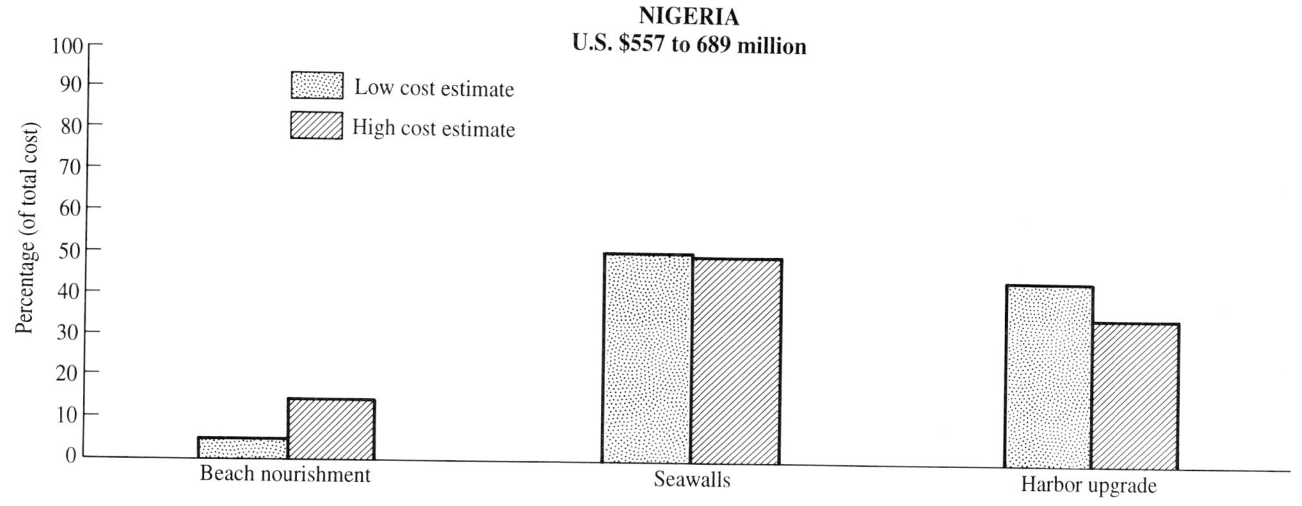

Figure 16 Constituent costs for Important Areas Protection with a 1 m rise in sea level for (1) beach nourishment; (2) seawalls; and (3) harbor upgrade. In Nigeria, the cost of island construction to protect oil production facilities is included under the cost for seawalls. (After Nicholls and Leatherman, 1995b.)

relative burden caused by protection requirements in response to sea-level rise will likely show significant variation at a national scale.

Land loss

Senegal and Nigeria show significant land loss with Important Areas Protection and a 1 m rise in sea level, these losses mainly comprising coastal wetlands. Uruguay shows the least losses for any of the countries studied.

Other impacts

Sea-level rise impacts in addition to land loss have not been investigated systematically but are discussed qualitatively.

Saltwater intrusion could have a major impact on many coastal cities as it would often affect the water supply. The aquifer beneath Dakar, Senegal, is already overexploited and additional water has to be piped over 200 km from the Senegal River at a higher unit cost than that for water from the aquifer (Dennis *et al.*, 1995a). Sea-level rise can only exacerbate these problems. The water supply to cities such as Shanghai (Wang *et al.*, 1995), Bombay and Calcutta is at risk from saltwater intrusion in the rivers. The provision of dependable water resources requires more investigation, and the associated infrastructure investment requires long lead times.

Increased flooding will be a problem in many areas with a rise in sea level, particularly in low-lying deltas. The efficiency of drainage will decrease and both surges (floods from the sea) and freshwater flooding from rainfall and high

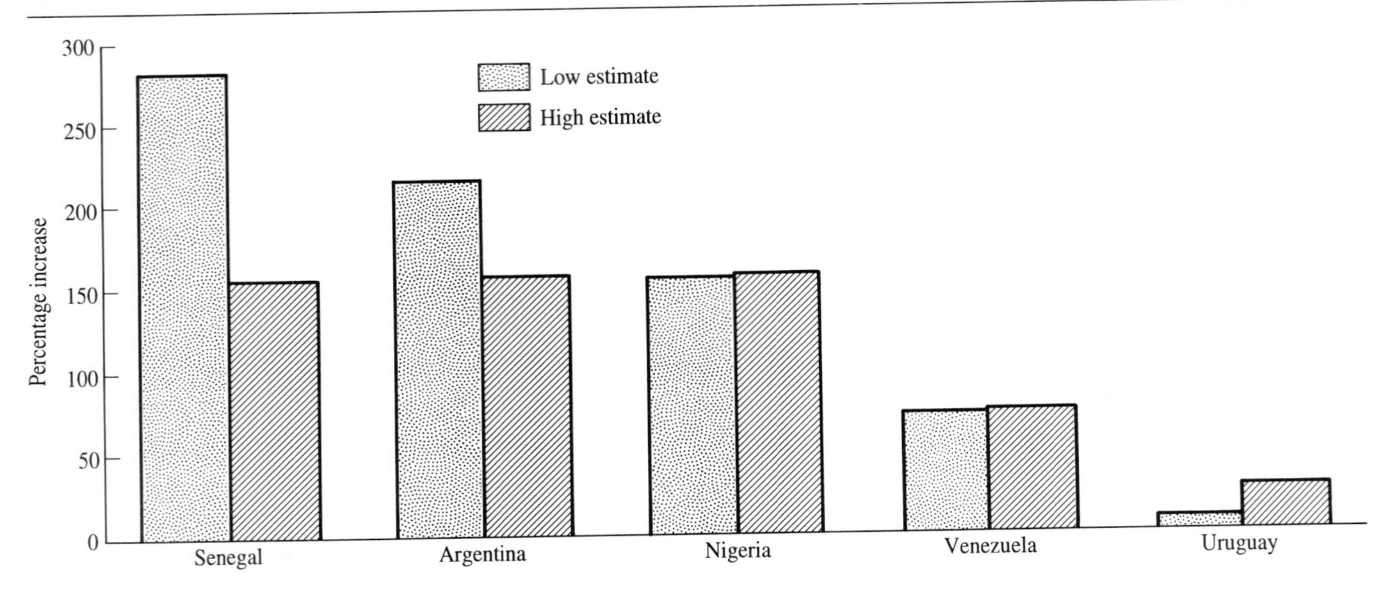

Figure 17 The percentage increase in costs from Important Areas Protection to Total Protection.

runoff will be exacerbated (e.g., Neves and Muehe, 1995). Bangladesh and Nigeria already suffer serious coastal flooding. In China, flood prevention structures exist along the rivers and deltas, but they will need to be elevated to accommodate higher river stages and sea levels.

DISCUSSION

A number of important points with general application can be drawn from these studies. The discussion that follows is based on a 1 m rise in sea level except where otherwise stated.

Deltaic problems and possible solutions

Deltas are more vulnerable to sea-level rise than other areas because they comprise extensive low-lying areas that tend to subside (e.g., Penland and Ramsey, 1990). Human activities, such as groundwater extraction, dam construction (causing downstream sediment starvation), and the construction of flood prevention structures within the delta, exacerbate this natural vulnerability (Milliman *et al.*, 1989). Thus, these areas are vulnerable to relative sea-level rise with or without climate change.

There is a wide range of risk for humans in deltaic settings. Some deltas such as the Ganges–Brahmaputra are densely populated (Huq *et al.*, 1995), whereas others such as the Orinoco in Venezuela have low populations (Volonté and Arismendi, 1995). Thus a wide range of responses is likely, ranging from conventional protection to retreat. The complexity of deltaic processes may necessitate novel approaches to the concept of protection, approaches that in many aspects are yet to be fully developed (for example, spraying of thin layers of sediment over living wetlands to increase elevation).

Integrated delta management, including the upstream catchment, would minimize the vulnerability of many deltaic areas to sea-level rise. In particular, the negative impacts of existing and proposed dams on sediment supply to downstream deltas should be considered. The possibility of sediment bypassing of dams should be evaluated where appropriate.

Where large sediment loads are carried by rivers, these materials could be utilized, at least theoretically, to raise land levels through deposition and hence to counter the adverse impacts of sea-level rise. This procedure has already been proven effective in the Yellow River Delta: there, the land surface adjacent to the river has been raised several meters by sediment deposited during controlled flooding (Han *et al.*, 1995b). This increase in elevation is more than sufficient to offset the IPCC sea-level-rise scenarios. Han *et al.* (1995b) advocate sediment management to combat sea-level rise in the Yellow River Delta.

Similar strategies may be appropriate in other deltaic areas. For instance, controlled diversions of the Mississippi River in coastal Louisiana is being advocated to provide both sediment and freshwater to preserve the delta marshes (Templet and Meyer-Arendt, 1988; Louisiana Coastal Wetlands Conservation and Restoration Task Force, 1993). In Bangladesh, the Ganges–Brahmaputra carries large quantities of sediment. Thus, sediment management is a realistic but as yet relatively unevaluated strategy for combating sea-level rise. However, the present \$5 to \$10 billion World Bank development plans to control river flooding in Bangladesh

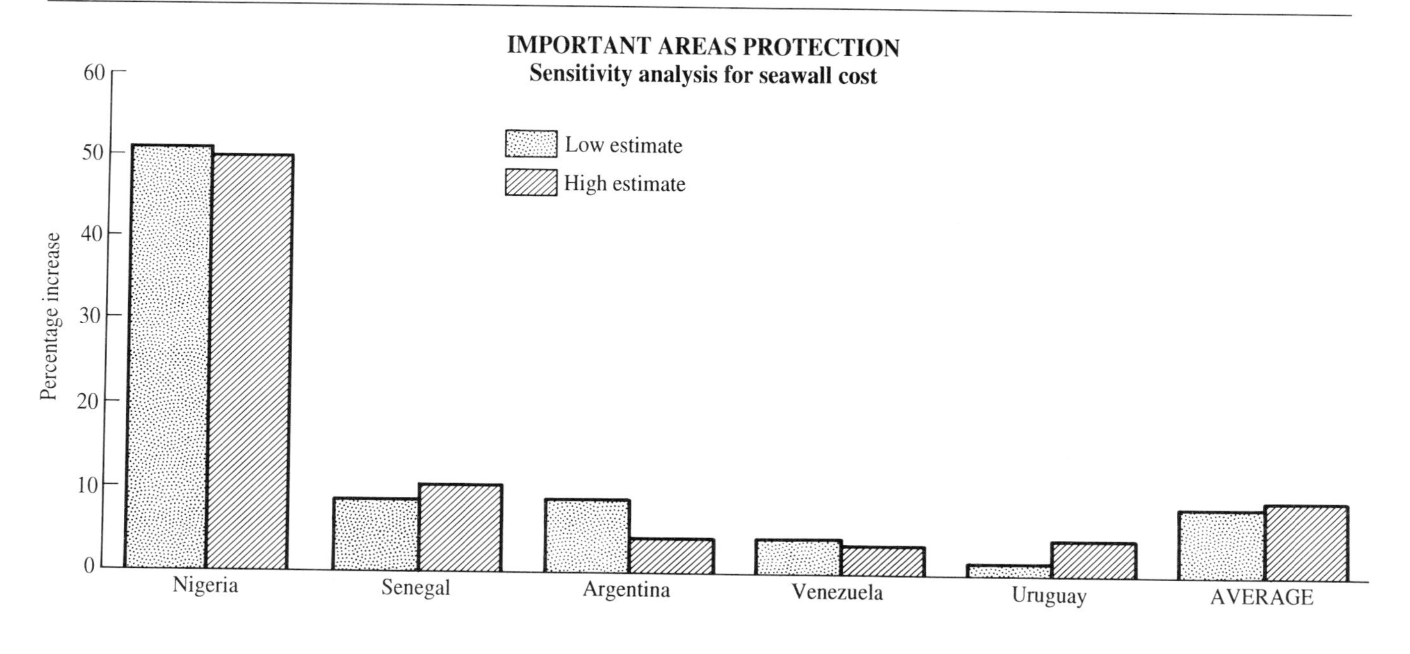

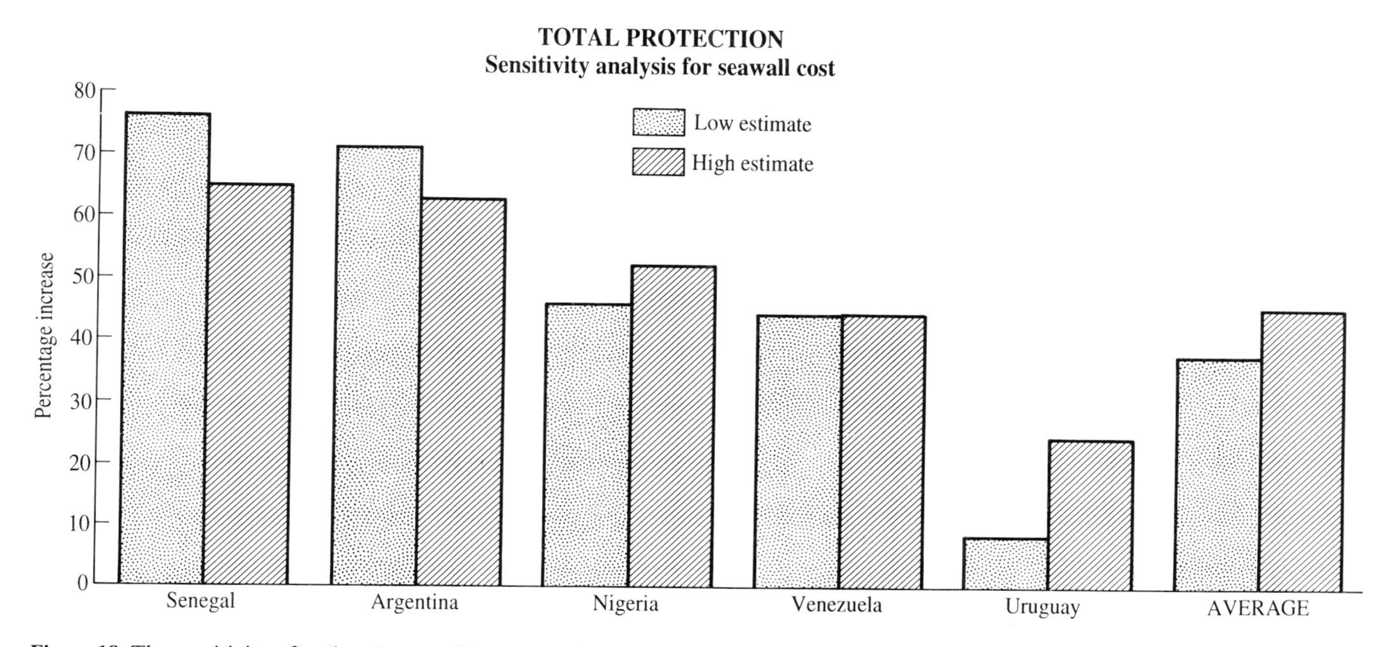

Figure 18 The sensitivity of national costs of Important Areas Protection and Total Protection to the doubling of seawall costs.

using levees would have the opposite effect, stopping sediment input to the delta plain (Chowdhury, 1991).

In some delta systems, such as the Nile, sediment management is unrealistic because the major source of sediment has already been removed by upstream dams (e.g., Milliman *et al.*, 1989). Sediment bypassing of the Aswan High Dam on the Nile is probably not technically feasible due to insufficient river flow to carry the sediment from the dam to the delta. Therefore, in this case, only conventional engineering solutions such as polderization of the delta are available.

Whatever strategies are adopted in deltaic areas, they will require long lead times and careful planning. Han *et al.* (1995b) note the vital need for institutional development and long-term planning so that the most effective responses to sea-level rise can be developed and applied in a timely fashion.

Coastal wetlands

Coastal wetlands are presently being lost at significant rates due to both human and natural changes, as shown in this study and by IPCC (1992). Given accelerated sea-level rise,

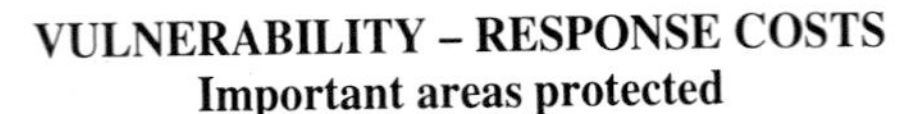

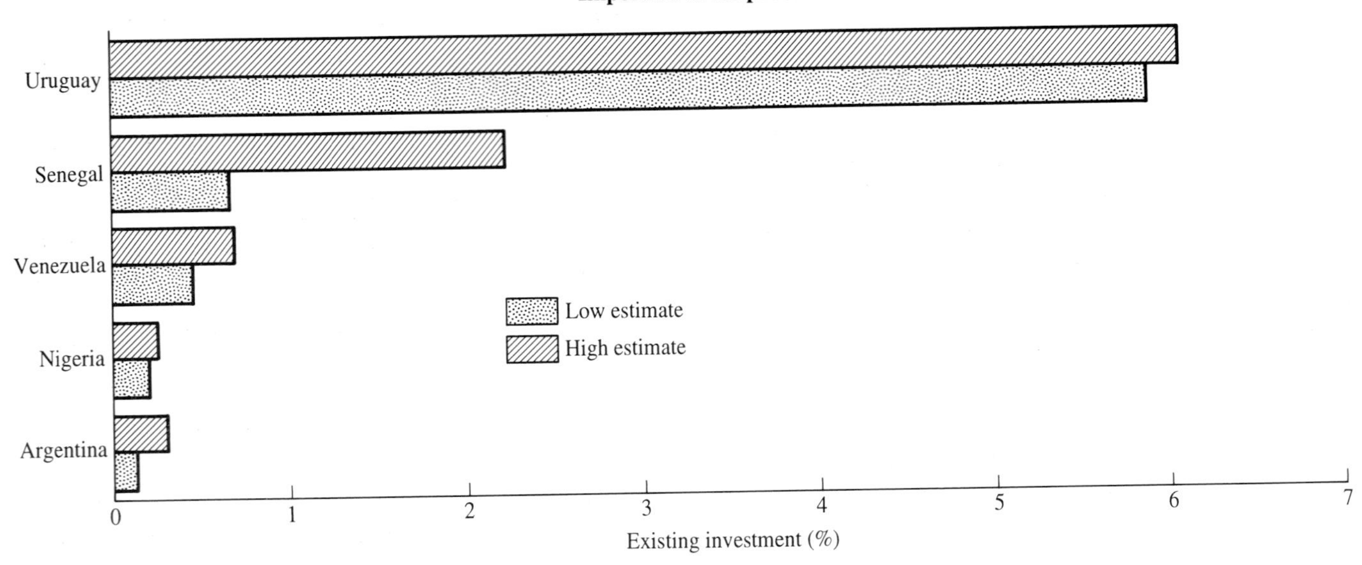

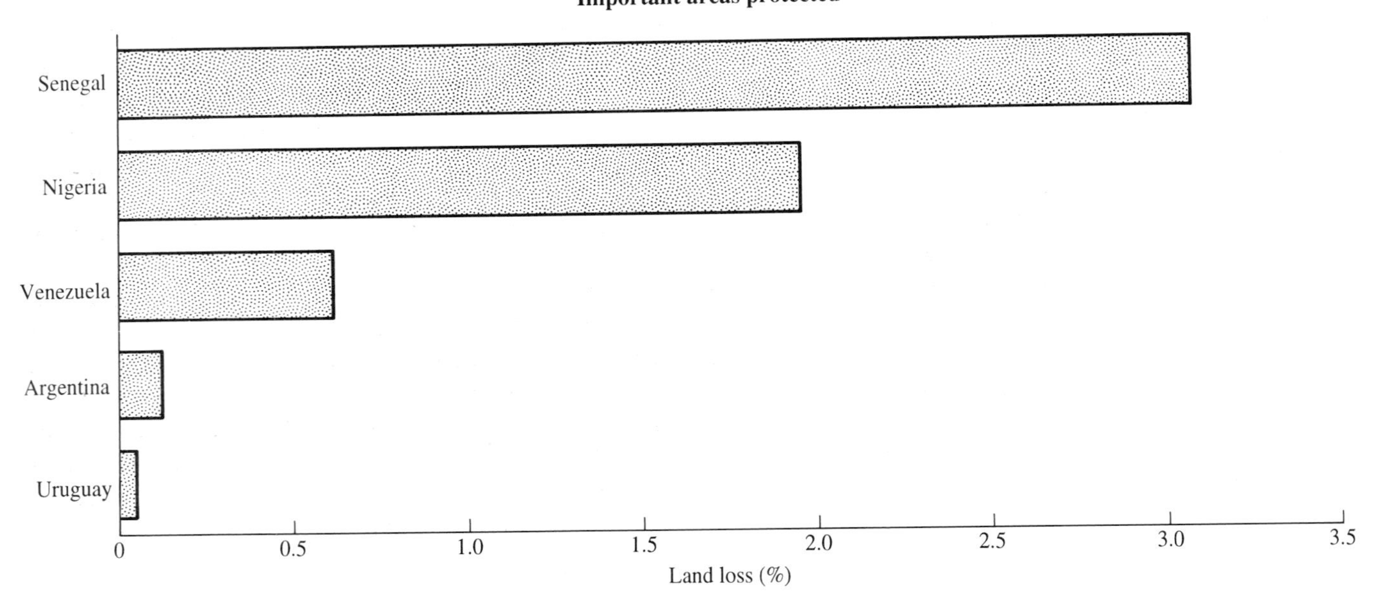

Figure 19 A two-part vulnerability profile for Important Areas Protection and a 1 m sea-level-rise scenario. The protection costs are assumed to occur uniformly over 50 years (2051 to 2100) and are expressed as a percentage of gross investment in 1990 (World Bank, 1992). The land loss is given as a percentage of total land area. (After Nicholls and Leatherman, 1995b.)

our projections of significant wetland loss will likely be repeated worldwide. Coastal wetlands are among the most productive environments on Earth, and their loss would have major implications for biodiversity of coastal ecosystems and coastal fisheries.

Options for preserving significant areas of existing wetlands appear to be limited, even for present rates of wetland loss. Many engineering approaches appear to be prohibitively expensive, although as already discussed in the context of deltas, sediment management may be possible in some locations. Lower cost approaches such as land use planning to assist wetland survival by allowing inland migration of wetlands under rising sea level are often advocated (Titus *et al.*, 1991; IPCC, 1992). However, the functional value of new wetlands may be reduced. If wetland migration is permitted or new wetlands are artificially created, will these ecosystems function physically, chemically, and ecologically like the counterparts that were destroyed? This is not well understood. Many coastal wetlands have existed for one or two millennia, and it is not clear if the new wetlands can truly replace them, at least in the short-term, when compared on a hectare-to-hectare basis (National Research Council, 1992). Thus, preservation of existing coastal wetlands should be favored, where possible.

Costs of beach nourishment

Nourishment of sandy beaches is likely to constitute a large component of protection costs. In the AVVA study countries, about 250 km of shoreline may require nourishment and the costs are more expensive per kilometer than those for building seawalls (Tables 4 and 10). Uruguay, which has relatively little land at risk, would have high protection costs because it is likely that beach nourishment would be needed in many areas (Volonté and Nicholls, 1995).

In the IPCC (1990) study, beach nourishment was found to be a smaller element of the estimated response costs than in the AVVA studies (Table 10). Two factors explain the difference. First, IPCC significantly underestimated – by a factor of two and a half – the length of shoreline for which beach nourishment was the most appropriate response. Second, the unit costs assumed by IPCC generally fell at the low end of our unit costs. In Uruguay, the nourishment costs we utilized are up to five times greater than those used by IPCC. Senegal has similar unit costs to Uruguay. In all cases, our higher costs are due to more realistic design of beach nourishment projects based on wave conditions and including longshore losses. The substantial and continuing growth in beach-based tourism worldwide makes it likely that future beach nourishment requirements will be much greater than presented here.

Table 10. *Beach nourishment for the 1-meter scenario – total length (in kilometers) and mean unit cost (in millions of dollars per kilometer) as measured by IPCC (1990) and the AVVA studies*

	IPCC (1990)		AVVA Studies	
Country	Length	Unit cost	Length	Unit cost
Argentina	40	6.75	69.2	3.6/12.9
Nigeria	25	7.88	6.0	4.5/16.3
Senegal	0	n.a.	21.5	8.8/33.2
Uruguay	25	6.76	84.3	—/33.9
Venezuela	10	9.00	70.5	5.2/12.3
Total	100		251.5	

Standard of protection

If sea level rises, many densely populated areas will lie below regular tidal inundation or even below mean sea level, and large populations would face catastrophic consequences in the event of failure of coastal defenses. This situation prevails today in the Netherlands. Half of this densely populated country is situated beneath present sea level (Peerbolte *et al.*, 1991). Coastal defenses are built to a high standard and can withstand an estimated 1 in 10,000 year flood! In Shanghai, China, which lies below existing sea level in places, coastal defenses are being upgraded to a 1 in 1,000 year standard (Wang *et al.*, 1995). In New Orleans, parts of which are over 3 meters below sea level, defenses are built to a 1 in 200 or 1 in 300 year standard (Hote, 1992). This standard of protection is under review in the aftermath of Hurricane Andrew, which battered Louisiana in August 1992.

Coastal defenses designed to higher standards of protection than previously existed would probably be prudent for many coastal cities and other densely populated areas given sea-level rise – despite the fact that the implementation of higher standards would raise the costs of protection. Huq *et al.* (1995) estimated that maintenance of the status quo in Bangladesh (protection against a 1 in 20 year flood) would cost $1 billion; however, estimates of costs to remove the risk of cyclone-produced floods are $10 billion (Turner *et al.*, 1990). A benchmark of a 1 in 1,000 year flood and storm event is often used as a standard for densely developed urban areas. Whatever standard is chosen, decisions on appropriate response strategies should explicitly consider the level of protection which is required. Turner *et al.* (1990) noted that a reliance on engineered flood-control structures has already increased the risk of catastrophic flooding. In effect, adaptation by protection still leaves large populations at risk from flooding from coastal storms and other events. This risk must be acknowledged and kept at an acceptably low level.

Building setbacks

In Uruguay, buildings are set back from the shoreline. Although the explicit purpose of this buffer was not to counter beach erosion, substantial shoreline recession can occur in many areas of Uruguay with minimal property losses, reducing the value at risk for small rises in sea level. (Coast-parallel roads would be affected in many locations in Uruguay – the implications of their loss were not considered). In Venezuela and Senegal, no formal building setbacks exist, and development close to the shoreline is common. Hence, the value at risk is relatively large for small amounts of sea-level rise. Thus, building setbacks, particularly if integrated with other coastal management policies, will reduce vulnerability to sea-level rise. Building setbacks can also help in the management of existing erosion hazards (e.g., National Research Council, 1990).

The appropriate setback is a site-specific decision that will depend on a number of factors. At the least, moving setbacks should be established based on present rates of shoreline change. Once established, such a mechanism would serve to combat an existing problem and also to reduce vulnerability to erosion impacts of accelerated sea-level rise. In addition, the plan could later be modified to deal explicitly with accelerated sea-level rise.

Long-term responses to sea-level rise

Based on these studies it might simply be concluded that we can adapt to sea-level rise. Significant economic growth may occur over the next few decades in all the countries studied, and the necessary investment resources may be available when required. This would certainly seem to be the case at a macroscale in developed countries, such as the United States (Titus *et al.*, 1991) and the Netherlands (Peerbolte *et al.*, 1991), although significant reduction of coastal wetlands would still be expected to occur. However, even if this conclusion is correct, which remains uncertain, other factors, which are discussed below, should be considered.

ANTICIPATORY RESPONSES

There is a great opportunity to plan effectively for sea-level rise (and other changes in the coastal zone) and hence reduce future costs. The impacts of accelerated sea-level rise will depend not only on the magnitude of global sea-level rise, but also on the human responses to that change (Vellinga and Leatherman, 1989). A 'do nothing' policy with no anticipatory adaptation may lead to unwise decisions, such as construction in areas vulnerable to sea-level rise. In contrast, a well-planned response that seeks to anticipate the physical impacts of sea-level rise in a timely fashion will minimize such mistakes and should result in less adverse impacts and lower costs for reactive responses such as protection. Such planning would be best implemented within the context of integrated coastal zone management as recommended by IPCC (1990, 1992). Long-term planning is required for anticipatory adaptation, though to date such planning is rare in the developing world. A notable exception is in Hong Kong. There, the West Kowloon reclamation is being constructed with a surface elevation 0.8 m above earlier reclamations to give a safety factor for sea-level rise (Yim, 1995). Future reclamations are expected to be raised to a similar degree in anticipation of sea-level rise. Yim (1995) recommends raising the older reclamations in Hong Kong when redevelopment allows. In general, however, the established planning horizons in the developing world are too short to deal with the problems of sea-level rise.

PROBLEMS OF IMPLEMENTATION

Even when appropriate engineering solutions are feasible and socioeconomic factors are favorable, it cannot be assumed that the solutions will be implemented. Venice, Italy, has experienced increased flooding from the Adriatic Sea due to a combination of local subsidence and global sea-level rise. An exceptional flood in November 1966 focused attention on this serious problem, and plans have been developed for a series of mobile gates to temporarily seal the lagoon from the sea when required (Zanda, 1991). However, they have not yet been constructed, and a number of conflicting uses of the Venice lagoon require resolution (Pirazolli, 1987). At the same time, Venice is steadily depopulating (due to a range of factors, including the flooding): from 175,000 inhabitants in 1951 to fewer than 80,000 inhabitants in 1989 (Zanda, 1991). The slow response to flooding in Venice is an excellent model of how not to deal with sea-level rise. The long lead times for responses involving major coastal engineering must be considered when examining adaptation.

EXTREME EVENTS

The role of extreme events, such as severe storms, on shaping human response to slower changes, such as sea-level rise, must be recognized. Around the North Sea in Europe much of the present flood protection was constructed only after the disastrous 1953 storm. The vulnerability of these areas to flooding had been increasing due to *relative* sea-level rise for a century or more: At London Bridge, extreme high waters have risen over 1 m since 1791 (Gilbert and Horner, 1984); but it took a disaster of major proportions and the loss of thousands of lives around the North Sea to promote action. Reconstruction after extreme events cannot be assumed – such events could equally serve as triggers for coastal abandonment.

Event-driven responses to accelerated sea-level rise should be avoided. Given current projections of accelerated sea-level rise, the increasing risks associated with extreme floods and storms can be predicted. Anticipating such changing risks emphasizes the importance of developing a long-term institutional viewpoint that can facilitate adaptation, including recognizing and solving problems before they become serious.

POLICY ISSUES

The results reported herein have some important policy implications concerning decisions in the coastal zone.

Integrated coastal zone management

The effects of sea-level rise presented here take into account the present pattern and level of development, which is almost certainly a low assumption due to the continuing trend of development and population growth in the coastal zone. The World Bank projects that global population will increase by 95 percent (above 1990 levels) by the year 2050 and predicts a further increase by 16 percent by the year 2100, giving a global population of over 12 billion (Bos and Bulatao, 1990). Thus, the effects we are describing will probably be significantly enhanced if no measures are undertaken to minimize the vulnerability of future development. Such measures would require careful planning of development in the coastal zone to an unprecedented degree. Many conflicting and competing demands already exist in the coastal zone – conflicts that would best be resolved within the framework of integrated coastal zone management, as recommended by IPCC (1990, 1992). As already noted, development of a long-term institutional viewpoint is an important requirement for an effective response to sea-level rise.

The timing of appropriate responses can be determined only through dynamic analyses of such uncertainties as sea-level rise, coastal development, and economic development. Certain responses, such as the construction of seawalls and beach nourishment, have short lead times so that implementation can be delayed until needed; other responses, such as land use planning, require implementation in the near future to provide maximum benefit. If we do not evaluate the full range of possible responses and make appropriate decisions today, we will leave future generations with a more restricted range of options for dealing with sea-level rise.

Delta catchment management

Deltas are naturally vulnerable to sea-level rise, and human activities such as groundwater withdrawal or changing sediment availability often enhance this vulnerability. Because decisions made large distances from the coast may adversely impact these sedimentary depositional areas, traditional coastal zone management may be ineffective within deltaic areas. Many of these human-induced changes, such as the construction of the Aswan Dam on the Nile, are for all practical purposes irreversible. The effect of human activities on deltaic systems will likely increase in the coming years and, if unregulated, will have increasingly severe impacts on deltas around the world. Therefore, coastal zone management plans for deltaic areas should include a management plan for the delta catchment.

A situation in Bangladesh illustrates the conflicting pressures in a deltaic environment. A major project utilizing levees is proposed by the World Bank to stop annual river flooding associated with the monsoon. The plan is controversial (Chowdhury, 1991), and while it offers many benefits, negative impacts are also apparent. If the levees prevent sediment from accumulating on the delta surface during floods, land subsidence will produce a relative rise of sea level.

This is already happening in the Mississippi River delta plain in Louisiana, where similar levees have been constructed (Day and Templet, 1989). In that area, sea level is rising in excess of 1 cm/yr (Penland and Ramsey, 1990), and annual land losses from 1983 to 1990 are estimated to have exceeded 52 km^2, which is the largest coastal land loss problem on the U.S. coast (Britsch and Dunbar, 1993). It is hoped that Bangladesh can learn the lessons of the Mississippi Delta and manage the problems of deltaic flooding in a manner that does not increase the future vulnerability of the country to sea-level rise.

Building setbacks

Shoreline recession is already a problem worldwide (Bird, 1985) and can be expected to continue. Development of shorelines should acknowledge this problem, and building setbacks based on existing erosion rates should be proposed. (For a discussion of proposed building setbacks in the United States, see, e.g., National Research Council [1990].) Establishing such building setbacks is a sensible policy with or without accelerated sea-level rise. As this study demonstrates, keeping development away from the shoreline reduces vulnerability to future erosion, whatever the cause.

The situation in Uruguay demonstrates that building setbacks and, more generally, the implementation of sensible land use planning in the coastal zone are important in reducing potential building losses to erosion.

Coastal-based tourism in particular should be considered, given the high potential costs of beach nourishment at such locations. Tourist infrastructure continues to be built close to the sea (Figure 14, Plate 7) even when undeveloped land is available in less vulnerable locations. Sensible land use planning could preserve a more natural coastal landscape and minimize the costs of beach nourishment.

Wetland value/preservation

Coastal wetlands do not appear to be valued in the developing world despite their vital contribution to fisheries resources and biodiversity. Significant wetland loss may occur without any climate change due to reclamation and other destruction. Sea-level rise could significantly increase these losses.

Policies that encourage wetland preservation should be encouraged, particularly those that preserve wetlands in situ, as existing wetlands appear to have a higher 'value' than artificially created wetlands (National Research Council, 1992). Such policies focus on maintaining the wetland elevation with respect to sea level and include wetland nourishment – the artificial provision of sediment to the wetland surface. These procedures are presently in their infancy. Policies to allow wetland migration under rising sea level should also be considered, but it must be recognized that these new coastal wetlands may not fully replace in size or function those that are lost.

FUTURE RESEARCH

This study has identified a number of basic research and data needs that must be met if we are to improve our understanding of vulnerability and possible responses to sea-level rise in developing countries.

Coastal data and coastal models

There are insufficient basic data to make truly quantitative estimates of the impacts of sea-level rise. Important information, such as rates of land subsidence, flood return periods, existing rates of land loss, and the availability of sediment, is often poorly known or completely unavailable. The long-term cost of response strategies is also difficult to assess quantitatively.

A focused plan that identifies the most critical information required for coastal management should be developed, rather than uncoordinated data collection with no integrated plan for data use.

Improved models of coastal evolution, incorporating the impacts of sea-level rise, are also required. In particular, the response of muddy coasts, including mangrove swamps, to sea-level rise is poorly understood. Such shorelines occur in Argentina (Dennis *et al.*, 1995b), Brazil (Muehe and Neves, 1995), and Malaysia (Midun and Lee, 1995), among other areas, and clearly require more research. The interrelationship of data collection and modeling to management goals should also be carefully considered.

Dynamic analyses

An analysis of the impacts of and possible responses to sea-level rise through time (for example, on a decadal basis), including the effects of economic growth and increased coastal development, is needed. Government officials need 'real-time' economic data and benefit-cost analyses in order to respond to sea-level-rise impacts. Also, such short (10-year) time scales are meaningful within existing planning procedures.

The most adverse impacts of sea-level rise are not expected to be experienced until the latter part of the next century, and their actual magnitude has a high uncertainty. Therefore, only 'no-regret' responses with a low cost are likely to be appropriate in the next few decades. Identifying and implementing such decisions in order to maximize our flexibility in the future are of key importance.

Analysis of vulnerable areas

It is important to continue to identify areas that are vulnerable to sea-level rise. It is already clear that certain classes of landforms are particularly vulnerable, such as deltas (already discussed) and small islands and, in particular, coral atolls (Pernetta, 1991; Roy and Connell, 1991). Any human activity in these areas is at risk to sea-level rise. However, our understanding of these vulnerable environments is poor, particularly given the important role of human modifications to the future behavior of these systems. Simplified generalizations about landform response to sea-level rise should be avoided in favor of process-oriented understanding. More detailed models that incorporate the full range of delta morphology and process are required to predict coastal evolution. A similar viewpoint can be argued for small islands (cf. Pernetta, 1991).

Other manifestations of climate change: The role of 'shocks'

The sensitivity of the coastal zone to other manifestations of climate change such as increased hurricane intensity and frequency need to be evaluated (Emmanuel, 1988). For instance, for the sandy coast in the Netherlands, it has been estimated that realistic adverse changes in wind and wave conditions combined with a 0.85 m rise in sea level would be 37 percent more costly than a 1.0 m rise in sea level with constant wind and wave conditions (Peerbolte *et al.*, 1991). Such analysis could usefully be coupled with consideration of the impact of 'shocks' or sudden change. We often treat the world as a slowly changing place, but in reality extreme events have a big role on both impacts and responses. Future storms or droughts may have large effects on decisions. Under such conditions, the poorest people in society are usually at greatest risk.

CONCLUSIONS

A 1 m rise in sea level could have a major impact on developing countries with coastal areas. Several vulnerable countries have been clearly identified: China, Bangladesh,

India, and Egypt. The AVVA studies have demonstrated that sea-level rise could cause significant problems in Uruguay and Senegal. The problems in Uruguay are particularly related to tourist-based developments along its coastline and to the high cost of beach nourishment. Allied with these conclusions is the recognition that certain environments are highly vulnerable to sea-level rise – notably coastal wetlands and deltas (and, from previous work, small islands). Land loss in deltas could have major impacts on agriculture, especially in the rice-growing countries of Asia. The worldwide vulnerability of coastal wetlands to land loss is of particular concern. Coastal wetlands are already being lost due to natural and human-induced factors, such as reclamation of mangrove swamps.

Responding to sea-level-rise impacts does not imply the need for panic, nor can it permit complacency. The more adverse effects of accelerated sea-level rise are not likely to be manifest for decades. However, our results do indicate that significant impacts and large costs likely lie ahead for the developing world due to sea-level rise. There is a need for continued studies of how best to manage sea-level-rise impacts and how best to implement appropriate low-cost responses in the next 10 years. The studies described in this chapter are a first step in this process.

The coastal zones of the world are being developed at an unprecedented rate (IPCC, 1990) and most of this development is taking place with little regard to the present problems of coastal land loss and storm flooding, let alone the risks of accelerated sea-level rise. Thus, vulnerability to sea-level rise is increasing with time. It is important to recognize that much could be done to reduce future vulnerability to sea-level rise by systematic data collection and the utilization of these data within the framework of integrated coastal zone management.

In practice, this means addressing existing coastal hazards and other problems within an integrated plan that maximizes flexibility of response to climate change. We should consider climate change today so that decisions in the coming decades do not increase the vulnerability of future generations to sea-level rise. Although accelerated sea-level rise is not yet certain, its associated risks may in some cases be sufficient to effect the decisions made today. Procedures to identify such situations should be developed and utilized as a priority. Development of the long-term institutional perspective required to manage sea-level rise, and global change in general, should be more strongly emphasized.

It is clear that the coastal zones of the world in the year 2100 will be very different from the existing conditions with or without climate change. We hope that these changes occur in a planned and sustainable manner and within the framework of robust policies that can cope with climate change and maintain the value of the coastal zone to all humankind.

ACKNOWLEDGMENTS

This work was funded by the Office of Policy Analysis of the U.S. Environmental Protection Agency (Joel Smith, project manager; Jim Titus, project officer). The reviews of Mark Byrnes of the Louisiana Geological Survey and Margaret Davidson of the South Carolina Sea Grant Consortium greatly improved an earlier draft of this chapter. We would like to thank all our principal collaborators, who are listed in Table 2 and without whom this study would not have been possible. We would also like to thank the following: for useful discussions on aerial video techniques, the staff of the Louisiana Geological Survey, particularly Shea Penland and Karen Westphal; for discussions on the protection options, Edward Fulford of Andrew Miller and Associates and Robert Hallermeier of Dewberry and Davis; and for discussions on coastal wetlands, Michael Kearney of the University of Maryland.

REFERENCES

Anonymous. 1990. Adaptive options and policy implications of sea-level rise and other impacts of global climate change. Miami Workshop Report to the Coastal Zone Management Subgroup of the Intergovernmental Panel on Climate Change. In *Changing Climate and the Coast*, Vol. 1, ed. J. G. Titus, 3–50. Washington, DC: U.S. Environmental Protection Agency.

Asthana, V. 1989. National assessment of effects of a possible sea-level rise and responses in India. Unpublished paper. University of Maryland, College Park.

Bird, E. C. F. 1985. *Coastline Changes – A Global Review*. Chichester: John Wiley – Interscience.

Bos, E., and R. A. Bulatao. 1990. *Projecting fertility for all countries*. Working Paper 500. Washington, DC: World Bank.

Britsch, L. D., and J. B. Dunbar. 1993. Land loss rates: Louisiana Coastal Plain. *Journal of Coastal Research* 9:324–38.

Broadus, J., J. Milliman, S. Edwards, D. Aubrey, and F. Gable. 1986. Rising sea level and damming of rivers: possible effects in Egypt and Bangladesh. In *Effects of Changes in Stratospheric Ozone and Global Change*, Vol. 4, ed. J. G. Titus, 165–89. Washington, DC: U.S. Environmental Protection Agency.

Carter, R. W. G. 1987. Man's response to sea-level change. In *Sea Surface Studies: A Global View*, ed. R. J. N. Devoy, 464–98. London: Croom Helm.

Chowdhury, J. U. 1991. Flood Action Plan: one-sided approach? *Bangladesh Environmental Newsletter* 2(2):1–3.

Davison, A. T., R. J. Nicholls, and S. P. Leatherman. 1992. Beach nourishment as a coastal management tool: an annotated bibliography on developments associated with the artificial nourishment of beaches. *Journal of Coastal Research* 8:984–1022.

Day, J. W., and P. H. Templet. 1989. Consequences of sea level rise: implications from the Mississippi delta. *Coastal Management* 17:241–57.

Dean, R. G. 1983. Principles of beach nourishment. In *Handbook of Coastal Processes and Erosion*, ed. P.D . Komar, 217–31. Boca Raton, Fla.: CRC Press.

Debusschere, K., S. Penland, K. A. Westphal, P. D. Reimer, and R. A. McBride. 1991. Aerial videotape mapping of coastal geomorphic changes. In *Proceedings Coastal Zone 91*, 370–90. New York: American Society of Civil Engineers.

Dennis, K. C., I. Niang, and R. J. Nicholls. 1995a. Sea-level rise and Senegal: potential impacts and consequences. *Journal of Coastal Research*, Special Issue No. 14:243–61.

Dennis, K. C., E. Schnack, F. Mouzo, and C. R. Orona. 1995b. Sea-level

rise and Argentina: potential impacts and consequences. *Journal of Coastal Research*, Special Issue No. 14:205–23.
Downs, L. L. 1993. *Historical shoreline analysis: Rockhold Creek to Solomons Island, Chesapeake Bay, Maryland.* M. A. Thesis, University of Maryland, College Park.
Douglas, B. C. 1991. Global sea-level rise. *Journal of Geophysical Research* 96(C4):6981–92.
Emmanuel, K. A. 1988. The dependence of hurricane intensity on climate. *Nature* 326:483–5.
El-Raey, M., S. Nasr, O. Frihy, S. Desouki, and Kh. Dewidar. 1995. Potential impacts of accelerated sea-level rise on Alexandria Governorate, Egypt. *Journal of Coastal Research*, Special Issue No. 14:190–204.
French, G. T., L. F. Awosika, and C. E. Ibe. 1995. Sea-level rise and Nigeria: potential impacts and consequences. *Journal of Coastal Research*, Special Issue No. 14:224–42.
Gilbert, S., and R. W. Horner. 1984. *The Thames Barrier*. London: Thomas Telford Ltd.
Gornitz, V., S. Lebedeff, and J. Hansen. 1982. Global sea-level trend in the past century. *Science* 215:1611–4.
Han, M., J. Hou, and L. Wu. 1995a. Potential impacts of sea-level rise on China's coastal environment and cities: a national assessment. *Journal of Coastal Research*, Special Issue No. 14:79–95.
Han, M., J. Hou, L. Wu, C. Liu, G. Zhao, and Z. Zhang. 1995b. Sea-level rise and the North China Coastal Plain: a preliminary assessment. *Journal of Coastal Research*, Special Issue No. 14:132–50.
Hands, E. B. 1983. The Great Lakes as a test model for profile responses to sea level changes. In *Handbook of Coastal Processes and Erosion*, ed. P. D. Komar, 167–89. Boca Raton, Fla.: CRC Press.
Holdgate, M. W., ed. 1989. *Climate Change: Meeting the Challenge*. Report to the British Commonwealth, London.
Hote, J. (U.S. Army Corps of Engineers, New Orleans District) 1992. Personal communication, September 8, 1992.
Huq, S., S. I. Ali, and A. Rahman. 1995. Sea-level rise and Bangladesh: a preliminary analysis. *Journal of Coastal Research*, Special Issue No. 14:44–53.
Intergovernmental Panel on Climate Change (IPCC). 1990. *Strategies for Adaption to Sea-Level Rise*. Report of the Coastal Zone Management Subgroup. Rijkswaterstatt, The Netherlands: IPCC Response Strategies Working Group.
Intergovernmental Panel on Climate Change (IPCC). 1992. *Global Climate Change and the Rising Challenge of the Sea*. Report of the Coastal Zone Management Subgroup. Rijkswaterstatt, The Netherlands: IPCC Response Strategies Working Group.
Jacobson, J. L. 1988. *Environmental Refugees: A Yardstick of Habitability*. Worldwatch Paper 86. Washington, DC: Worldwatch Institute.
Kana, T. W., J. Michel, M. O. Hayes, and J. R. Jensen. 1984. The physical impact of sea-level rise in the area of Charleston, South Carolina. In *Greenhouse Effect and Sea-Level Rise: A Challenge for This Generation*, 105–50. New York: Van Nostrand Reinhold.
Leatherman, S. P. 1984. Coastal geomorphic responses to sea-level rise: Galveston Bay, Texas. In *Greenhouse Effect and Sea-Level Rise: A Challenge for This Generation*, 151–78. New York: Van Nostrand Reinhold.
Leatherman, S. P. 1991. Coasts and beaches. In *The Heritage of Engineering Geology: The First Hundred Years*, ed. G. A. Kiersch, 183–200. Geological Society of America, Centennial Special Issue Vol. 3.
Leatherman, S. P., R. J. Nicholls, and K. C. Dennis. 1995. Aerial videotape-assisted vulnerability analysis: a cost-effective approach to assess sea-level rise impacts. *Journal of Coastal Research*, Special Issue No. 14:15–25.
Louisiana Coastal Wetlands Conservation and Restoration Task Force. 1993. *Louisiana Coastal Wetlands Restoration Plan*. June 1993 draft. New Orleans District, U.S. Army Corps of Engineers.
Midun, Z., and S. Lee. 1995. Implications of a greenhouse-induced sea-level rise: a national assessment for Malaysia. *Journal of Coastal Research*, Special Issue No. 14:96–115.
Milliman, J. D., J. M. Broadus, and F. Gable. 1989. Environmental and economic implications of rising sea level and subsiding deltas: the Nile and Bengal examples. *Ambio* 18:340–5.
Muehe, D., and C. F. Neves. 1994. Implications of sea level rise on the Brazilian coast: A preliminary assessment. *Journal of Coastal Research*, Special Issue No. 14. (In press).
National Research Council. 1987. *Responding to Changes in Sea Level: Engineering Implications*. Washington, DC: National Academy Press.
National Research Council. 1990. *Managing Coastal Erosion*. Washington, DC: National Academic Press.
National Research Council. 1992. *Restoration of Aquatic Ecosystems*. Washington, DC: National Academic Press.
Neves, C. F., and D. Muehe. 1995. Potential impacts of sea-level rise on the metropolitan region of Recife, Brazil. *Journal of Coastal Research*, Special Issue No. 14.
Nicholls, R. J., and S. P. Leatherman, eds. 1995a. The potential impacts of accelerated sea-level rise on developing countries. *Journal of Coastal Research*, Special Issue No. 14.
Nicholls, R. J., and S. P. Leatherman. 1995b. The Implications of Accelerated Sea-Level Rise and Developing Countries: A Discussion. *Journal of Coastal Research*, Special Issue No. 14:303–23.
Nicholls, R. J., K. C. Dennis, C. R. Volonté, and S.P. Leatherman. 1993. Methods and problems in assessing the impacts of accelerated sea-level rise. In *The World At-Risk: Natural Hazards and Climate Change*, ed. R. Bras, AIP Conference Proceedings, #277, 193–205. New York: American Institute of Physics.
Nicholls, R. J., S. P. Leatherman, K. C. Dennis, and C. R. Volonté. 1995. Impacts of sea-level rise: qualitative and quantitative assessments. *Journal of Coastal Research*, Special Issue No. 14:26–43.
Organization of American States. 1991. *Primer on Natural Hazard Management in Integrated Regional Development Planning*. Washington, DC: Organization of American States.
Peerbolte, E. B., J. G. de Ronde, L. P. M. de Vrees, B. Mann, and G. Baarse. 1991. *Impact of Sea-Level Rise on Society: A Case Study of the Netherlands.* Delft: Delft Hydraulics.
Penland, S., and K. E. Ramsey. 1990. Relative sea-level rise in Louisiana and the Gulf of Mexico: 1908–1988. *Journal of Coastal Research* 6:323–42.
Pernetta, J. C. 1991. Cities on oceanic islands: A case study of Male, capital of the Republic of the Maldives. In *Impact of Sea-Level Rise on Cities and Regions: Proceedings of "Cities on Water," Venice, December 11–13, 1989*. Venice: Marsilio Editori, 169–82.
Pernetta, J., and G. Sestini. 1989. *The Maldives and the Impact of Expected Climatic Changes*. Regional Sea Reports and Studies No. 104. Nairobi, Kenya: United Nations Environmental Programme.
Pirazolli, P. A. 1987. Recent sea-level changes and related engineering problems in the Lagoon of Venice (Italy). *Progress in Oceanography* 18:323–46.
Roy, P., and J. Connell. 1991. Climatic change and the future of Atoll States. *Journal of Coastal Research* 7:1057–75.
Smith, J. B., and D. A. Tirpak. 1989. *The Potential Effects of Global Climate Change on the United States*. Report to Congress. Washington, DC: U.S. Environmental Protection Agency.
Stevenson, J. C., L. G. Ward, and M. S. Kearney. 1986. Vertical accretion in marshes with varying rates of sea level rise. In *Estuarine Variability*, 241–58. New York: Academic Press.
Stive, M. J. F., R. J. Nicholls, and H. J. De Vriend. 1991. Sea-level rise and shore nourishment: a discussion. *Coastal Engineering* 16:147–63.
Templet, P. H., and K. J. Meyer-Arendt. 1988. Louisiana wetland loss: regional water management approach to the problem. *Environmental Management* 12:181–92.
Titus, J. G., ed. 1990. *Changing Climate and the Coast*, 2 vols. Washington, DC: U.S. Environmental Protection Agency.
Titus, J. G., S. P. Leatherman, C. H. Everts, D. L. Kriebel, and R. G. Dean. 1985. *Potential Impacts of Sea-Level Rise on the Beach at Ocean City, Maryland.* Washington, DC: U.S. Environmental Protection Agency.
Titus, J. G., R. A. Park, S. P. Leatherman, J. R. Weggel, M. S. Green, P. W. Mausel, S. Brown, C. Gaunt, M. Trehan, and G. Yohe. 1991. Greenhouse effect and sea-level rise: potential loss of land and the cost of holding back the sea. *Coastal Management* 19:171–204.
Turner, R. K., P. M. Kelly, and R. C. Kay. 1990. *Cities at Risk*. London: BNA International Inc.
United Nations Population Fund (UNFPA). 1991. *Population, Resources and the Environment: The Critical Challenges*. New York: UNFPA.
Vellinga, P., and S. P. Leatherman. 1989. Sea-level rise: consequences and policies. *Climatic Change* 15:175–89.
Volonté, C.R., and J. Arismendi. 1995. Sea-level rise and Venezuela: potential impacts and responses. *Journal of Coastal Research*, Special Issue No. 14:285–302.
Volonté, C. R. and R. J. Nicholls. 1995. Sea-level rise and Uruguay:

potential impacts and responses. *Journal of Coastal Research*, Special Issue No. 14:262–84.
Wang, B., S. Chen, K. Zhang, and J. Shen. 1995. Potential impacts of sea-level rise on the Shanghai area. *Journal of Coastal Research*, Special Issue No. 14:151–66.
Warrick, R. A., and H. Oerlemans. 1990. Sea-level rise. In *Climate Change: The IPCC Scientific Assessment*, eds. J. T. Houghton, G. J. Jenkins, and J. J. Ephraums, 257–81. Cambridge: Cambridge University Press.
Weggel, J. R. 1986. Economics of beach nourishment under scenario of rising sea level. *Journal of Waterway, Port, Coastal and Ocean Engineering, ASCE* 112:418–27.
Wigley, T. M. L., and S. C. B. Raper. 1992. Implications for climate and sea level of revised IPCC emissions scenarios. *Nature* 357:293–300.
Wind, H. G., ed. 1987. *Impact of Sea Level Rise on Society*. Rotterdam: A. A. Balkema.
World Bank. 1992. *World Development Report 1992. Development and the Environment*. New York: Oxford University Press.
Yim, W. W.-S. 1995. Implications of sea-level rise for Victoria Harbour, Hong Kong. *Journal of Coastal Research*, Special Issue No. 14:167–89.
Yohe, G. 1990. The cost of not holding back the sea: toward a national sample of economic vulnerability. *Coastal Management* 18:403–31.
Zanda, L. 1991. The case of Venice. In *Impact of Sea-Level Rise on Cities and Regions: Proceedings of "Cities on Water," Venice, December 11–13, 1989*. Venice: Marsilio Editori, 51–9.

5 Human Health

LAURENCE S. KALKSTEIN

Department Of Geography

University Of Delaware

GUANRI TAN

Department of Atmospheric Sciences

Zhongshan University

Guangzhou, People's Republic of China

SUMMARY

The objective of this study was to estimate the potential impact of a global warming on human health. Potential changes in heat-related mortality were estimated for three countries: The People's Republic of China, Canada, and Egypt. In addition, a framework for analyzing two vector-borne diseases that may spread in a warmer world was developed.

The following conclusions were made:

- Based on various global warming scenarios, heat-related mortality could rise significantly in many places around the world, especially in developing countries, such as China and Egypt. It is possible that in such countries the population may not be capable of physiologically adapting to the increasing warmth. However, this study did not address urban infrastructure changes that may mitigate the impact of global warming on mortality.
- The areas that are presently most sensitive to rises in heat-related mortality are those with intense but irregular heat waves. Locales with generally constant heat and little day-to-day climate variability seem less vulnerable, and lower death rates from heat-induced mortality assuming a global warming are estimated for those regions. Thus, significant increases in mortality are a function of greater climate variability as well as increases in temperature.
- There appears to be a poleward boundary of heat-related impacts, which was determined in Canada. A large-scale warming might shift this boundary further north, and cities presently less vulnerable to heat-induced mortality might become more sensitive.
- The elderly appear to be disproportionately stressed, and death rates from heat-related mortality will likely increase most rapidly for this age group.
- Relationships between winter mortality and weather are much weaker than relationships between summer mortality and weather in all the evaluated cities. Thus, it is not expected that any potential decreases in winter mortality will offset the larger increases expected in heat-related, summer mortality.
- It is feasible that simple weather/health watch-warning systems can be constructed for developing country and domestic cities most vulnerable to heat-related mortality. People are often unaware that dangerous weather conditions exist which might contribute to heightened mortality rates; a watch-warning system could be constructed to advise people when stressful weather conditions are imminent.

Some of the more specific findings include the following:

- Heat-stress-related mortality is presently an important cause of death in China, especially in the mid-latitude city of Shanghai. The impact is somewhat less in the tropical city of Guangzhou.
- Present-day death rates from heat-related mortality in Shanghai are over two times higher than those in New York City, one of the most heat-sensitive cities in the United States.
- There is evidence that the Chinese will be less likely than the populations of the other study areas to acclimatize to the increased warming under $2 \times CO_2$ (double baseline CO_2 concentration) conditions. Thus, heat-related mortality in China could rise significantly under such conditions.
- Heat-related death rates in Shanghai were estimated to rise from the present 6 per 100,000 to about 45 per 100,000 population during an average summer under the GFDL (Geophysical Fluid Dynamics Laboratory) scenario. This mortality rate is over 7 times the present death rate, and it is by far the highest for any city evaluated in this study.

Guangzhou rates were estimated to rise less significantly.

- Of 10 Canadian cities evaluated, only three (Toronto, Montreal, and Ottawa) yielded significant summer mortality relationships. It appears that a northern boundary of heat-related impacts can be determined in Canada.
- Under $2 \times CO_2$ scenarios, heat-related mortality rose significantly in Toronto and Montreal, even when it was assumed that the population would acclimatize as expected. For example, assuming acclimatization, the Montreal death rate was estimated with the GISS (Goddard Institute for Space Studies) scenario to be over 8 per 100,000 population, which is very close to New York's acclimatized rate. This is equivalent to 218 extra deaths attributed to heat during an average summer.
- In Cairo, Egypt, mortality currently rises linearly with temperature. This differs from mortality trends in all other cities evaluated, where specific threshold temperatures were noted.
- Cairo currently has a higher heat-related mortality rate than U.S. and Canadian cities and a lower rate than Shanghai.
- Acclimatization to increased warmth could be minimal in Cairo, and increasing warmth could contribute to much higher mortality rates there. For example, under the GISS scenario, Cairo's estimated heat-related mortality rate was almost 18 per 100,000 population. This rate is slightly higher than New York's and Montreal's, but it is considerably lower than Shanghai's.
- Two vector-borne infectious diseases, onchocerciasis and malaria, were selected for intense evaluation. Sites in several west African countries have already been selected for the onchocerciasis study, and the blackfly vector population has been monitored at these sites for up to 18 years. In addition, meteorological data and streamflow measurements are readily available. Considering that the blackfly's life cycle and reproductive rate are closely related to the magnitude of streamflow, a water budget approach appears warranted.

INTRODUCTION

There has been a growing concern in both the medical and climatological communities that a significant global warming could create major international health problems. Research recently completed in the United States and described in several reports indicates that a global warming could have a major impact on the number of heat-related deaths in that country (Smith and Tirpak, 1989; Kalkstein and Giannini, 1991; Kalkstein, 1991b). Considering that most people in the United States have some access to air conditioning – which might mitigate the impact of global warming on human health – the impact of heat-related mortality in other regions of the world might be more extreme. This is especially true of Third World countries, where most of the population is more exposed to the vagaries of the weather. An additional major health impact of a long-term climatic change might involve the spread of debilitating infectious diseases, whose vectors might migrate poleward if the climate warms. It is not surprising, then, that the World Meteorological Organization (WMO) considers the international health ramifications of a global warming to be among the most pressing problems for the upcoming century (WMO, 1986), and the World Health Organization (WHO) has expressed similar concern (WHO, 1990).

The goal of this report is to summarize research recently completed under Environmental Protection Agency (EPA) auspices on the potential impact of global warming on human health. First, a discussion of potential changes in heat-related mortality will be presented for three countries: The People's Republic of China (PRC), Canada, and Egypt. Second, a framework for analyzing two vector-borne diseases that may spread in a warmer world will be discussed.

LITERATURE REVIEW

Interest in studies on the impact of climate on human health has dramatically increased, and a number of comprehensive reports summarizing most of this research have appeared in the last few years. Among them are White and Hertz-Picciotto (1985), Kalkstein and Valimont (1987), Ewan *et al.* (1990), and WHO (1990). The first two of these reports dealt primarily with historical climate/health relationships and thus shed little light on projected health problems under various global warming scenarios. However, Ewan *et al.* devoted much of their discussion to heat-stress problems (citing much EPA-funded research), infectious disease transmission, and potential mitigating strategies. The WHO report discussed both direct impacts of climate on human health (heat-stress-related illnesses) and indirect impacts (transmission of infectious diseases). Both documents provided ample evidence to suggest that human health may be severely impaired if the Earth warms over the next century.

The Intergovernmental Panel on Climate Change (IPCC) Working Group II Report (Tegart *et al.*, 1990), recognized the potential impact of increased heat-stress-induced mortality but did not expound upon it. Similarly, it gave insufficient treatment to the problem of vector-borne infectious disease spread. Much of the IPCC health evaluation concentrated on UV-B radiation exposure and related problems (e.g., skin cancers). A recent Electric Power Research Institute (EPRI) evaluation of the IPCC report (EPRI, 1991) criticized the IPCC for the perfunctory treatment it gave to the potentially major topics of heat-stress-related mortality and the spread of infectious diseases. In addition, the EPRI review pointed out that the IPCC neglected to evaluate regions where health might actually *improve* if a global warming takes place.

A large majority of the research on the relationships between global warming and health has dealt with heat stress, whereas little research has concentrated on infectious disease problems. Kalkstein (1988, 1989a, 1991a) has written specifically on domestic heat-stress/mortality impacts of a global warming and more generally on climate change and public health problems (Kalkstein, 1989b, 1990, 1991b). In the area of infectious disease, Haile (1989) developed computer simulations evaluating potential patterns of vector-borne disease transmission in a warmer world. However, the most innovative work on infectious disease spread and global warming (and probably the only other attempt to date to do such research) was that done by Dobson and Carper (1988), who evaluated the potential spread of trypanosomiasis (sleeping sickness) from west to central Africa given a 2°C warming scenario. None of the infectious disease work reported in the literature used sophisticated climatic modeling. The discussion of the methodologies used in the present study is, in part, an attempt to improve the climatological component within infectious disease studies.

METHODOLOGY

The structure of the study

The heat-stress/mortality component of this research expanded upon the previous domestic research (Kalkstein, 1989a) by assessing the role of extreme weather events on mortality. This was partially accomplished with the use of a new 'synoptic climatological approach,' which evaluates weather *situations* rather than individual weather *elements*. Several target countries that may be inordinately affected if the globe warms as expected were identified and selected for study. These countries and the international researchers who assisted in the evaluations are listed below.

- Canada (Data were provided by Environment Canada's Canadian Climate Centre. The research is being performed by the author at the University of Delaware and is directed by Abdel R. Maarouf, Biometeorologist, Bioclimate Adaptation Division, Canadian Climate Centre.)
- China (The research director is Guanri Tan, Chairperson, Department of Atmospheric Sciences, Zhongshan University, Guangzhou, People's Republic of China.)
- Egypt (Data were provided by Rifky Faris, Department of Community, Environmental, and Occupational Medicine, Ain Shams University, Cairo. The research is being performed by the author at the University of Delaware and is directed by Karl A. Western, Assistant Director for International Research, National Institute of Allergy and Infectious Diseases [NIAID], National Institutes of Health [NIH].)

The infectious disease project will be directed by the NIAID at NIH, under the direction of Karl Western. In November 1990, NIH and EPA organized a workshop on the identification of infectious vector-borne diseases, available datasets, and target regions that may be susceptible to changes in disease incidence if the globe warms. As a result of that workshop, two infectious diseases were identified for study:

- onchocerciasis (river blindness), to be evaluated in west Africa (Data were provided by A. Seketeli, Assistant Director, WHO Onchocerciasis Control Program [OCP], Ouagadougou, Burkina Faso.)
- malaria, to be evaluated in west or east Africa (Data sources have been identified by Karl Western at NIH, and the United Nations Environment Program [UNEP] has also agreed to provide data and participate in an advisory role.)

Procedure for the heat-stress/mortality study

Detailed mortality data bases were available for all three evaluated countries. For Canada, daily mortality sums were provided for the following 10 cities and their adjacent metropolitan areas for the years 1958 to 1988: Calgary, Edmonton, Halifax, Montreal, Ottawa, Quebec, St. John, Toronto, Vancouver, and Winnipeg. These data were grouped by age (less than 1 year, greater than 65 years, all ages) and cause of death (weather-related, including all respiratory causes, influenza, stroke/cerebrovascular, injury, and heat stroke/heat stress; and all causes). There is conflicting evidence in the literature about the validity of evaluating specifically weather-related causes of death. Many researchers continue to utilize total mortality figures in their analyses, as deaths from a surprisingly large number of causes appear to escalate with more extreme weather conditions (Applegate *et al.*, 1981; Jones *et al.*, 1982). In an attempt to circumvent this apparent disagreement among researchers, weather-related and all-causes categories were evaluated separately in this study.

For China, mortality data were obtained for Shanghai, which has a mid-latitude climate, and Guangzhou (Canton), which has a subtropical climate. Each city contains a central mortuary that receives virtually all corpses; a record is kept for each individual. Mortality data were tabulated daily for summer (June, July, and August) and winter (December, January, and February) over a 10-year period from 1980 through 1989. Specific causes of death were not noted, but age groups of greater than 65 years old and total deaths were evaluated.

For Egypt, mortality data were extracted for Cairo for a five-year period from 1981 through 1985. Daily totals were tabulated in the same manner as for the Chinese data, although for Cairo data on neither cause of death nor age at death were available.

Two procedures were used to ascertain historical weather/mortality relationships, which must be determined before employing climate scenarios. The first involved the identification of 'threshold temperatures,' which represent the temperature at each location beyond which mortality significantly increases (Kalkstein and Davis, 1989). The second depended upon the identification of offensive synoptic situations, or specific 'air masses' that appear to be associated with particularly high mortality totals (Kalkstein, 1991a).

In the first procedure, the threshold temperature is calculated objectively by measuring the dissimilarity of mortality rates above and below a given temperature. The mean mortality is calculated for all days above and for all days below the temperature selected. Then the sum of the squared mean deviations is computed for each group, yielding two variance measures. Finally the sum of these two variances, or the total sum of squares (TSS), is calculated for this temperature. This procedure is then repeated at one degree increments, producing an array of TSS values. The temperature with the *smallest* TSS is chosen as the threshold temperature, since this represents the point where between-group variances are maximized and within-group variances are minimized. If there is no change in these variances across the one degree increments for a given locale, it is assumed that no threshold temperature exists, and the temperature–mortality relationship for this particular location is considered insignificant. The threshold temperature in Shanghai for the summer, for example, is 34°C (Figure 1); mortality increases dramatically at temperatures above this level. This procedure is conducted both for the summer and for the winter. In winter, the threshold temperature represents the temperature *below* which mortality increases.

Once the threshold temperature has been established, an 'all regression' procedure is used to determine which combination of weather elements produces weather/mortality models with the highest coefficient of determination (R^2) for days beyond the threshold temperature (Draper and Smith, 1981). The variables used in the regression include two non-meteorological elements, 'time' and 'day in sequence' (Table 1). The time variable determines whether an above-threshold day occurs early or late within the summer season and evaluates short-term and intra-seasonal acclimatization response. Day in sequence notes how a particular day above the threshold is positioned within a consecutive-day sequence. The algorithms developed through this regression analysis can be employed to estimate number of deaths or death rates attributed to weather. According to Box and Wetz (1973), an algorithm may be considered a worthwhile predictor if the actual F-ratio is larger than some multiple of the critical F; the magnitude of this multiple is dependent on the degrees of freedom. Only algorithms meeting the Box and Wetz criterion were utilized for estimation; if an algorithm for a city did not meet that criterion, it was deemed that weather did not affect mortality for that city.

Table 1. *Weather variables used in the mortality study*

Maximum Temperature (MAXT)
Minimum Temperature (MINT)
Maximum Dewpoint (MAXTD)
Minimum Dewpoint (MINTD)
Cooling Degree Hours (CDH): Summer only[a]
Heating Degree Hours (HDH): Winter only[b]
3 AM Visibility (VISAM)
3 PM Visibility (VISPM)
3 AM Windspeed (WNDAM)
3 PM Windspeed (WNDPM)
Mean 10 AM to 4 PM Cloud Cover (CLD)
Time
Day in Sequence

Notes:

[a] CDH represents a measure of the day's warmth, and is calculated as follows:

$$\mathrm{CDH} = \sum_{i=1}^{N} (T - T_s), \text{ where } T > T_s$$

T represents the hourly temperature and N represents total hours with temperature above T_s.

T_s is the temperature one standard deviation above the mean daily maximum in the summer.

[b] HDH represents a measure of the day's coldness, and is calculated as follows:

$$\mathrm{HDH} = \sum_{i=1}^{M} (T_w - T), \text{ where } T < T_w$$

T represents the hourly temperature and M represents total hours with the temperature below T_w.

T_w is the temperature one standard deviation below the mean daily maximum in the winter.

The synoptic procedure involves an entirely different approach and, as mentioned, is an attempt to identify discrete 'air masses' within a specific locale. Air masses are bodies of air that contain distinctive meteorological characteristics; all days under the influence of a particular air mass are thus similar in meteorological character (Perry, 1983). Examples of air masses include 'continental polar,' which is characterized by cold, dry, and generally clear conditions, and 'maritime tropical,' which is distinguished by hot, humid, and partly cloudy conditions as well as a high probability of thunderstorms. When evaluating the impact of an air mass upon an environmental variable such as human mortality, it is assumed that the variable responds to a set of meteorological elements working *in concert*, rather than individually. Thus, the synoptic approach assumes that humans respond to the entire 'umbrella of air' that surrounds them, rather than to individual weather elements, such as air temperature. The synoptic approach therefore permits an evaluation of synergistic relationships among weather elements; it assumes that the combined impact of several elements is greater than the sum of the individual impacts of each element.

An automated air mass-based synoptic climatological index was developed for each city (except Cairo, due to meteorological data constraints). This air mass identification procedure, known as the temporal synoptic index (TSI), is designed to categorize days into meteorologically homogeneous groups, or synoptic categories (Kalkstein, 1991a). Each day is defined in terms of six readily available meteorological elements: air temperature, dewpoint temperature, total cloud cover, sea-level air pressure, wind speed, and wind direction. These elements are measured four times daily, and the 24 variables obtained represent the basis for categorization. (Refer to Kalkstein, Tan, and Skindlov [1987] for the procedural framework for TSI.)

The mean daily mortality for each synoptic category, along with the standard deviation, is then determined to ascertain whether particular categories have distinctively high or low mortality. Potential lag times are accounted for by evaluating the daily synoptic category on the day of the deaths as well as one, two, and three days prior to the day of the deaths. Daily mortality is then sorted from highest to lowest during the period of record to determine whether certain synoptic categories (deemed 'offensive') are prevalent near the top or bottom. To evaluate the impact of within-category variations in meteorology for the offensive categories, a stepwise multiple regression analysis is performed on all days within the offensive category utilizing the variables listed in Table 1. The algorithms developed through this procedure are used as predictors of future mortality if they pass the Box and Wetz criterion.

After historical relationships for the target cities were established by these two approaches, the next step was to estimate changes in mortality that might occur with climatic warming. This study used a variety of general circulation model (GCM) scenarios as well as arbitrary scenarios to develop these estimates. The number of days above the temperature threshold and within offensive synoptic situations was established using the scenarios, and the algorithms developed through the historical analysis were used to estimate mortality rates or totals.

When measuring the impact of global warming on future mortality, the question of acclimatization must be considered. If the globe warms, will people within each locale respond to weather as they do today, or will their reactions be similar to those of people who presently live in hotter climates and are acclimatized to heat? A procedure was developed to account for such acclimatization by evaluating mortality responses in a city during heat waves in cool and hot summers. An assumption was made that if people acclimatize, they will respond to heat in a more extreme fashion during cooler summers, when heat waves are infrequent, than during hotter summers, when heat waves are a common occurrence. This assumption was based on previous U.S. research indicating that acclimatization to hot weather can occur very rapidly in some places, often within one season (Kalkstein, 1991b; Rotton, 1983). In addition, recent research in the medical community on the role of heat shock proteins, which are synthesized in the body as a response to environmental stresses such as temperature change, also implies that acclimatization may occur quite rapidly (Born *et al.*, 1990). Thus, a relationship was established for each evaluated city between the number of summer days above the threshold (or, for the synoptic approach, within an offensive air mass) and the total mortality for each of these hot days. For example, if more people die *per hot day* during a cooler summer, and fewer die per hot day during a warmer summer, it may be inferred that some degree of acclimatization has occurred during the warmer summer. Thus, it is probable that such a population might also acclimatize to a slow, long-term warming, such as that which would be associated with a human-induced climatic change.

If mortality per hot day in a certain city was significantly and inversely related to the number of days above the threshold temperature over a series of summers, it was assumed that the population of that city could acclimatize to the warmer conditions expected under the $2 \times CO_2$ conditions. For such cities, estimates of acclimatized mortality under the various scenarios were calculated by using the slope of the regression line expressed by the scenarios to estimate diminished mortality during the hotter summers.

It is important to note that these acclimatization estimates account for only physiological and short-term behavioral adjustments, not infrastructure changes such as major architectural improvements in urban areas that may be made in response to the increased warmth. It is virtually impossible to predict how city structure might change under global warming conditions, and no research involving such infrastructure changes has ever been attempted. Nevertheless, the acclimatization procedure outlined here may yield the most realistic estimates, as cultural or social adjustments, such as architectural changes, may lag far behind the physiological adjustments of the human body, especially in developing countries (Kalkstein, 1991b).

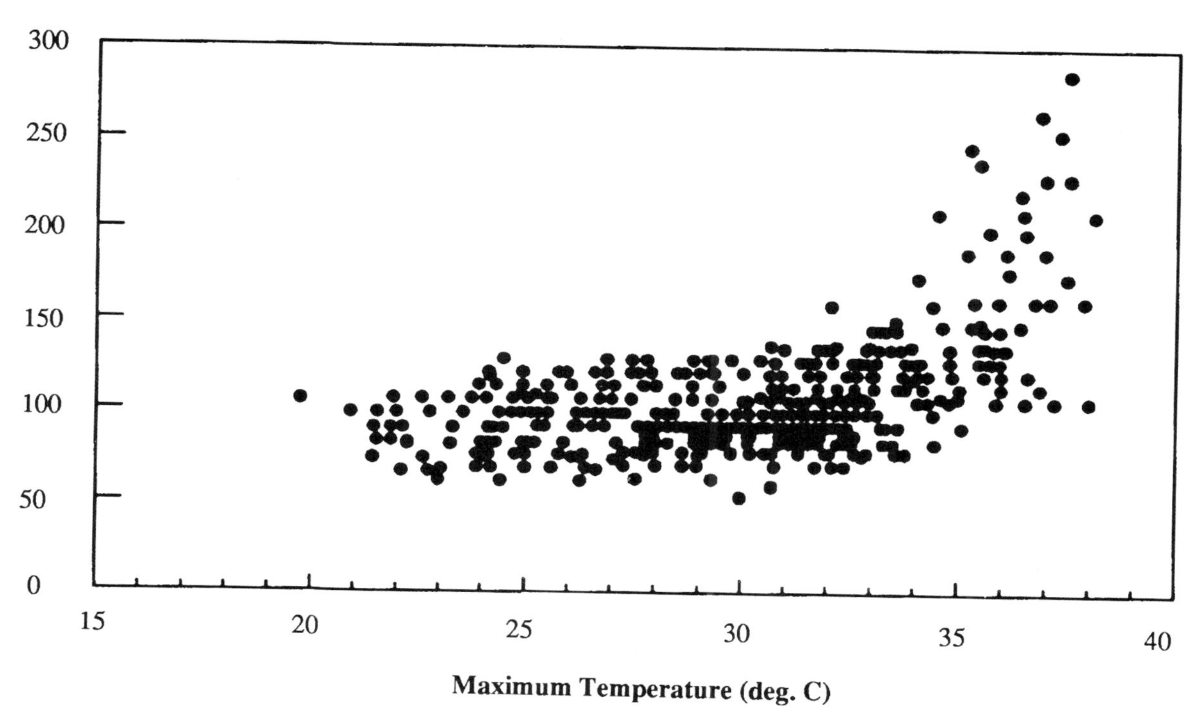

Figure 1 Relationship between maximum temperature and mortality: Shanghai.

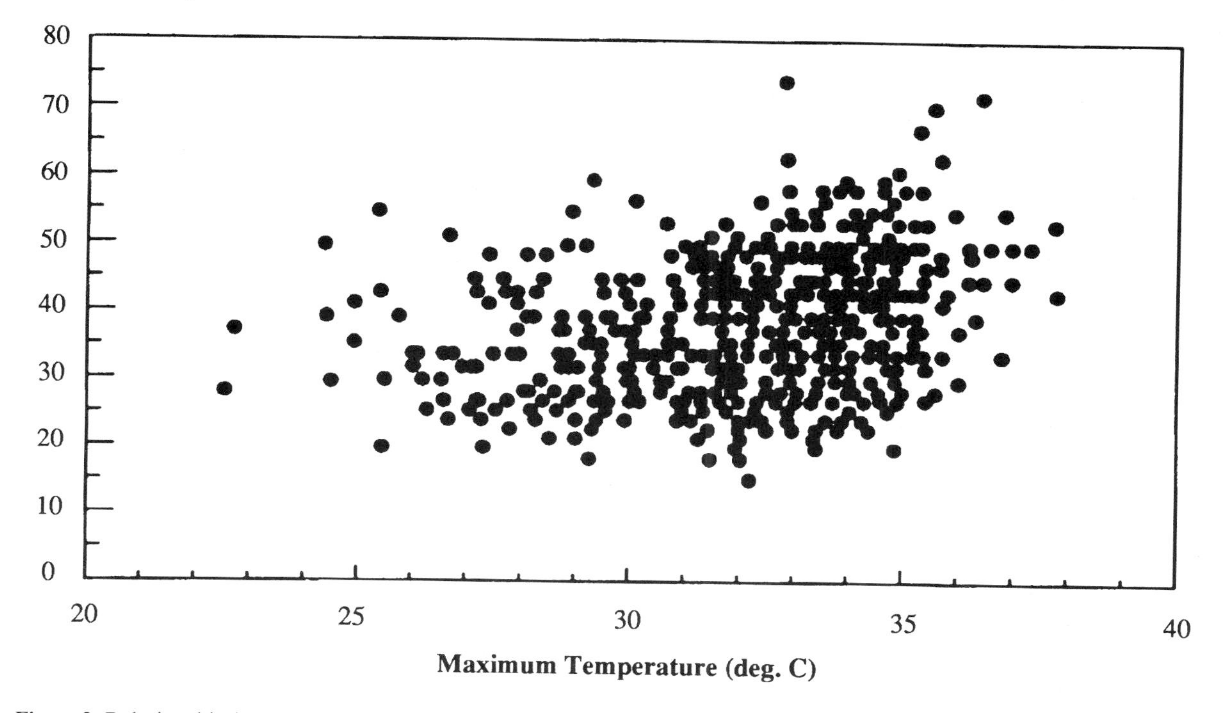

Figure 2 Relationship between maximum temperature and mortality: Guangzhou.

Limitations of the procedures used

The procedures used in this study to develop heat stress/mortality estimates suffer from two major shortcomings. First, the impact of air pollution, which may be closely related to weather, is ignored. Second, the acclimatization procedure is weakened by our limited knowledge of human responses to large changes in climate.

The air pollution limitation is underscored by the work of Schwartz and Dockery (1992), which determined that human health in Los Angeles is markedly impacted by variations in air pollution concentration. However, our previous studies of 10 U.S. cities (Kalkstein, 1991a) indicated that daily mortality fluctuations are much more sensitive to weather than to pollution concentration. Those synoptic categories that appeared to possess the highest mortality rates were the hottest and most oppressive, whereas their associated pollution loads did not appear to be unusually high. Nevertheless, when one considers the potential for high concentrations of fine particles and ozone in cities like Cairo and Shanghai, it becomes clear that air pollution impacts must be integrated into any future research.

The acclimatization issue is much more difficult to evaluate. Previous studies have relied on 'analog cities' (which are locales currently possessing climates similar to the climate estimated for the target city using the warming scenarios) to determine how mortality rates of a city might respond under different climate change scenarios. (Refer to Kalkstein [1989a] for a procedure for determining analog cities.) Unfortunately, analog cities cannot duplicate the demographics or urban structure of the target city, which greatly confines their use. For this reason, the analog city approach was abandoned for this evaluation. However, the acclimatization procedure utilized here suffers from a different shortcoming: it assumes that if no interseasonal acclimatization is noted, there will be little acclimatization expected under long-term global warming conditions. In spite of the fact that humans appear to acclimatize rapidly to hot weather, there is no available research to determine acclimatization rates over periods of several decades, which represents the time period of a long-term, human-induced global warming. In addition, our acclimatized results are indications of sensitivities assuming current infrastructure and little change in the behavior of the population.

RESULTS

Chinese mortality

The evaluation of Shanghai and Guangzhou strongly suggested the following:

- Heat-stress is presently an important cause of death in China.
- The impact is greater in the mid-latitude city (Shanghai) than in the tropical city (Guangzhou).
- Winter mortality increases only slightly in these cities during extreme weather.
- There is evidence that the Chinese will not acclimatize to the increased warming induced by 2 × CO_2 concentrations, and under these conditions heat-stress-related mortality could rise significantly.

Although the threshold temperatures for Guangzhou and Shanghai were very similar (Figures 1 and 2), the rise in mortality above the threshold was much more dramatic for Shanghai. This finding is consistent with research conducted in the United States that indicated that northern and midwestern cities, which experience severe but non-constant heat, are more sensitive to heat-stress-related mortality than southern cities, in which extreme heat is routine (Kalkstein and Davis, 1989).

In Shanghai, days with hot afternoon temperatures, low wind speeds, and low humidity were associated with the greatest mortality increases. In fact, a regression model relating statistically significant climate variables with mortality for days above the threshold temperature yielded an R^2 of 0.740. For Guangzhou, hot afternoon temperatures and very warm nights produced the greatest mortality, with wind speed and humidity being insignificant. In addition, hot weather early in the summer season in Guangzhou produced greater mortality than similarly hot conditions later in the season, a result consistent with U.S. findings. However, the R^2 for the Guangzhou regression model (0.195), although statistically significant, was considerably lower than that for the Shanghai model.

Direct mortality comparisons between Shanghai and Guangzhou indicated significant differences in response between the two cities (Table 2; Figure 3). Although the number of days above the threshold temperature was greater for Guangzhou during an average summer, total deaths and death rates on hot days were considerably greater for Shanghai. In fact, the average death rate for Shanghai on days above the threshold (1.972 per 100,000 population) was 59 percent greater than for Guangzhou (1.240 per 100,000). During an average summer, heat-stress-related deaths were three times higher at Shanghai than Guangzhou (418 vs. 135), representing a 50 percent higher death rate. In addition, the Shanghai heat-stress-related death rate was 132 percent higher than that of New York, which is one of the most sensitive U.S. cities. During the worst year during the study period in Shanghai, heat-stress-related deaths totaled 1,140, which approximates the average number of heat-stress-related deaths in 15 of the largest U.S. cities combined (Kalkstein, 1989a).

Table 2. *Summer mortality analysis: Shanghai & Guangzhou*

Category	Shanghai	Guangzhou
Daily deaths in an average summer	103.3/1.524[a]	36.6/1.118
Threshold (maximum) temperature	34°C	34°C
Days meeting threshold per year	12	25
Percentage of days over threshold	13%	28%
Daily deaths on hot days	133.6/1.972	40.6/1.240
Heat-related deaths in an average summer[b]	418/6.169	135/4.126
Heat-related deaths in the hottest summer	1,140/16.825	216/6.602

Notes:
[a] Actual deaths/death rate per 100,000 population.
[b] The difference between the computed number of deaths by the regression on days above the threshold and the average of the regular days.

The synoptic evaluation for the two cities revealed two potentially offensive air masses for Shanghai (Table 3). Air mass 102, characterized by hot, clear, and dry conditions, was associated with a mean daily mortality that was about 9 percent above the total mean. However, although this air mass occurred on only 14 percent of summer days, it was present over one-third of the time on the 50 highest mortality days recorded in Shanghai during the sample period. Air mass 103, associated with sultry maritime tropical conditions, appeared to be even more stressful. This air mass was present on 60 percent of the top 50 mortality days, which is remarkably high considering that its normal summer frequency was less than 18 percent. Thus, the combination of these two apparently stressful air masses, which occurred less than 33 percent of the time during summer, accounted for 47 of the top 50 mortality days in Shanghai (94 percent). With the GISS $2 \times CO_2$ scenario, it was predicted that these air masses will occur much more frequently. For example, air mass 103 is predicted to almost double in frequency, from 17.9 percent to 32.7 percent, using the GISS scenario. This may affect heat-stress-related mortality in Shanghai very negatively.

The synoptic evaluation for Guangzhou suggested that no particular air mass increased mortality significantly. This more benign relationship in Guangzhou is consistent with results from the threshold temperature study.

The winter evaluation indicated that extreme cold had little impact on daily mortality for either city. Although mean daily mortality was slightly higher in winter than in summer in Shanghai, no threshold temperature could be determined for winter in that city. Thus, no mortality peaks were achieved at low temperatures, and the slightly higher general winter mortality was attributed to influenza and other infectious diseases not directly related to cold temperatures. For Guangzhou, a threshold temperature of 12°C was determined, but daily mortality on days below this threshold was only marginally higher than the overall mean. Thus, potential changes in winter weather under various global warming scenarios should have an insignificant impact on weather-related mortality.

Estimates of summer mortality changes with global warming in Shanghai and Guangzhou were made using five different climate change scenarios. Separate estimates were derived using the threshold temperature approach (for both cities) and the synoptic climatological methodology (for Shanghai only). The increased frequency of offensive air masses was determined by running a TSI on the five climate change scenarios.

It appears that climate change could have a very large negative impact on heat-stress-related mortality, especially for Shanghai (Figures 4 and 5). Assuming no acclimatization and using the threshold approach, heat-stress-related mortality for the city under the UKMO scenario exceeded 4,800 deaths, which represents a tenfold increase over present conditions. This value is 3 times the estimated $2 \times CO_2$ heat-stress-related mortality for New York City. Estimates using the GFDL, GISS, and two arbitrary scenarios were somewhat smaller but still considerably higher than for any U.S. city. Even the estimated values for Guangzhou were considerably higher than those for U.S. cities, with the GISS $2 \times CO_2$ scenario yielding an estimated 1,569 extra deaths.

Data from this study indicate that acclimatization will not be a major factor in either Chinese city. When the acclimatization procedure described previously was used, there was no difference in mortality response to heat whether the summer was very hot or comparatively cool. Thus, people responded similarly to heat whether it was very frequent or infrequent. For this reason, it is suggested that physiological acclimatization to increasing warmth may very well not take place. On the surface, the weaker mortality response at Guangzhou as compared to Shanghai contradicts this contention and seems to support the notion that some degree of acclimatization might take place. However, it is likely that acclimatization at Guangzhou was a function of the small climatic *variability* that occurs there in summer, rather than a function of the intensity of the heat itself. The GCMs and arbitrary scenarios utilized in this study assumed that climatic variability will not change in either city. If this is the case, Shanghai's summer climate will become warmer, but the variability will not be reduced. The implication would then be that intense but irregular heat waves will continue to occur at Shanghai, lessening the probability of population acclimatization.

The scenario results when the synoptic air-mass-based analysis was used were even more extreme (Figure 5). When a TSI was developed for Shanghai from the GISS $2 \times CO_2$ scenario, the number of unacclimatized heat-stress-related deaths exceeded 5,000 during an average summer. This was

Table 3. *Offensive air masses in summer for Shanghai*

Air mass type	102	103
Weather	Hottest in the season, light winds from SE, little cloudiness, comparatively high atmospheric pressure (Subtropical Anti-cyclone dominant)	Almost as hot as 102, very light wind from SE at night and from SW by daytime, mostly cloudy, low atmospheric pressure (Maritime Tropical Air Mass)
Mean daily mortality	113.9	116.9
Percent above overall mean	8.8%	11.2%
Percent of all summer days	14.2%	17.9%
Days in top 50 highest mortality days/percent	17/34.0%	30/60.0%
Percent frequency of air mass under GISS 2 × CO_2 scenario	18.4%	32.7%

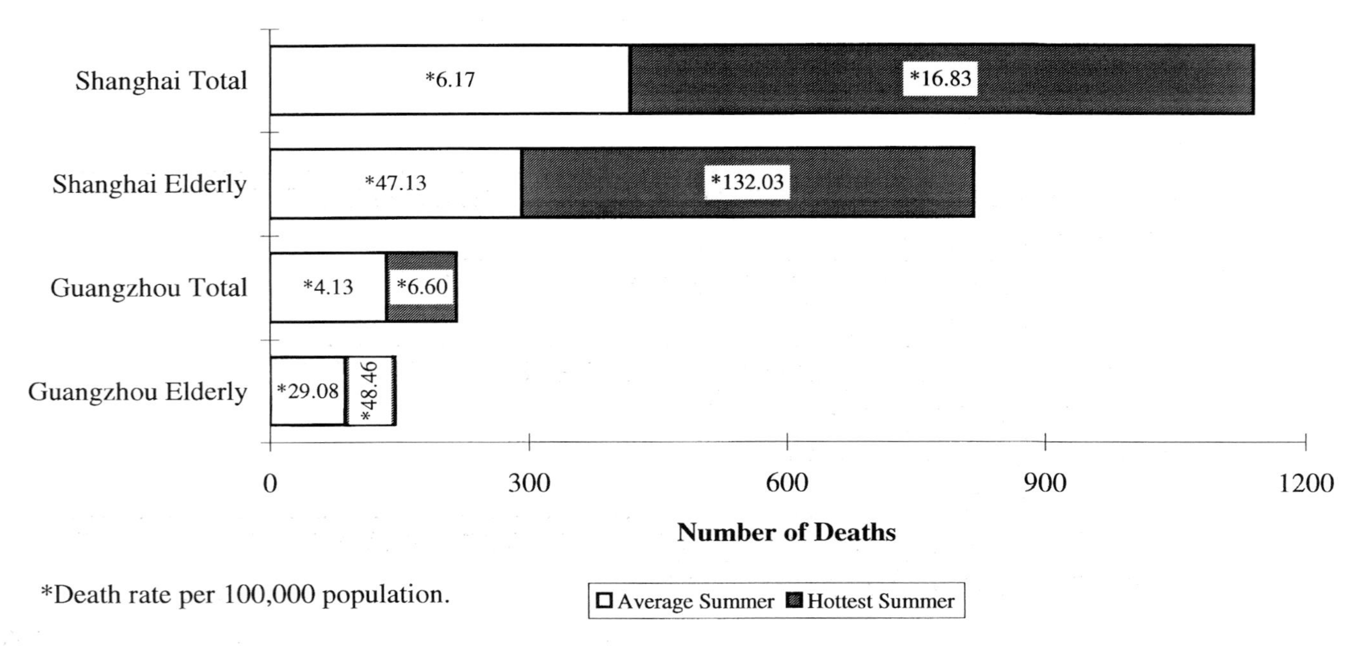

Figure 3 Summer mortality analysis for Shanghai and Guangzhou.

based on the greater frequency of the two offensive air masses expected in a warmer world. As might be expected, the elderly will constitute the largest proportion of this mortality.

Egyptian mortality

Some of the key findings of the Egyptian evaluation were the following:

- Unlike other places evaluated, mortality in Cairo presently rises in a linear fashion as temperature increases. Thus, there is no particular threshold temperature despite a strong mortality/temperature relationship.
- Egyptian heat-stress/mortality relationships are stronger than those found for the United States and Canada but less dramatic than those found for Shanghai.
- Acclimatization to increased warmth will probably be minimal in Egypt, and increasing warmth could contribute to much higher mortality rates.

An evaluation of maximum temperature/mortality rela-

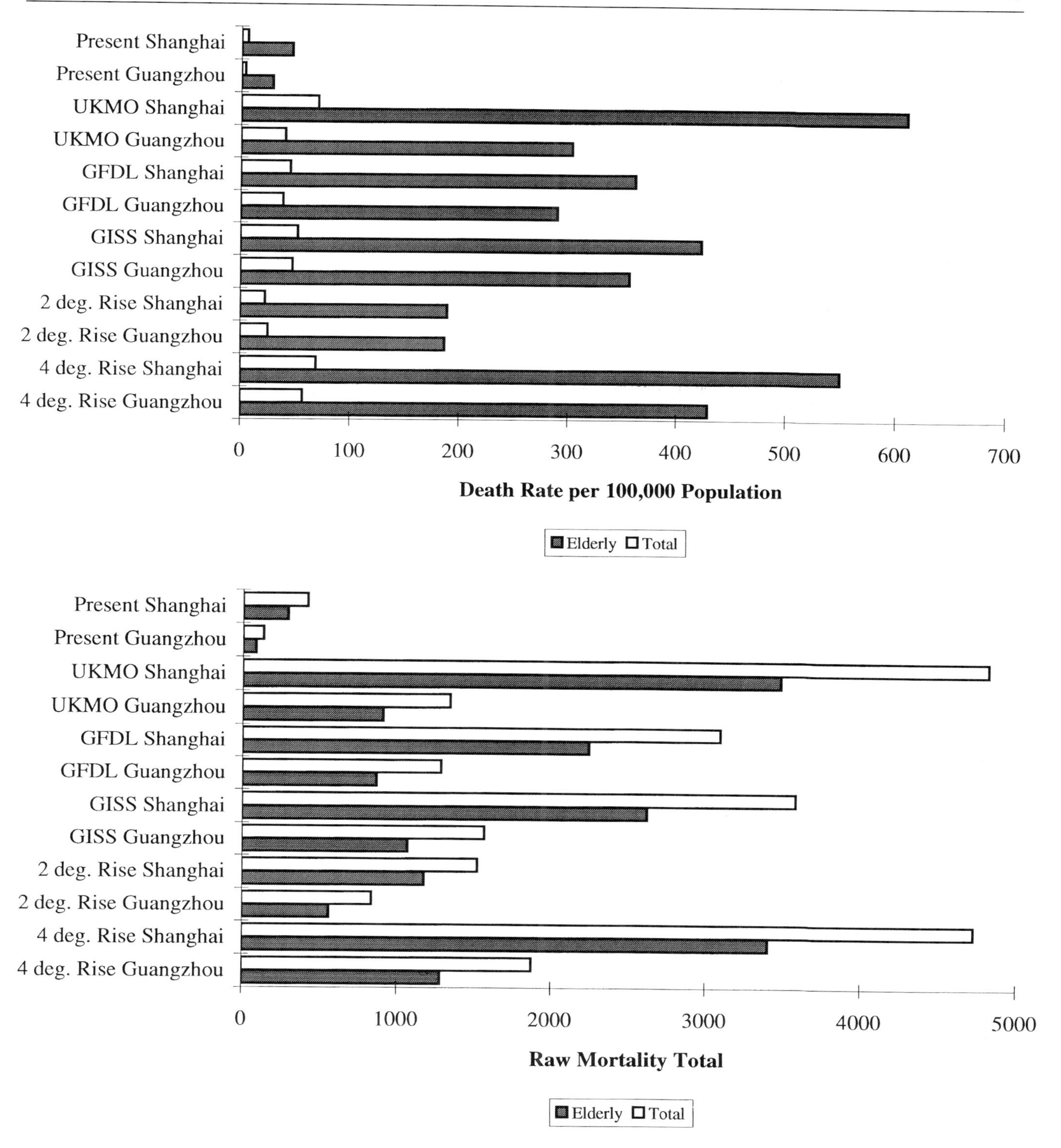

Figure 4 Estimates of future Chinese mortality in summer: threshold approach.

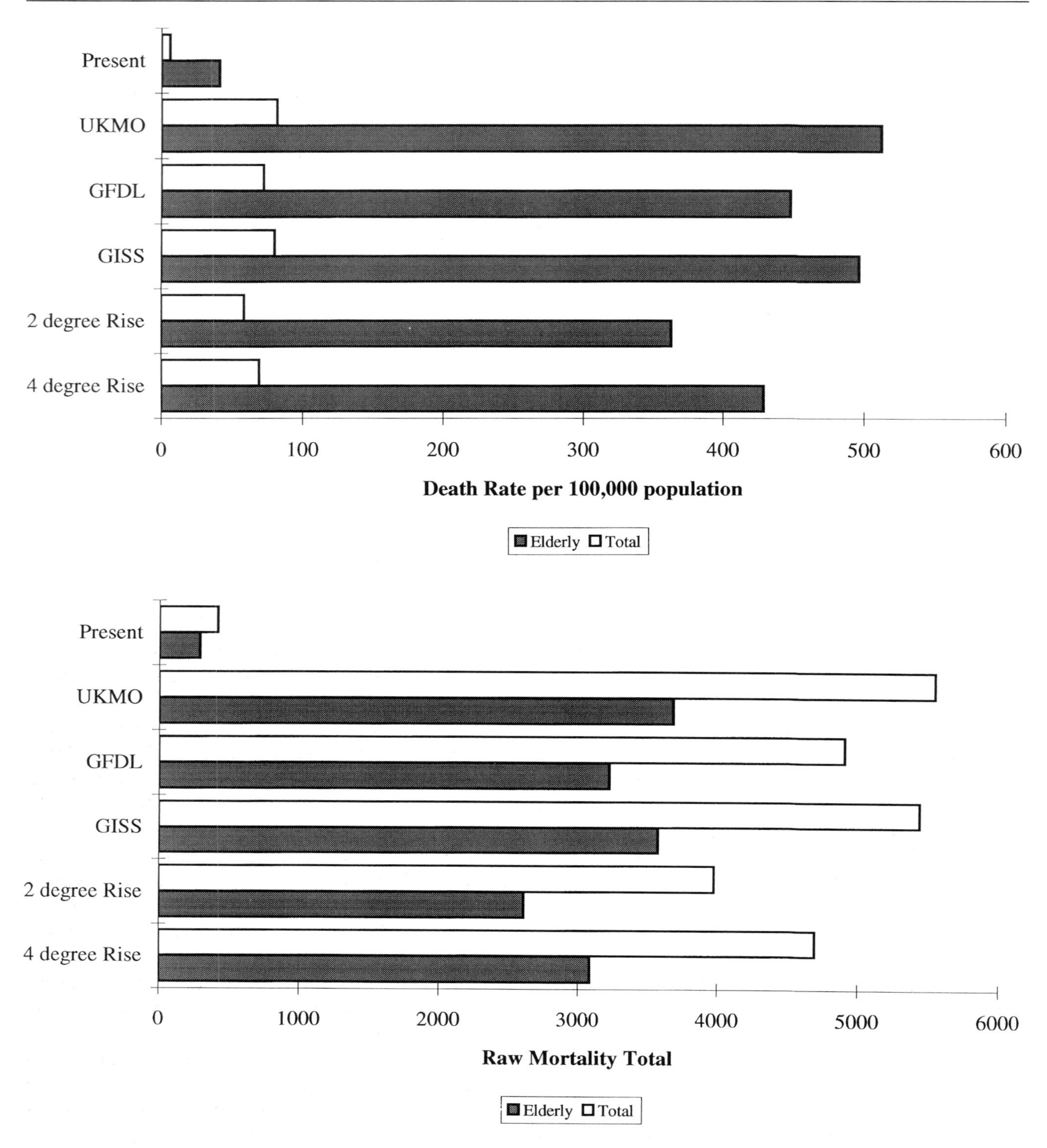

Figure 5 Estimates of future Shanghai mortality in summer; synoptic approach.

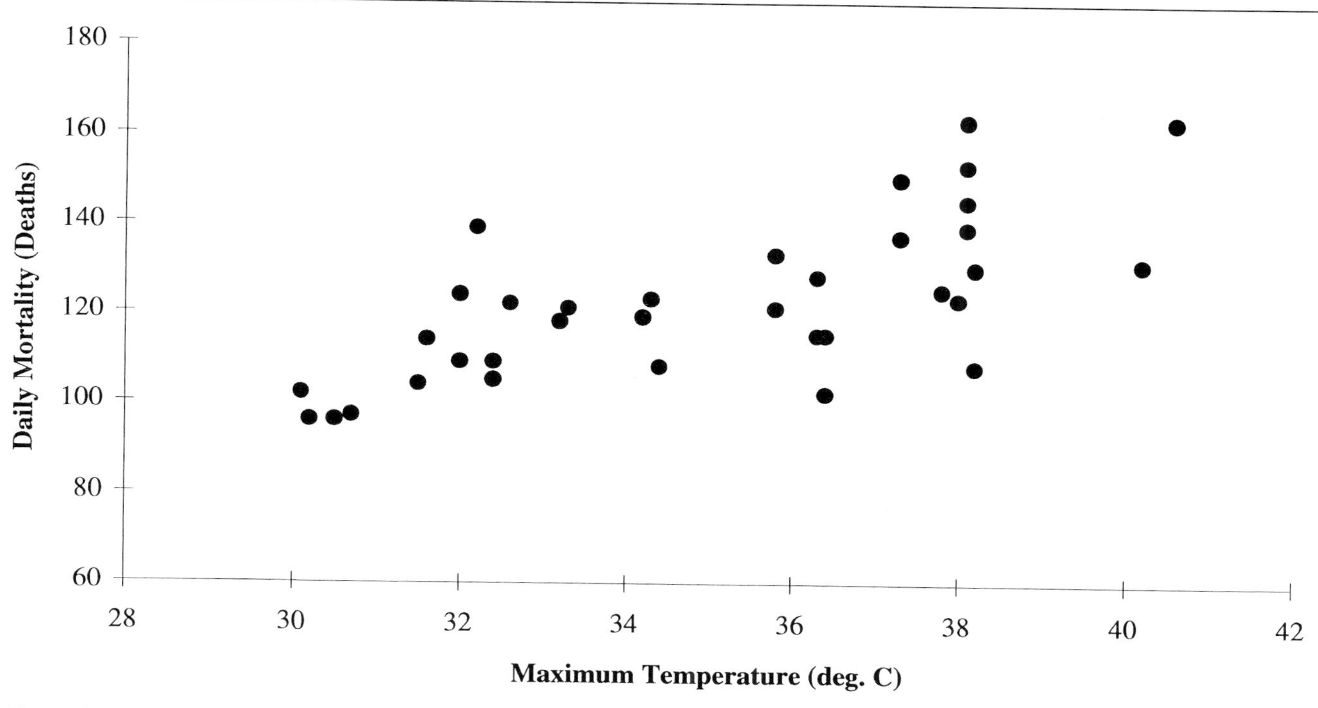

Figure 6 Relationship between maximum temperature and mortality: Cairo, Egypt: 1982.

tionships in summer for Cairo for the period of 1981 to 1985 indicated that mortality from all causes increased quite rapidly as the temperature increased (Figure 6). The relationship between these variables led to two interesting findings: (1) there was a one-day lag between weather and mortality response, and (2) a threshold temperature could not be determined for Cairo, even though it was obvious that a strong temperature/mortality relationship existed. The lag response is similar to results found for all the countries evaluated, developed or developing, and strongly suggests that people respond rapidly to stressful weather. However, the threshold result is unique to Egypt only, as all other cities with a strong weather/mortality relationship exhibited a distinctive threshold temperature. Additionally, Cairo's response differs from results obtained for Phoenix, Arizona, which has a summer climate even more extreme than Cairo's. In Phoenix, no significant weather/mortality relationship was uncovered, possibly because of the population's easy access to air conditioning. However, the role of air conditioning in mitigating heat-related mortality is especially complex, and there is disagreement in the literature as to whether air conditioning is helpful at all (Kalkstein, 1988). The EPA is presently funding a project specifically designed to determine the impact of air conditioning access on heat-related mortality, and results will be forthcoming shortly.

Although weather variables other than maximum temperature were unavailable for this particular evaluation, several other interesting features were noted in the Cairo study. First, the timing of hot weather through the season had little impact in Cairo. Thus, a very hot day in August would contribute to mortality just as strongly as a similar day in June. This is dissimilar to results for most other locales, where heat early in the season had a much greater impact than if it occurred later. Second, consecutive hot days did not seem to exacerbate heat-related mortality. This also differs from Chinese, Canadian, and U.S. results, where the impact of heat increased with increasing duration of heat. Third, the impact of hot weather on heat-related mortality was similar in Cairo for hot and 'cool' seasons. This is similar to the Chinese results but differs from the Canadian and implies that acclimatization may not be a major issue in Cairo.

Present-day heat-related mortality in Cairo numbers almost 300 individuals during an average summer (Figure 7). The death rate of 4.45 per 100,000 is higher than that for any U.S. or Canadian city evaluated (New York's is 2.67 per 100,000) but is lower than Shanghai's (6.17 per 100,000). Present-day heat-related mortality in Cairo during the hottest year was estimated with available data (1981) to approach 400 people, but it should be noted that the 5-year period evaluated did not include any extremely hot years. Thus, during certain summers in Cairo, it can be expected that heat-related mortality would be much higher.

Estimates utilizing the climate scenarios indicated that Cairo may be vulnerable to very large increases in heat-related mortality if the globe warms (Figure 7). Analyses based on the GCM scenarios showed that average summer mortality totals may exceed 1,000; such death rates would be considerably higher than those for all evaluated U.S. and

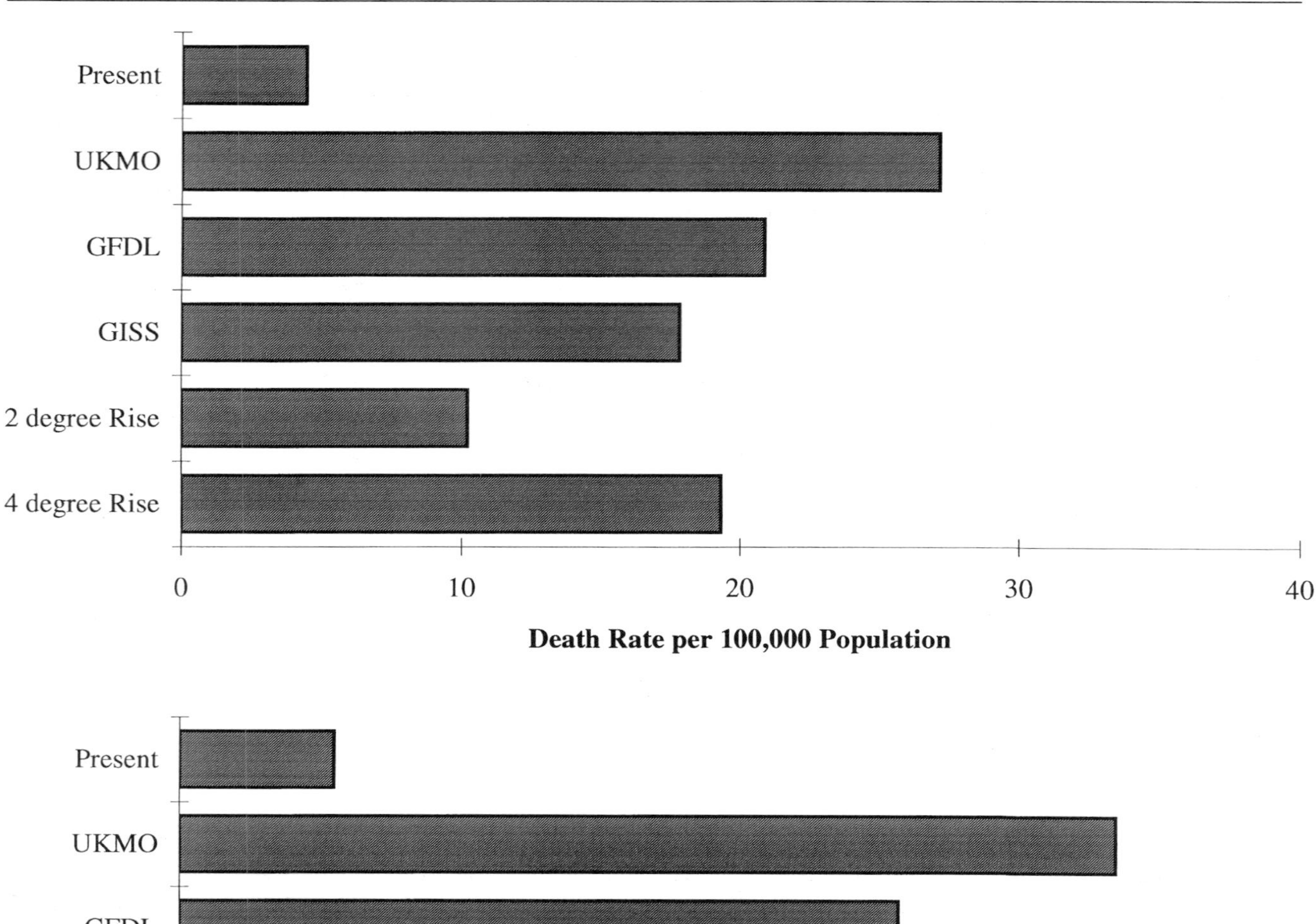

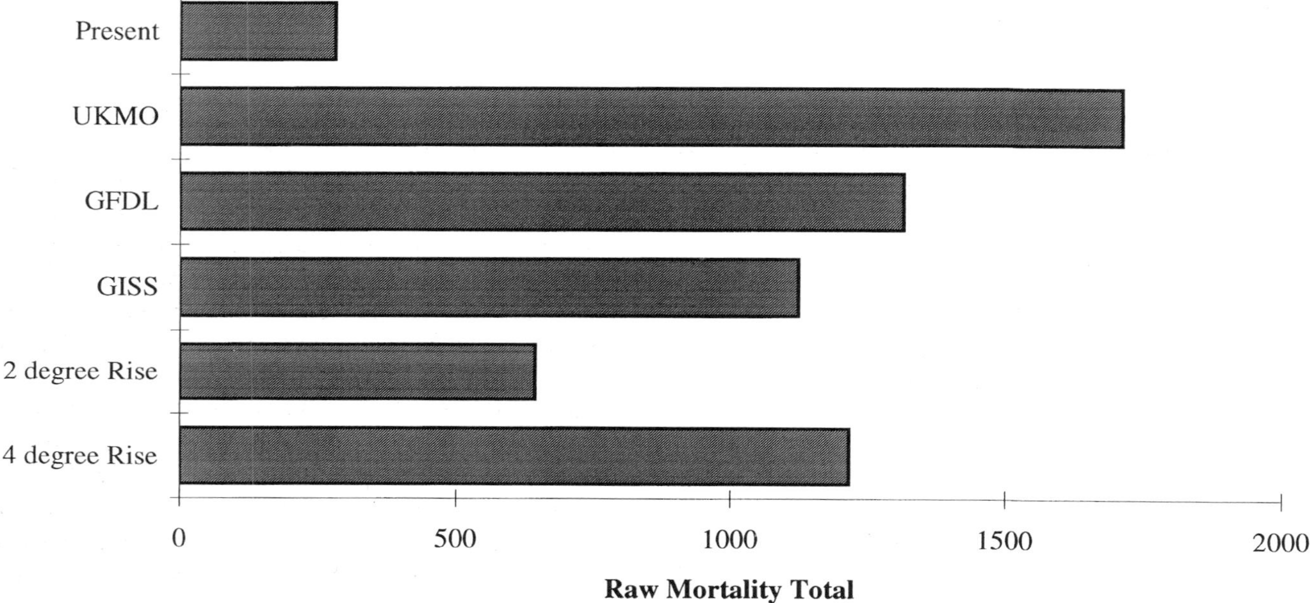

Figure 7 Estimates of future Cairo mortality in summer: threshold approach.

Canadian cities. However, rates and totals were considerably lower than those estimated for Shanghai.

A comparison of Cairo's heat-related mortality rates with five other major cities in the United States, China, and Canada revealed the city's comparative sensitivity to hot weather (Table 4). Much like the Chinese cities, acclimatization may not occur in Cairo. This projection is based on the lack of a differential mortality response during hot and cool summers. Thus, it is possible that Cairo's heat-related mortality rate will increase by three or four fold under the various scenarios. However, considering the extreme heat predicted for Cairo using the GCM scenarios, these values might overestimate mortality. It is possible that the *consistency* of hot weather, as expressed by a low variation in day-to-day maximum temperatures, may help mitigate heat-related mortality. However, it is also possible that the rising temperatures might place Cairo's summer weather outside the range of human tolerance; if this is the case, mortality rates could

Table 4. *A comparison of heat-related mortality rates for six large cities*

	Total	UKMO		GFDL		GISS		2° Rise		4° Rise	
		No Acc.	Acc.	No Acc.	Acc.	No Acc.	Acc.	No Acc.	Acc.	No Acc.	Acc.
Montreal	69[a]	1,135	652	1,070	656	430	218	195	87	418	211
	2.62[b]	43.16	24.79	40.69	24.95	16.35	8.29	7.42	3.31	15.90	8.02
Toronto	19	885	0	441	6	251	3	78	1	238	2
	0.73	33.95	0	16.92	0.23	9.63	0.12	2.99	0.04	9.13	0.08
Shanghai	418	4,826	n/a	3,098	n/a	3,587	n/a	1,562	n/a	4,730	n/a
	6.17	71.23	n/a	45.72	n/a	52.94	n/a	23.05	n/a	69.81	n/a
Guangzhou	135	1,344	n/a	1,288	n/a	1,569	n/a	838	n/a	1,876	n/a
	4.13	41.08	n/a	39.37	n/a	47.96	n/a	25.61	n/a	57.34	n/a
New York	320	n/a	n/a	n/a	n/a	1,743	880	577	388	775	647
	2.67	n/a	n/a	n/a	n/a	14.53	7.33	4.81	3.23	6.46	5.39
Cairo	281	1,713	n/a	1,317	n/a	1,125	n/a	645	n/a	1,218	n/a
	4.46	27.17	n/a	20.89	n/a	17.84	n/a	10.23	n/a	19.32	n/a

Notes:
Acc. = Acclimatization
[a] Raw mortality data.
[b] Death rate per 100,000 population.

be considerably higher. Unfortunately no study to this point has delineated the upper limit of human tolerance to heat, rendering this statement highly speculative.

Canadian mortality

The Canadian mortality study suggested the following:

- Of 10 Canadian cities evaluated, only three (Toronto, Montreal, and Ottawa) yielded significant summer mortality relationships.
- None of the cities showed any winter relationships.
- It appears that a northern boundary of heat-related impacts can be determined in Canada.
- It is estimated that several hundred extra deaths will occur in Toronto and Montreal under 2 × CO_2 conditions, even if the population acclimatizes.

Although there appeared to be a significant relationship between summer weather and mortality in Canada, there is no doubt that the overall impact was much less than in China. When the threshold temperature procedure was used, only 2 of the 10 cities evaluated (Toronto and Montreal) demonstrated a strong relationship. The synoptic approach uncovered significant summer relationships for three cities: Toronto, Montreal, and Ottawa. These results suggest that the other Canadian cities did not exhibit the intensity of heat waves found in China and the United States, although it was surprising that results from Winnipeg, Calgary, and Edmonton, which suffered from periodic heat waves, were not significant. The lack of a relationship for these cities might be attributed to the continental, rather than maritime, source of hot air, which may be less stressful. It is also possible that the small size of these cities contributed to a noisier mortality data base; with a smaller data base, non-weather induced factors, such as a major traffic accident, can have a major impact on mortality totals. This was also a problem in studies of smaller midwestern and western U.S. cities (Kalkstein, 1988).

Threshold temperatures for Montreal (29°C) and Toronto (33°C) were considerably lower than those for Chinese, and most U.S., cities. It appears that Montreal and Toronto represent the present-day northern limit for heat-related mortality, as air masses from tropical sources do intrude occasionally into southern Canada. An evaluation of variables affecting mortality above the threshold temperature indicated strong similarities between Montreal and Toronto (Table 5). For both cities, the consecutive-day variable was directly related to mortality above the threshold, indicating that a long string of stressful days produces an adverse physiological response. For both cities, the time variable was inversely related, implying that heat waves early in the season were more damaging than those late in the season; this corresponds with U.S. results and implies intra-seasonal acclimatization (Kalkstein, 1988). For Montreal, there was a direct relationship between minimum dewpoint and mortality, which underscores the importance of hot, humid air masses. One apparent difference between the two cities

Table 5. *Regression analysis for days above threshold for Montreal and Toronto: summer*

Montreal (0 day lag, threshold temperature = 29°C)

Step	Variable	Coefficient	Partial R^2 (%)	Model R^2 (%)	F	Prob > F
	Intercept	11.5015			0.89	0.3456
1	Cons. Days	1.4972	8.81	8.81	18.17	0.0001
2	Time	−0.1315	4.86	13.67	33.35	0.0001
3	MinDPT	0.7529	5.83	19.50	19.28	0.0001
4	MaxT	0.9358	1.05	20.55	4.99	0.0260

Toronto (1 day lag, threshold temperature = 33°C)

Step	Variable	Coefficient	Partial R^2 (%)	Model R^2 (%)	F	Prob > F
	Intercept	28.0129			16.62	0.0001
1	Cons. Days	4.0598	18.10	18.10	11.65	0.0011
2	MinT	1.0680	6.64	24.74	7.40	0.0083
3	CC1400	−0.7068	4.81	29.55	4.18	0.0449
4	Time	−0.0581	1.12	30.67	1.08	0.3028

relates to the lag time between offensive weather and mortality response; there was a one-day lag for Toronto, but there was no lag for Montreal. For both cities, the coefficient of determination was sufficiently high to pass the Box and Wetz test, indicating that the algorithm developed from the analyses may be used for predictive purposes.

The synoptic evaluation uncovered offensive air masses for Montreal, Toronto, and Ottawa (Table 6). The stressful air masses had similar characteristics for all three cities: very warm air (by Canadian standards), high dewpoint, moderate to strong southwesterly winds, and anticyclonic control. Thus, it appears that maritime tropical air masses, which are noted for high relative humidity, are important in Canada, whereas a more continental air mass, which is hot but possesses lower humidity, is of greater importance in Shanghai.

Estimates of present-day mortality using the threshold and synoptic approaches indicated that several hundred extra deaths can be attributed to stressfully hot weather during warm summers (Figures 8 and 9). These values were estimated to be more modest during average summers. Values derived independently using the threshold and synoptic approaches were strikingly similar. These totals were comparable to heat-stress-related mortality totals for moderately sensitive U.S. cities, such as Los Angeles and Minneapolis (Kalkstein, 1989a).

Death rates for Toronto and Montreal were much lower than for either Chinese city. Rates for Montreal were slightly lower than New York's but still averaged over two per

Table 6. *Characteristics of offensive air masses: Montreal, Ottawa, and Toronto*

	Mean value		
Variable	Montreal	Ottawa	Toronto
Midnight			
Temperature (°C)	22.1	22.9	20.8
Dewpoint (°C)	18.2	19.3	17.0
Windspeed & direction	strong, SW	strong, SW	weak, WSW
Cloud cover (tenths)	6	5	2
6 a.m.			
Temperature (°C)	21.1	21.5	19.5
Dewpoint (°C)	18.1	19.3	16.7
Windspeed & direction	strong, SW	strong, SW	weak, WSW
Cloud cover (tenths)	7	6	3
Noon			
Temperature (°C)	26.2	29.8	30.1
Dewpoint (°C)	18.9	20.6	18.6
Windspeed & direction	very strong, SW	strong, WSW	strong, SW
Cloud cover (tenths)	6	6	4
6 p.m.			
Temperature (°C)	24.9	28.4	29.2
Dewpoint (°C)	17.1	20.1	18.5
Windspeed & direction	strong, WSW	strong, WSW	strong, SW
Cloud cover (tenths)	6	6	4

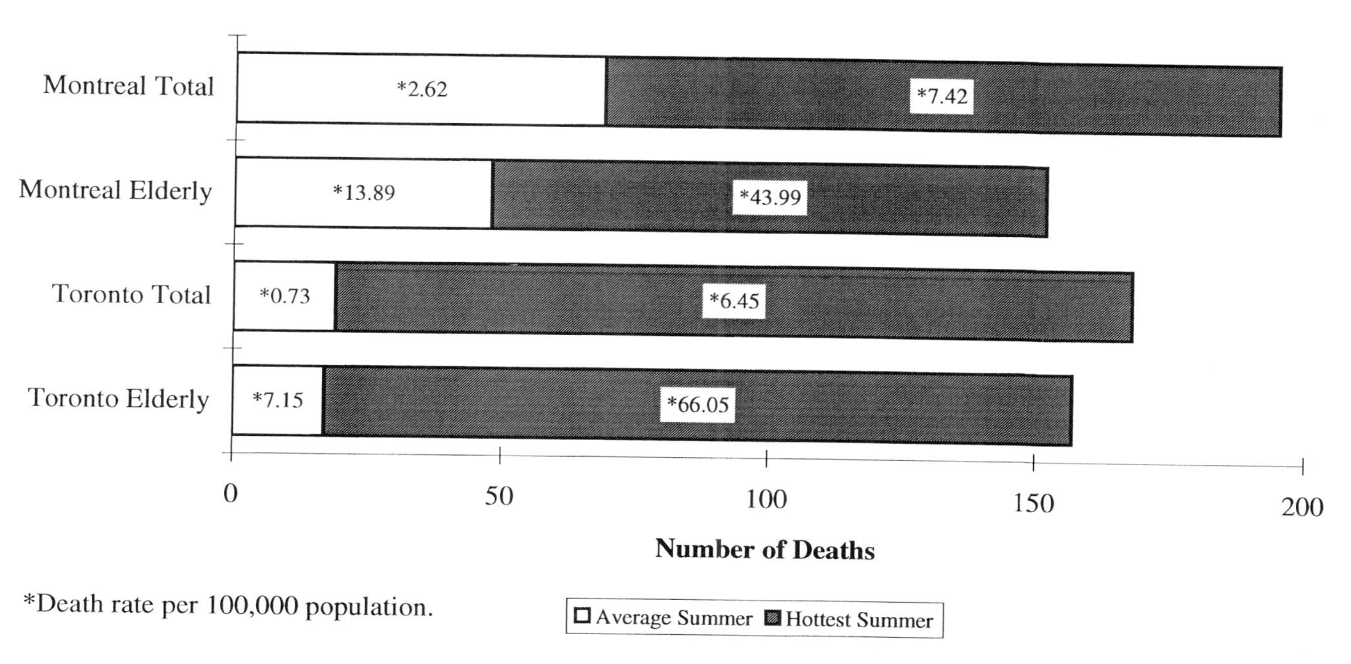

Figure 8 Summer mortality analysis for Montreal and Toronto: threshold procedure.

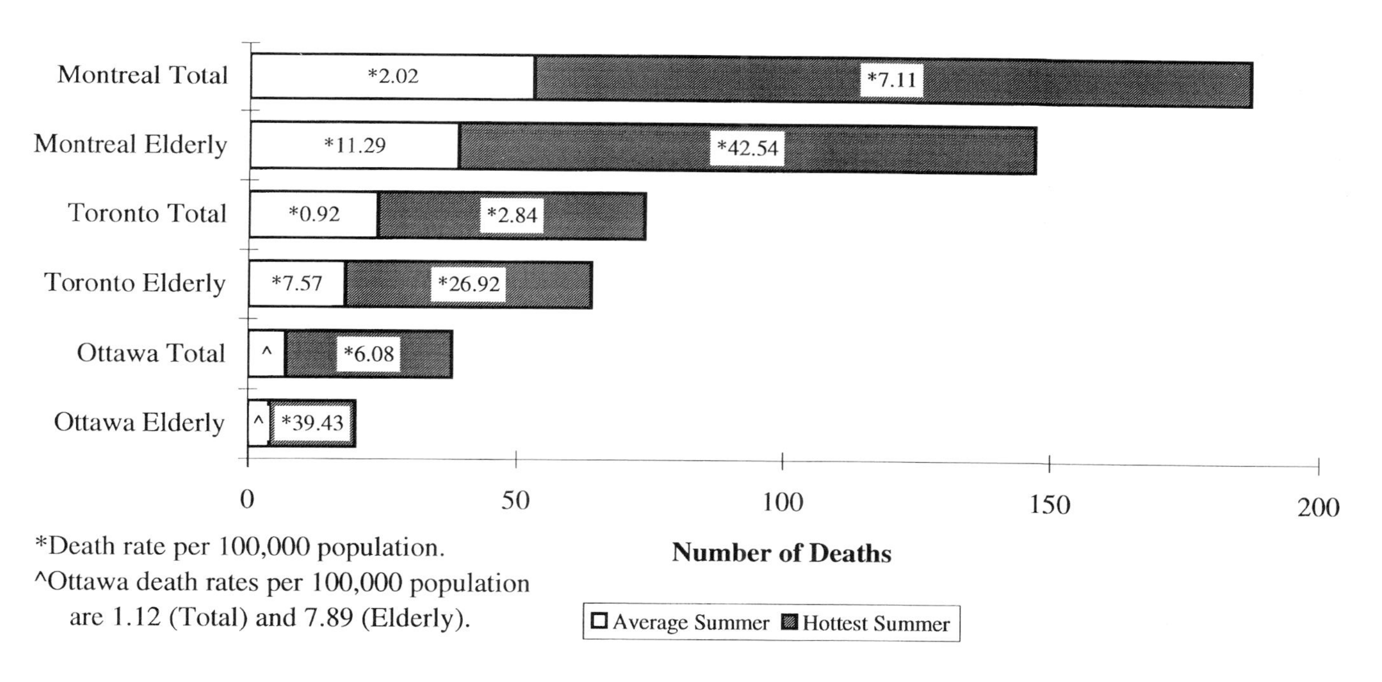

Figure 9 Summer mortality analysis for Montreal, Toronto, and Ottawa: synoptic procedure.

100,000 population during an average summer, indicating moderate sensitivity. These rates more than tripled during the warmest summers. However, under average summer conditions, Toronto's heat stress-related mortality rates were among the lowest for mid-latitude North American cities.

There are strong indications that global warming could significantly increase heat-related mortality, even with acclimatization (Figures 10 and 11). Estimates assuming no acclimatization were particularly large, and exceeded 1,000 extra deaths at Montreal using the threshold procedure for the GFDL and UKMO scenarios. These values were similar to results for highly sensitive U.S. cities, such as New York (Kalkstein, 1989a). In addition, assuming no acclimatization, death rates at Montreal would be considerably higher than those of New York. The prime reason for the marked increases for Toronto and Montreal is that those cities will likely have a much greater frequency of days above the threshold temperature under $2 \times CO_2$ conditions. The estimates assuming acclimatization were quite high for Montreal but were considerably lower for Toronto. The differential can be attributed to the acclimatization procedure. The mortality rate for Toronto was projected to be much lower during hot summers than during cool summers. Acclimatization is less of an issue in Montreal, where heat-induced mortality rates were projected to be similar during hot and cool summers.

Infectious diseases

Approximately 25 million people suffer from onchocerciasis, which leads to painful skin lesions and often blindness. It has been responsible for huge population movements in west Africa, where villages near rivers have long been vulnerable to large blackfly (*Simulium* spp.) populations. The blackfly transmits the disease to humans through bites that inject worm larvae (microfilaria) into the body (WHO, 1985). The onchocerciasis evaluation has commenced, and 15 study sites have been selected in Benin, Togo, Ghana, Senegal, Mali, and Burkina Faso. Since onchocerciasis is a chronic disease resulting from repeated bites of infected blackflies, the study design is based on the blackfly vector population rather than humans afflicted with the disease. The goal of the onchocerciasis project is to determine the climate-limiting factors of the vector for the purpose of estimating future blackfly populations and determining potential shifts in the range of the insect under the various global warming scenarios. Continuous measurements of blackfly populations have been made (usually twice weekly) at the study sites (Figure 12).

The study sites represent numerous environments in west Africa, which is important because there are several subspecies of the vector. The savanna subspecies is most closely associated with the most dangerous aspects of the disease, such as blindness (WHO, 1985). The sites in Mali, Benin, and Burkina Faso are associated with the savanna subspecies, whereas the others are primarily forest sites. At least one high-quality meteorological station has been selected, with the assistance of the Burkina Faso National Meteorological Office, near each group of insect population sites. No meteorological station is more than 50 km from an insect site. In addition, since the reproduction of the blackfly is largely dependent upon streamflow, gauging stations have also been selected adjacent to the insect sites.

The insect data are being subdivided into categories to determine the potential of the insects to spread onchocerciasis. The most basic category is total number of females captured (only the female blackfly bites humans). Most captured females are dissected to obtain data on number of parous (egg-laying) females, number infected with the microfilaria that are associated with the disease, number of females with microfilaria in the head region (where they can be transmitted to humans through a bite), and number of microfilaria found in the head region. It is necessary to separate microfilaria counts from insect counts, as it is possible that climatic-limiting factors for the vector and the infective agent are different.

A climatic water budget approach, which partitions the use of available precipitation within the environment, provides streamflow estimates and will be one of the necessary methodologies for determining insect/climate relationships. Various water budget procedures are available that can be applied directly to GCM data, such as the Thornthwaite/Mather water budget (Mather, 1978). However, a synoptic evaluation or some other approach that evaluates thermal and other atmospheric variables will also be necessary to determine climatic-limiting factors that are not directly related to streamflow.

The malaria project is less well-defined at this time, although data sources for west and central Africa have been identified by NIAID, which is co-researching the infectious disease project. It appears that the dependent variable in the malaria study will be the measurement of infection within the mosquito, and it is possible that a sporozoitic index will be used.

POLICY IMPLICATIONS

The alarming estimates of increased heat-related mortality and the potential spread of important vector-borne diseases under $2 \times CO_2$ conditions suggest that anticipatory responses should be considered to mitigate these problems. The following are possible mitigating actions:

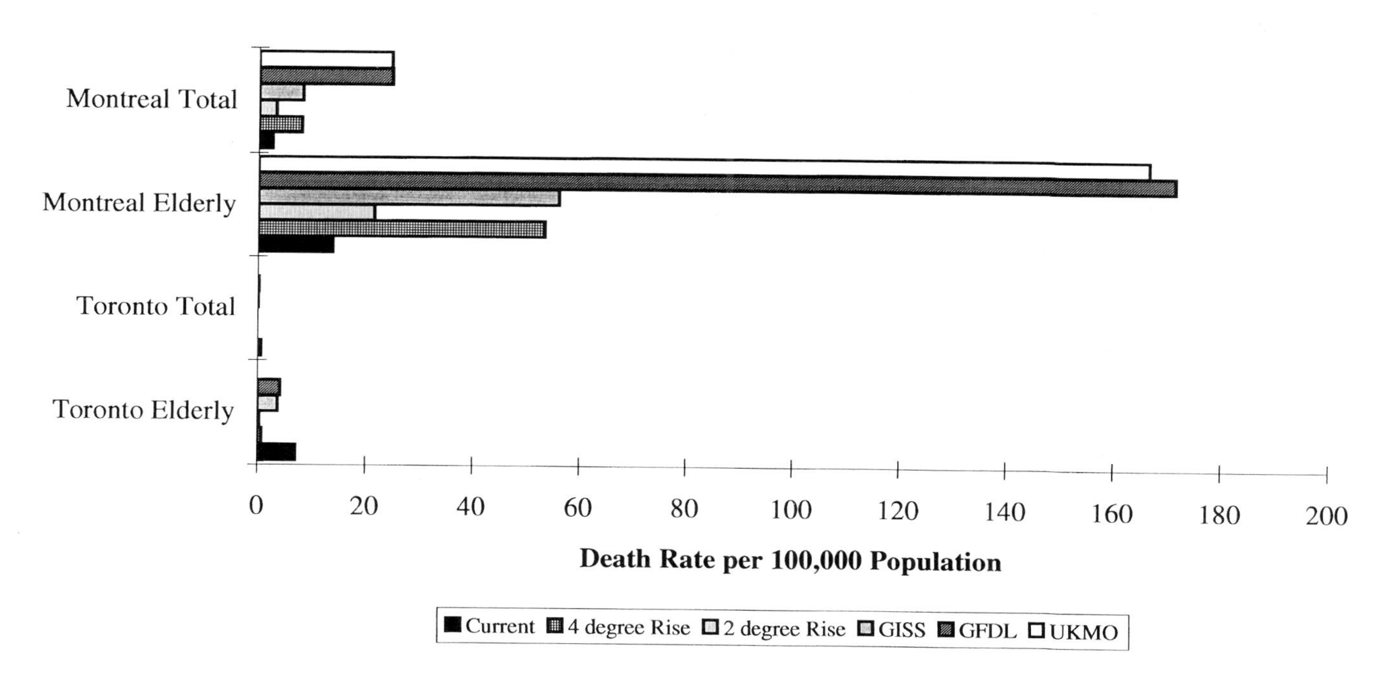

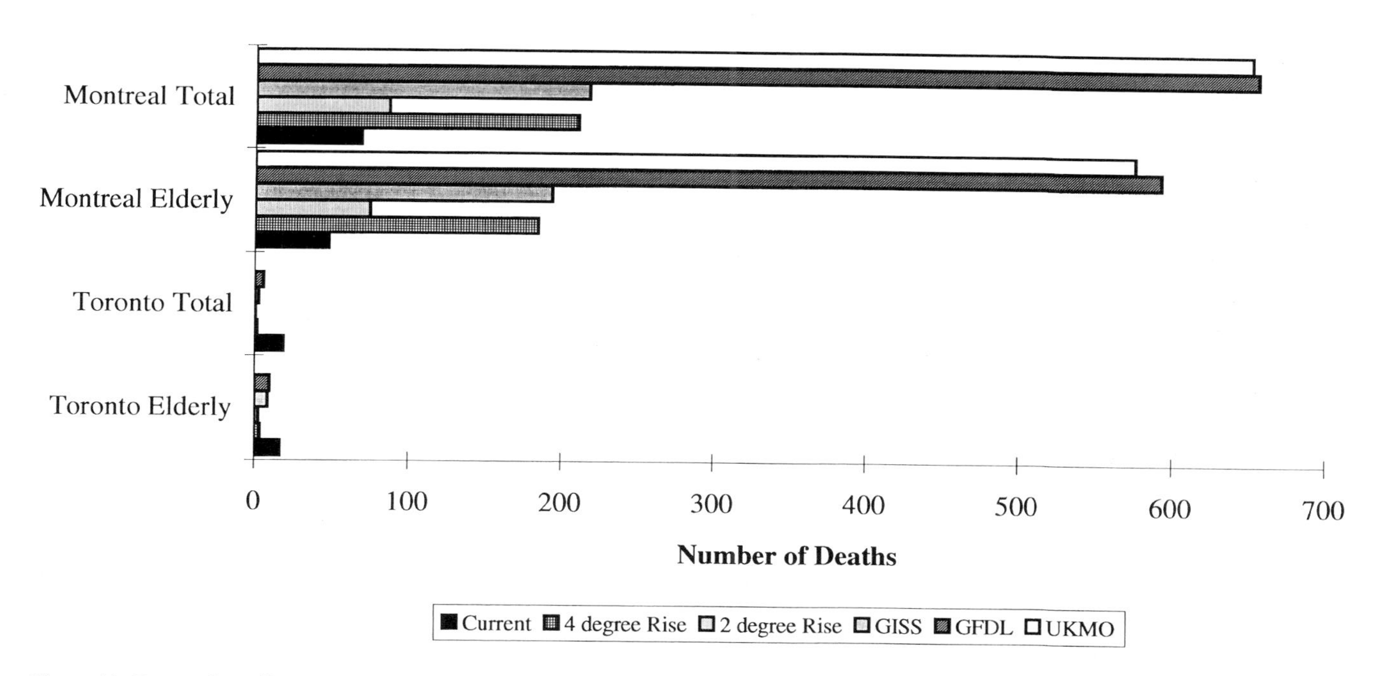

Figure 10 Future Canadian mortality estimates for summer due to weather: threshold approach.

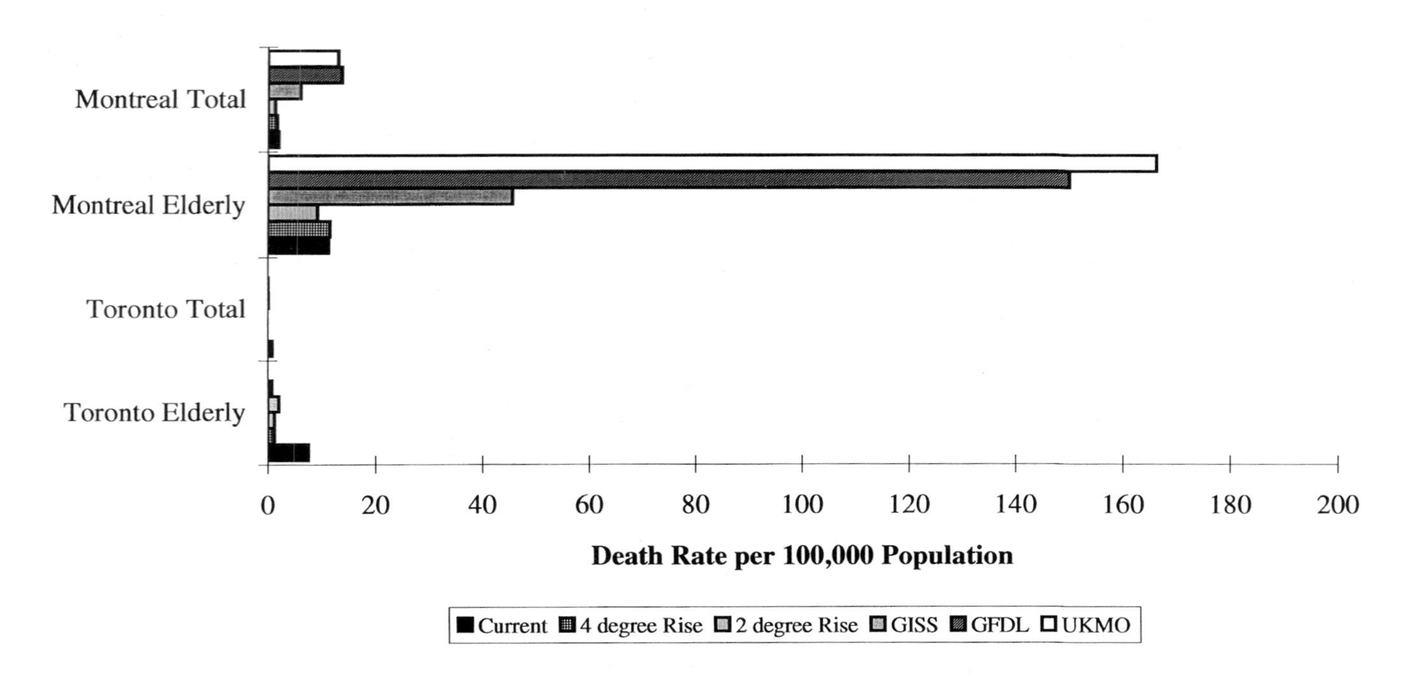

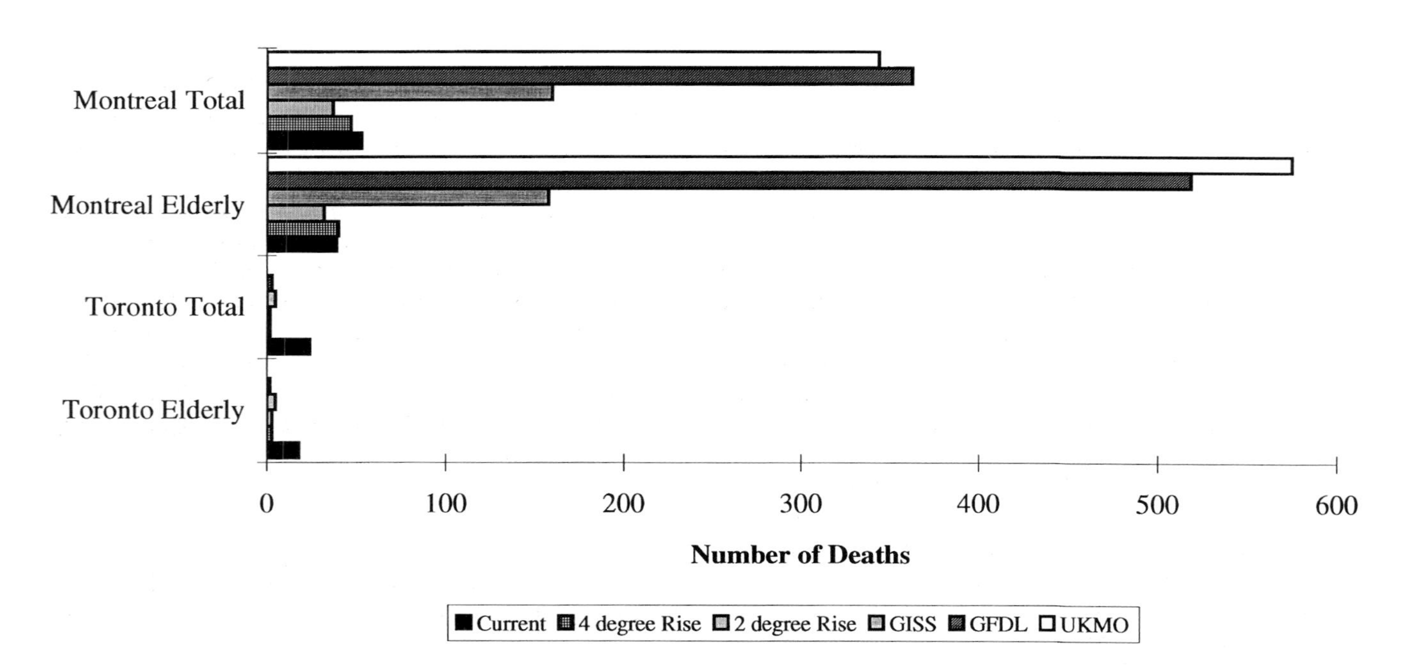

Figure 11 Future Canadian mortality estimates for summer due to weather: synoptic approach.

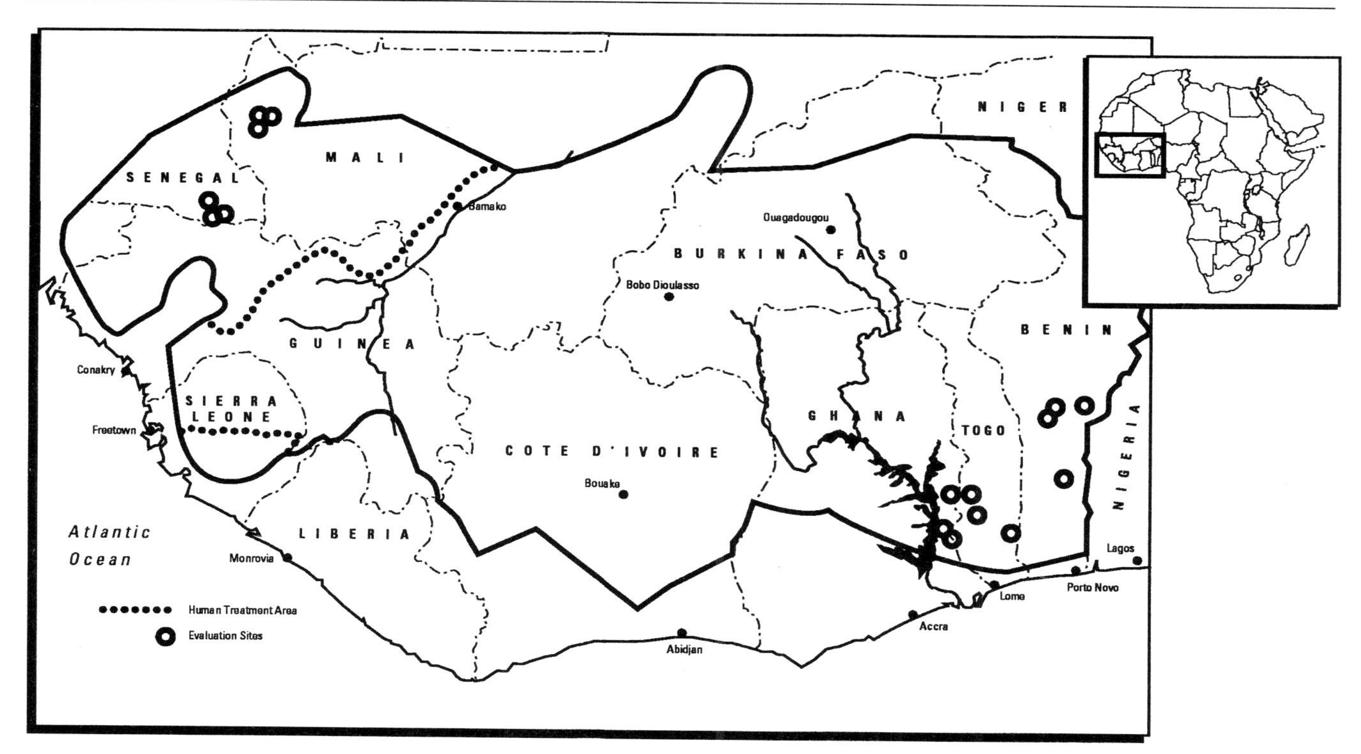

Source: World Health Organization

Figure 12 Onchocerciasis study area.

- Development of a weather/health watch–warning system. People are often unaware that dangerous weather conditions exist, which might contribute to heightened mortality rates. A watch–warning system is necessary (much like the systems presently available to warn of hurricanes and severe snowstorms) to advise people when stressful weather conditions are imminent. A cooperative effort is already under way between the Centers for Disease Control and the state of Missouri to develop such a system for that state. Simple weather stress algorithms could be developed for such population centers as Cairo, Toronto, and Shanghai.
- Improvement of public health procedures. With the potential of certain vector-borne diseases spreading into new areas, an effort should be made to provide medicines, immunization programs, and more intensive care to affected individuals, as diseases such as onchocerciasis and malaria are curable or controllable.
- Improved vector control. An integrated pest management program of insect control, which includes pesticide application, should be designed for malaria to lessen the populations of disease-carrying organisms and to promote vector eradication techniques. Such a program has been developed for onchocerciasis in several west African countries and has led to the resettlement of lands formerly uninhabitable due to the existence of infected blackflies (WHO, 1985). However, the onchocerciasis program must be continued beyond its 10-year life span (presently scheduled to end in 1995), as blackflies will migrate to previously treated areas once pesticide application has ceased. In addition, a monitoring program is necessary to determine any negative environmental impacts from long-term pesticide application.
- Improved surveillance systems. At present, data on the incidence and spread of malaria are especially poor in developing countries of Africa and Latin America. Improved surveillance programs are necessary to determine the potential for these diseases to spread from sources of infection. The success of the Onchocerciasis Control Program, which is centered on intense surveillance of the blackfly, underscores the need for such a program to help control malaria.

RECOMMENDATIONS FOR FUTURE RESEARCH

There are numerous uncertainties surrounding the estimates presented in this report. For example, how might pollution interact with weather to alter heat-stress-induced mortality? Can an acclimatization procedure be developed that describes more accurately the potential social and infrastructure changes that would be associated with global warming? How will changes in the range of important vector-borne diseases contribute to human migration patterns? The following studies are suggested to address these and other unanswered questions:

- The potential impact of pollution on heat-stress-related mortality in Egypt and China. For both of these developing countries, it has been assumed that changes in mortality are dependent upon climate alone. It is quite likely that mortality is very sensitive to the synergistic impacts of climate and pollution concentration. A present study funded for the United States will evaluate this interaction in detail, and Environment Canada is considering funding a similar study for that country.
- The impact of acclimatization on heat-stress-related mortality. Results here suggest that humans might acclimatize to increased heat, especially in Canada and the United States. However, acclimatization impacts in developing countries are more nebulous, and social scientists from those countries should be included in a future study to determine potential infrastructure changes that could alter the mortality results.
- The impact of infectious disease spread on demographics. This study would concentrate on migration patterns, changes in population age distributions, and other demographic impacts that may occur if certain infectious diseases migrate to other developed or developing nations.
- The impact of global warming on other infectious diseases. The present study concentrated on only two diseases, but other vector-borne ailments, such as yellow fever, dengue fever, schistosomiasis, and trypanosomiasis, kill millions of people in developing nations and may expand in range if the globe warms. With the success of climate/disease models for onchocerciasis and malaria, an effort should be made to uncover climate/disease relationships for other equally dangerous infectious diseases. In addition, the National Academy of Science's Institute of Medicine recently published a report on emerging infections that are becoming increasingly important within the United States (Institute of Medicine, 1992). Study of the potential impact of climate change on these emerging infections may lessen their prevalence in decades to come.

ACKNOWLEDGMENTS

We would like to thank a number of people who contributed significantly to this paper. University of Delaware graduate students Hengchun Ye, David Barthel, Karen Smoyer, and Shouquan Cheng deserve special thanks for their technical support and assistance in manipulating very large data sets. Dr. Karl Western, National Institutes of Health, and Joel Smith, RCG/Hagler, Bailly, Inc., reviewed the manuscript and provided numerous helpful suggestions. Joan Hahn developed the graphics and tables, and offered considerable moral support to assure the successful completion of this work. This research was supported by a cooperative agreement (CR817693) with the Climate Change Division of the U.S. Environmental Protection Agency, and I greatly appreciate the continued enthusiasm of that agency about climate change/human health issues.

REFERENCES

Applegate, W. B., J. W. Runyan, Jr., L. Brasfield, M. L. Williams, C. Konigsverg, and C. Fouche. 1981. Analysis of the 1980 heat wave in Memphis. *Journal of the American Geriatrics Society* 29:337–42.

Born, W., M. P. Happ, A. Dallas, C. Reardon, R. Kubo, T. Shinnick, P. Brennan, and R. O'Brien. 1990. Recognition of heat shock proteins and cell functions. *Immunology Today* 11:40–3.

Box, G. E. P., and J. Wetz. 1973. *Criteria for Judging Adequacy of Estimation by an Approximating Response Function*. Technical Report No. 9. Madison, Wisc.: University of Wisconsin Statistics Department.

Dobson, A., and R. Carper. 1988. Global warming and potential changes in host–parasite and disease–vector relationships. *Proceedings of the Conference on the Consequences of Global Warming for Biodiversity*. New Haven, Conn.: Yale University Press.

Draper, N., and H. Smith. 1981. *Applied Regression Analysis*, 2d Edition. New York: John Wiley.

Electric Power Research Institute (EPRI). 1991. *Reviews of the Intergovernmental Panel on Climate Change Working Group Reports*. Palo Alto, Calif.: EPRI.

Ewan, C., E. Bryant, and D. Calvert. 1990. *Health Implications of Long Term Climate Change, Vol. 1, Effects and Responses*. Canberra: Australian National Health and Medical Research Council.

Haile, D. G. 1989. Computer simulation of the effects of changes in weather patterns on vector-borne disease transmission. In *The Potential Effects of Global Climate Change on the United States: Appendix G – Health*, eds. J. B. Smith and D. A. Tirpak, 3–1 to 3–48. Washington, DC: U.S. Environmental Protection Agency.

Institute of Medicine. 1992. *Emerging Infections: Microbial Threats to Health in the United States*. Special report edited by J. Lederberg, R. E. Shope, and S. C. Oaks, Jr. Washington, DC: National Academy Press.

Jones, T. S., *et al.* 1982. Morbidity and mortality associated with the July 1980 heat wave in St. Louis and Kansas City. *Journal of the American Medical Association* 247:3327–30.

Kalkstein, L. S. 1988. The impacts of predicted climate change on human mortality. *Publications in Climatology* 41:1–127.

Kalkstein, L. S. 1989a. The impact of CO_2 and trace gas-induced climate changes upon human mortality. In *The Potential Effects of Global Climate Change on the United States: Appendix G – Health*, eds. J. B. Smith and D. A. Tirpak, 1–1 to 1–34. Washington, DC: U.S. Environmental Protection Agency.

Kalkstein, L. S. 1989b. Global Climate Change and Human Health: What Are the Potential Impacts? *Medical and Health Annual, Encyclopedia Britannica*, 118–33.

Kalkstein, L. S. 1990. A proposed global warming/health initiative. *Environmental Impact Assessment Review* 10:383–92.

Kalkstein, L. S. 1991a. A new approach to evaluate the impact of climate upon human mortality. *Environmental Health Perspectives* 96:145–50.

Kalkstein, L. S. 1991b. The potential impact of global warming: climate change and human mortality. In *Global Warming and Human Mortality*, ed. R. L. Wyman, 216–23. New York: Routledge, Chapman and Hall.

Kalkstein, L. S., G. Tan, and J. Skindlov. 1987. An evaluation of objective clustering procedures for use in synoptic climatological classification. *Journal of Climate and Applied Meteorology* 26:717–30.

Kalkstein, L. S., and K. M. Valimont. 1987. Climate Effects on Human Health. In *Potential Effects of Future Climate Changes on Forests and Vegetation, Agriculture, Water Resources, and Human Health*, ed. D. Tirpak, 122–52. EPA Science and Advisory Committee Monograph #2538.

Kalkstein, L. S., and R. E. Davis. 1989. Weather and human mortality: An evaluation of demographic and inter-regional responses in the United States. *Annals of the Association of American Geographers* 79:44–64.

Kalkstein, L. S., and S. H. Giannini. 1991. Global Warming and Human Health: Does a Real Threat Exist? *The New Biologist* 3(8):727–8.
Mather, J. R. 1978. *The Climatic Water Budget in Environmental Analysis*, 239. Lexington, Mass.: Lexington Books, D. C. Heath & Co.
Perry, A. 1983. Growth points in synoptic climatology. *Progress in Physical Geography* 7:91–6.
Rotton, J. 1983. Angry, sad, happy? Blame the weather. *U.S. News and World Report* 95:52–3.
Schwartz, J., and D. Dockery. 1992. Particulate Air Pollution and Daily Mortality in Steubenville, Ohio. *American Journal of Epidemiology* 135(1):12–19.
Smith, J. B., and D. A. Tirpak, eds. 1989. *The Potential Effects of Global Climate Change on the United States*. Report to Congress. EPA 230-05-89-050. Washington, DC: U.S. Environmental Protection Agency, Office of Policy, Planning, and Evaluation.
Tegart, W. J. McG., G. W. Sheldon, and D. C. Griffiths. 1990. *Climate Change: The IPCC Impacts Assessment*. Special report of the Intergovernmental Panel on Climate Change. Canberra: Australian Government Publishing Service.
White, M. R., and I. Hertz-Picciotto. 1985. Human health: Analysis of climate related to health. In *Characterization of Information Requirements for Studies of CO_2 Effects: Water Resources, Agriculture, Fisheries, Forests and Human Health*, ed. M. R. White, 172–205. Washington, DC: Department of Energy.
World Health Organization (WHO). 1985. Review of the work of the Onchocerciasis Control Programme in the Volta River Basin Area from 1974 to 1984. In *Ten Years of Onchocerciasis Control in West Africa*. Geneva: WHO.
World Health Organization (WHO). 1990. *Potential health effects of climatic change*. Report of a WHO Task Group. Geneva: WHO.
World Meteorological Organization (WMO). 1986. *Urban climatology and its applications with special regard to tropical areas*. WMO No. 652.

6 Global Forests

T. M. SMITH, P. N. HALPIN, and H. H. SHUGART
Department of Environmental Sciences
University of Virginia

C. M. SECRETT
International Institute for Environment and Development

INTRODUCTION

Potential changes in the Earth's climate as a result of rising atmospheric concentrations of greenhouse gases would have a major impact on the current distribution and abundance of forest ecosystems. Forest ecosystems represent an important resource not only for timber, fuelwood production, and recreation, but as centers of biodiversity. It is estimated that tropical rainforests are home to more than two-thirds of the estimated 30 to 50 million species on Earth (Erwin 1988; Lugo 1988). In addition, forests are by far the largest component of terrestrial carbon storage, currently covering 38 percent of the land surface and containing 58 percent of the estimated total terrestrial carbon.

Consideration of the potential impacts of global climate change on forest ecosystems must also take into account changing patterns of human land use. Current fragmentation of forest cover will influence the response of forests to changing climate patterns, presenting barriers to long-term forest migration.

Recent studies have demonstrated the sensitivity of forest ecosystems to climate disturbance at various spatial scales ranging from site-specific (see, e.g., Solomon 1986; Bonan *et al.* 1990; Smith *et al.* 1990, 1992a; Urban and Shugart 1989) to global (Emanuel *et al.* 1985; Smith *et al.* 1990, 1992a,b). In addition, paleo-ecological studies have shown the long-term response of forest distribution to past changes in global climate patterns (see, e.g., Davis 1976; Peteet 1987; Webb 1987). However, no comprehensive study has been undertaken to address the potential impacts of climate change on forested systems at regional and global scales under an array of possible climate change scenarios. In recognition of this need for a comprehensive study, the Global Systems Analysis Program (GSAP) at the Department of Environmental Sciences of the University of Virginia, in cooperation with the United States Environmental Protection Agency, initiated this assessment of the implications of climate change for global forests.

OBJECTIVES AND APPROACH

The project had three main objectives: (1) to estimate the impacts of a climate change on the distribution of global vegetation, (2) to more specifically estimate the environmental impacts of a climate change on boreal and tropical forest systems, and (3) to identify and evaluate adaptive strategies for reducing the vulnerability of these forested ecosystems to climate change. We focused on these two forest zones because they represent contrasting cases of the interaction of human and biological constraints on the potential response of forest ecosystems to climate change. The boreal forest zone is the major source of softwood production and represents a zone of minimal human impact in terms of urban and agricultural land use at a global scale. In contrast, the tropical mesic forests of the world are declining at an annual rate of 1–3.5 percent as a result of conflicting land-use pressures for urban and agricultural development. The current global distribution of tropical forests represents less than 50 percent of its potential distribution. The interaction between land-use pressures and possible climate change presents a dual threat to remaining forest cover, with potentially severe consequences on patterns of biodiversity and possible feedback to the regional and global climate system (Shukla and Mintz 1982; Sellers 1987).

To meet the above objectives we evaluated the response of forests at two scales: (1) the boreal and tropical forest zones in the context of the global distribution of major terrestrial ecosystems, and (2) regional analyses based on case studies in specific countries. The global analyses provided an assessment of potential changes in the boreal and tropical forest zones within a global context, including spatial changes in the global distribution of major biome types (e.g., grassland or tundra to forest). The regional analyses allowed for the evaluation of transient dynamics associated with these transitions, as well as estimates of changes in forest composition, productivity, and diversity. The results of these analyses were then used to evaluate potential impacts and adaptive responses for timber and fuelwood production, protection of natural areas (i.e., biodiversity), and the global carbon cycle.

The project was international in scope, involving the collaborative efforts of scientists and institutions on five continents.

GENERAL DESCRIPTION OF MODELS —

A wide range of methodologies was employed, but they can be generally classified into three groups: (1) bioclimatic models, which correlate the distribution of vegetation with climate patterns, (2) physical/biophysical models of evapotranspiration, and (3) demographic models that simulate the dynamics of forest stands (ca. 1 hectare) and are able to track changes in species composition and forest structure. We will first discuss these three classes of models in general terms and will later refer to their specific applications in sections relating to global and regional analyses.

Climate–vegetation classification

Perhaps the simplest models for relating vegetation pattern to climate at a global scale are climate–vegetation classification models. These bioclimatic models (von Humbolt 1867; Grisebach 1838; Koppen 1900, 1918, 1936; Thornthwaite 1931, 1933, 1948; Holdridge 1949, 1959, 1967; Troll and Paffen 1964; Box 1978, 1981; Prentice 1990) relate the distribution of vegetation (e.g., biomes) or plant types to biologically important features of the climate (e.g., temperature and precipitation) to define *eco-climatic zones*. The eco-climatic zones therefore define the climate conditions associated with a given vegetation or ecosystem type (e.g., boreal forest). These models are essentially climate classifications defined by the large-scale distribution of vegetation. Bioclimatic classification models have a history of application in simulating the distribution of vegetation under changed climate conditions, both for past climatic conditions associated with the last glacial maximum (Manabe and Stouffer 1980; Hansen *et al.* 1984; Prentice and Fung 1990) and for predictions of future climate patterns under conditions of doubled CO_2 (Emanuel *et al.* 1985; Prentice and Fung 1990; Smith *et al.* 1992a,b).

HOLDRIDGE LIFE ZONE CLASSIFICATION

The bioclimatic model we chose to focus on for our global analyses was the Holdridge Life Zone Classification (Holdridge 1967). The features of the Holdridge Classification are summarized in Figure 1 (Plate 8). The life zones are depicted by a series of hexagons formed by intersecting intervals of climate variables on logarithmic axes in a triangular coordinate system. Two variables, average biotemperature and average annual precipitation, determine the classification. Average biotemperature is the average temperature over a year with the unit temperature values (i.e., daily, weekly, or monthly) that are used in computing the average set to 0°C if they are less than or equal to 0°C. In the Holdridge Diagram (Figure 1, Plate 8)), identical axes for average annual precipitation form two sides of an equilateral triangle. A logarithmic axis for the potential evapotranspiration (PET) ratio (effective humidity) forms the third side of the triangle, and an axis for mean annual biotemperature is oriented perpendicular to the base. By marking equal intervals on these logarithmic axes, hexagons are formed that designate the Holdridge Life Zones. Each life zone is named to indicate a vegetation association and represents the climate conditions or eco-climatic zone associated with that ecosystem or vegetation type. The terms *life zone* and *eco-climatic zone* will therefore be used synonymously throughout the discussion of results.

Potential evapotranspiration is the amount of water that would be released to the atmosphere under conditions of sufficient but not excessive available water throughout the growing season. The potential evapotranspiration ratio is the quotient of PET and average annual precipitation. Holdridge (1959) assumed, on the basis of studies of several ecosystems, that PET is proportional to biotemperature. The PET ratio in the Holdridge Classification is therefore dependent on the two primary variables, annual precipitation and biotemperature.

One additional division in the Holdridge System is based on the occurrence of killing frost. This division is along a critical temperature line that divides hexagons between 12°C and 24°C into warm temperate and subtropical zones. This line is adjusted to reflect regional conditions. The complete Holdridge Classification at this level includes 37 life zones.

The Holdridge model, like all bioclimatic models, is essentially a correlation between climate and vegetation patterns. Although it provides a convenient description of the association between features of the vegetation and regional climate, the model implicitly assumes that vegetation patterns are effectively at equilibrium with current climate patterns. Although this may be an essentially valid assumption under current conditions, it presents a problem in the application of this class of models to the climate change scenarios investigated in this study. In many cases, the time scale associated with the changes in climate (i.e., 70–100 years) is much faster than the processes required for vegetation to respond (e.g., long-distance migration of species) to these new patterns. Therefore, we restricted our use of this model to investigate the implications of the climate change scenarios on the distribution of eco-climatic zones associated with the vegetation or ecosystem types. *This method provided a means of evaluating the potential for a geographical area to sustain a particular vegetation type, not for evaluating whether a vegetation type would occur there within the time frame associated with the climate change scenario.* We could then examine the processes and possible time frames associated with the implied transitions. Despite this restrictive use, the methodology provided a valuable tool for assessing the implications of climate change on the current distribution of vegetation and the potential for change in vegetation patterns.

Models of evapotranspiration

Evapotranspiration is the flux of water from a terrestrial surface resulting from both evaporation and transpiration. A number of models have been developed to estimate evapotranspiration. In general these models vary in the degree of complexity with which transpiration is estimated. For the purposes of this study we chose to use the model developed by Priestley and Taylor (1972).

The Priestley–Taylor (Priestley and Taylor 1972) model of evapotranspiration differs from more biophysically complex models (e.g., the Penman–Monteith model [Penman 1948; Monteith 1973]) in its simplifying assumptions regarding the process of transpiration. In the present analyses the model was applied on a daily basis, using daily temperature, irradiance, and precipitation values to provide daily estimates of soil water deficits (soil texture and available water capacity values were taken from a Food and Agriculture Organization [FAO] soils database). The Priestley–Taylor model is very similar to the Penman–Monteith model, but it employs a simplifying assumption that allows regional scale estimates. The assumption is that given adequate soil moisture supply, atmospheric humidity and evaporation approach equilibrium over relatively large areas. This equilibrium is independent of wind speed and canopy properties and is controlled primarily by net radiation.

The Priestley–Taylor model is used for estimating both potential (PET) and actual (AET) patterns of evapotranspiration, reported on both a monthly and an annual basis. The ratio of AET/PET represents an index of plant moisture stress (Cramer and Prentice 1992), reflecting the ratio of supply to demand.

Demographic process models

The use of vegetation–climate classification models to evaluate plant response to climate change implicitly assumes a time scale sufficient for vegetation migration and eventual equilibrium of vegetation to the new, 'changed' climate patterns. However, simulating the temporal response of vegetation to changing climate conditions requires the explicit consideration of plant demographic processes. There are numerous models of vegetation dynamics that simulate the demographics of plant populations (Shugart and West 1980). One such class of demographic process models is 'gap models.' Forest gap models simulate the establishment, growth, and mortality of individual plants on a plot scaled to the maximum size of the plant species being simulated (Shugart 1984). These models have been developed for a wide range of forest and grassland ecosystems. Although the models differ in their inclusion of processes that may be important in the dynamics of the particular site being simulated (e.g., hurricane disturbance, flooding), all gap models share a common set of characteristics and demographic processes.

Each individual plant is modeled as a unique entity with respect to the processes of establishment, growth, and mortality. This allows the model to track species- and size-specific demographic behaviors. The model structure includes two features important to a dynamic description of vegetation: (1) the response of the individual plant to the prevailing environmental conditions, and (2) the way the individual plant modifies those environmental conditions (i.e., the feedback between vegetation structure/composition and the environment). Gap models have been applied to examine the response of forested systems to climate changes, both to reconstruct prehistoric Quaternary forests (Solomon *et al.* 1980, 1981; Solomon and Shugart 1984; Solomon and Webb 1985; Bonan and Hayden 1990; Bonan *et al.* 1990) and to project possible consequences of future climate change (Solomon *et al.* 1984; Solomon 1986; Pastor and Post 1988; Bonan *et al.* 1990; Urban and Shugart 1989; Overpeck *et al.* 1990; Smith *et al.* 1992a).

In contrast to the two modeling approaches discussed in the earlier sections, the gap model approach is a high resolution model in that it can predict species composition, vegetation structure and associated productivity, and standing biomass through time; however, it is limited in spatial scale. That is, it is limited in the spatial extent to which the results can be extrapolated. The reason for this limitation is that the information required to parameterize/initialize a model that can address changes in these features of the vegetation (e.g., *species* composition and productivity through time) relates to site-specific features such as topographic position, soil characteristics, land-use history and disturbance, and present vegetation structure, all of which may vary over short distances.

GLOBAL ANALYSIS

Global distribution of vegetation

The expected current distributions of Holdridge Life Zones (eco-climatic zones) were mapped using a climate data base of mean monthly precipitation and temperature at a 0.5° × 0.5° (latitude and longitude) resolution (Leemans and Cramer 1990). Simulations of current (i.e., $1 \times CO_2$) and $2 \times CO_2$climates from four GCMs (Table 1 in Chapter 1; see this chapter also for further description of GCM scenarios) were used to construct climate change scenarios. Changes in mean monthly precipitation and temperature were calculated for each GCM scenario for each computational grid element by taking the difference between simulated current and $2 \times CO_2$ climates. These data from each GCM were interpolated to 0.5° resolution, and changes in monthly precipitation and temperature were then applied to the global climate data base to provide a change scenario. The altered data bases corresponding to each of the four GCM scenarios were then used to reclassify the grid cells (0.5°) using the Holdridge Life Zone Classification.

Maps of the distribution of life zones (eco-climatic zones) under current climate and climate change scenarios based on the four GCMs are presented in Figure 2(a–e) (Plate 9). All four climate change scenarios investigated showed a significant shift in the distribution of eco-climatic zones (i.e., life zones). Changes in the areal coverage of major biome types associated with these climates are shown in Table 1. These biome types are further aggregates of the Holdridge Life Zones presented in Figure 1 (Plate 8), and the zones comprising the types are defined in Table 1.

There was a general qualitative agreement among scenarios in the directions of change. The extent of tundra and desert decreased, and that of grasslands and forests increased under all four scenarios. Despite the agreement in increased forest cover, the scenarios differed in the degree to which the increase was attributable to mesic (i.e., moist or wet) and dry forest components. Mesic forest cover increased under the GISS and OSU scenarios but decreased in the GFDL and UKMO scenarios. These predicted decreases in mesic forest by GFDL and UKMO were offset by larger increases in dry forest. Thus forest cover increased overall.

The changes in coverage of the biome types presented in Table 1 were the outcome of a dynamic process of spatial changes in the climate pattern and associated spatial changes in the distribution of life zones. These spatial dynamics can be described as a matrix of transitions between biome types (Table 2). Rows of the matrix show transitions from the designated biome type (i.e., aggregated life zone) to the type specified in the column headings. The diagonal elements show the area occupied by the biome type under current climate, which does not change (to another biome type) under the new climate conditions. Therefore, the sum of the elements in each row is the current coverage for that biome type, while the sum of the elements in each column is the coverage for that type under the changed climate conditions.

Table 1. *Changes in the areal coverage ($km^2 \times 10^4$) of major biome-types* under current and changed climate conditions*

	Current	OSU	GFDL	GISS	UKMO
Tundra	939	−302	−515	−314	−573
Desert	3699	−619	−630	−962	−980
Grassland	1923	380	969	694	810
Dry forest	1816	4	608	487	1296
Mesic forest	5172	561	−402	120	−519

Notes:

* Tundra: polar dry tundra, polar moist tundra, polar wet tundra, polar rain tundra

Desert: polar desert, boreal desert, cool temperate desert, warm temperate desert, subtropical desert, subtropical desert bush, tropical desert, tropical desert bush

Grassland: cool temperate steppe, warm temperate thorn steppe, subtropical thorn steppe, tropical thorn steppe, tropical very dry forest

Dry forest: warm temperate dry forest, subtropical dry forest, tropical dry forest

Mesic forest: moist, wet and rain forest for boreal, cool temperate, warm temperate, subtropical and tropical temperature zones

The decline in tundra observed under all scenarios was primarily due to a shift from tundra to mesic forest. This transition was a result of the warming at higher latitudes and the subsequent northward movement of boreal forest zone into the areas now occupied by wet tundra. A second major vector of change in the tundra region was the desertification of areas where warming and/or decreases in precipitation occur.

The decrease in the global extent of desert seen in all four scenarios was a function of the shift from desert to tundra in the higher latitudes and from desert to grassland in the temperate and tropical regions. Furthermore, there was a significant conversion from desert to mesic forest under the GFDL and UKMO scenarios. These shifts occurred in the northern latitudes where cold desert/dry tundra zones increased in both temperature and precipitation.

The increased cover of grassland under all scenarios was a function of both shifts from desert to grassland with increased precipitation in areas of the temperate and tropical

Table 2. *Transition matrices for the four GCM-based climate change scenarios. Matrices show the changes in coverage between biome types. Values are in km² × 10⁴ (T= Tundra; D= Desert; G= Grassland; DF= Dry Forest; MF= Mesic Forest)*

	To: T	D	G	DF	MF	Total
OSU						
From: T	189.3	111.3			638.3	938.6
D	447.9	2869.6	372.7		8.9	3699.1
G		60.0	1565.2	295.8	2.1	1923.1
DF			59.3	1282.3	474.1	1815.7
MF		14.6	306.1	242.3	4609.2	5172.2
Total	637.2	3055.5	2303.3	1820.4	5732.6	
GFDL						
From: T	18.5	177.6	29.2	0.2	713.3	938.8
D	405.2	2784.1	337.8	0.3	171.8	3699.2
G		67.9	1663.4	181.0	10.8	1923.1
DF			141.8	1490.8	183.2	1815.8
MF		9.2	720.6	751.9	3690.6	5172.3
Total	423.7	3038.8	2892.8	2424.2	4769.7	
GISS						
From: T	167.4	114.5	2.1		655.0	939.0
D	456.7	2559.5	617.0	5.4	60.5	3699.1
G		33.8	1645.3	198.2	45.7	1923.0
DF		152.9	1427.2	235.6		1815.7
MF		5.3	200.0	672.0	4295.2	5172.5
Total	624.1	2713.1	2617.3	2302.8	5292.0	
UKMO						
From: T	20.5	110.3	35.3	1.2	771.6	938.9
D	343.9	2505.1	566.9	4.1	278.7	3698.7
G	0.5	37.8	1520.4	353.8	10.5	1923.0
DF	0.3	5.8	211.7	1374.7	223.3	1815.8
MF	0.3	25.7	399.0	1378.3	3369.0	5172.3
Total	365.5	2684.7	2733.3	3112.1	4653.1	

regions and the transition of dry and mesic forests to grassland as a result of drying in all forested regions.

The extent of dry forest increased with increasing precipitation in grassland regions and with increased temperatures and/or decreased precipitation in mesic forests. The latter transition occurred primarily in the subtropical and tropical regions. This transition was the largest in the UKMO scenario, resulting in a doubling of the global extent of dry forest.

The major transition toward mesic forest was the shift from tundra to boreal forest discussed earlier. This transition was followed in importance by the shift from dry to mesic forest, primarily in the subtropical and tropical regions. The extent of the transition from tundra to boreal forest was similar (ranging from 638.3 × 10³ to 771.6 × 10³ km²) for the four scenarios. The major difference between the scenarios in the predicted areal coverage of mesic forest was the degree of mesic forest decline associated with drying in the subtropical and tropical regions (i.e., the shift to dry forest discussed earlier).

Changes in areal coverage of the five major mesic forest types defining the mesic forest biome type are presented in Figure 3. The types are defined by summing over the moist, wet, and rainforest life zones for each temperature zone in the Holdridge Life Zone Classification (Figure 1). The histograms present the areas for each forest type under current climate and the four climate change scenarios.

In general, there was a decrease in the area associated with the boreal, warm temperate, and subtropical forest types (i.e., eco-climatic zones). In contrast, there was an increase in the area associated with cool temperate and tropical forest. The largest increase was in the eco-climatic zone associated with tropical forest. The areal cover of the tropical forest zone increased dramatically under all scenarios, ranging from 87 percent for the UKMO scenario to 159 percent for OSU.

The black portion of each of the histograms in Figure 3 represents the area of that forest type that is stable (i.e., that experiences no shift in eco-climatic zone) under the given GCM scenario. Although there was a net increase in the cool temperate eco-climatic zone, there was a dramatic shift in the spatial distribution of this forest type. The percentage area of the current cool temperate forest that remained stable under the four scenarios ranged from only 19.3 percent (UKMO) to a high of 66.3 percent (OSU). In contrast, the tropical forest zone was relatively stable, ranging from 77.3 percent (OSU) to 89.8 percent (GFDL).

The histograms in Figure 3 can be interpreted in a number of ways. The difference between the land area for each forest type under current climate and the area remaining stable under the four scenarios is the area of that forest type that undergoes a shift in eco-climatic zone (Table 3). The difference between the stable land area (i.e., black portion) and the new total under the scenarios is the area that is currently in some other eco-climatic zone but is predicted to change to that forest type. If the eco-climatic zones are categorized into mesic forest and nonmesic forest, these two categories of change represent three possible classes of transition: (1) from one mesic forest type to another (e.g., boreal to cool temperate forest), (2) from a nonmesic forest biome type to mesic forest (e.g., grassland to forest), and (3) from a mesic forest to a nonmesic forest type (e.g., forest to grassland). These three classes of change can then be discussed in terms of the processes and time scales associated with the potential changes in vegetation type implied.

1 TRANSITIONS BETWEEN FOREST TYPES

The major transitions between mesic forest types are shown in Table 3. These shifts represent a poleward movement of

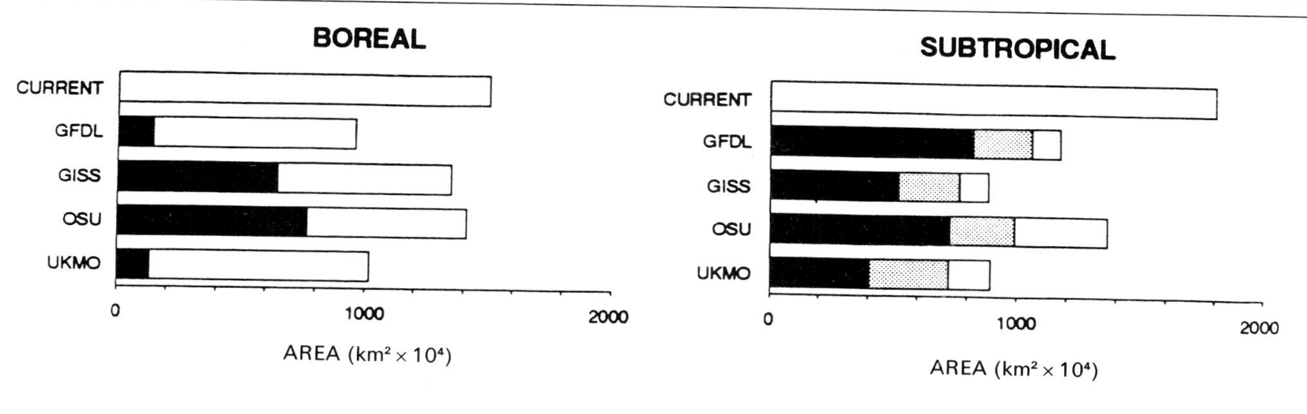

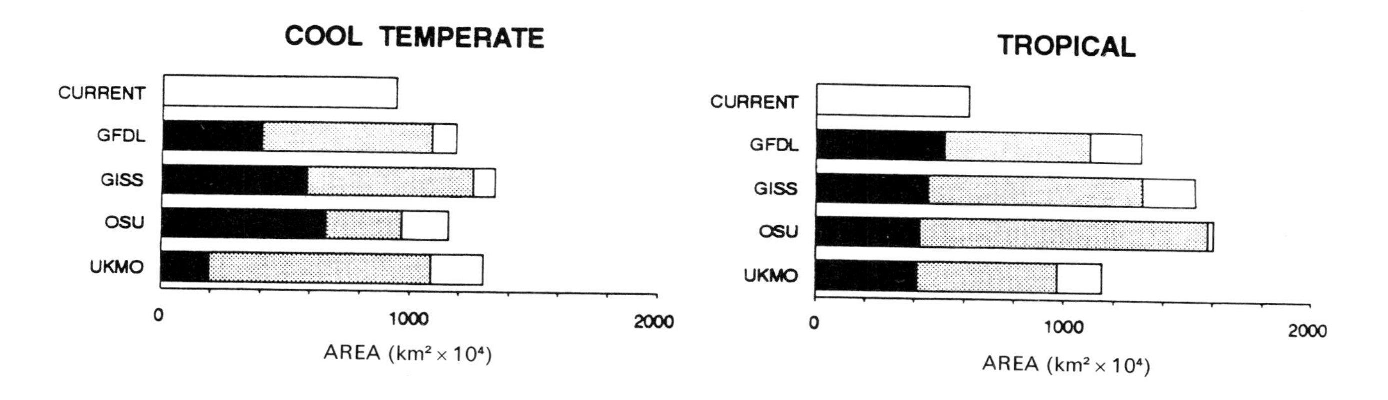

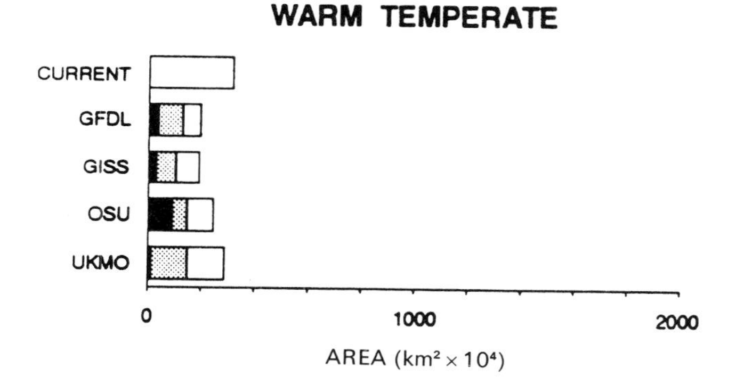

Figure 3 Areal coverage of mesic forest types under current climate and four climate change scenarios. Black portion of each histogram represents the area that is currently occupied by that forest type and remains so under the changed climate conditions (i.e., stable).

forest zones. This northward shift is due to the overall warming predicted by the GCMs and the subsequent expansion of the tropical zone as defined by biotemperature in the Holdridge model (Figure 1, Plate 8). Of the areas predicted to change to tropical forest under the climate change scenarios, 74 percent to 98 percent are currently occupied by another mesic forest type (e.g. warm temperate, subtropical).

There is a great deal of uncertainty in defining the processes and time frame associated with these shifts in forest type. Subtropical forests are generally lower in diversity than their tropical counterparts; however, there is a great deal of taxonomic overlap in these two categories of forest. The transition from subtropical to tropical forest would be limited by the ability of tropical forest species to disperse and become established in the existent subtropical forest. These tree species are long-lived, and the process of dispersal and establishment would undoubtedly require a period far beyond the time frame of the climate change scenarios (i.e., 70–100 years). In addition, the current fragmentation of forest cover in the tropics would hinder this transition by presenting a barrier to forest migration.

The transition from warm temperate to subtropical forest

Table 3. *Land area shifting eco-climatic zone from one forest type to another. Values are expressed in* $km^2 \times 10^4$

Forest Type	Scenario			
	GFDL	GISS	OSU	UKMO
Boreal				
↓	0.6	669.0	471.8	947.4
Cool temperate				
↓	161.4	161.0	141.6	292.4
Warm Temperate				
↓	229.5	232.2	208.9	262.7
Subtropical				
↓	655.7	867.8	847.6	655.7
Tropical				

is marked by the presence of broadleaf evergreen species. The biogeographic division between temperate and subtropical is related to the occurrence (and frequency) of frosts (i.e., winter temperatures below 0–5°C). Many of the subtropical species are not frost resistant, and therefore their ability to disperse northward would be dependent on the rates of warming and dispersal. Like the subtropical–tropical transition, this shift would require a long time relative to the time frame of the scenarios.

In contrast to the temperate–tropical transition, the cool temperate–warm temperate forest types are less taxonomically distinct. Many of the same species dominate these two forest types, and the transition is more subtle. This shared species pool would enable the transition to proceed more quickly, with changes in patterns of species dominance possibly occurring within the 70–100 year time frame associated with the scenarios.

The geographical transition from cool temperate to boreal forest is characterized by interspersed patches of the two forest types, with distributions within the transition zone dependent on edaphic features such as soils and topography (Pastor and Mladenoff 1992). The dominant species of the cool temperate forest (i.e., northern hardwoods) are primarily deciduous, whereas the boreal forest zone is dominated by coniferous species. Numerous climatic features have been related to the boundary between these two forest types, including the occurrence of minimum temperatures below −40°C (Arris and Eagleson 1989).

The expansion of temperate forest into the current boreal forest zone with increasing temperatures would be associated with a range of processes influencing the dispersal and establishment of northern deciduous species, including changes in patterns of nutrient cycling (Pastor and Post 1986, 1988) and rates of disturbance to existent forest (Overpeck *et al.* 1990). These processes would require a period of time longer than that outlined in the scenarios.

Table 4. *Total land area shifting eco-climatic zone from nonforest biome to mesic forest. Values are expressed in* $km^2 \times 10^4$

From Nonforest to:	Scenario			
	GFDL	GISS	OSU	UKMO
Boreal	819.9	700.3	644.9	881.9
Cool temperate	65.0	59.7	4.3	146.5
Warm temperate	0.9	1.2	3.6	2.1
Subtropical	71.1	102.7	261.5	69.6
Tropical	122.4	132.8	208.9	122.4

2 TRANSITIONS FROM NONFOREST BIOMES TO FOREST

The land areas involved in this class of transition under the four scenarios are presented in Table 4. In the context of the ecosystem types represented in the maps of Figure 2 (Plate 9), the major transitions are (1) steppe to dry forest, (2) tundra to boreal and cool temperate forest, (3) polar desert to boreal forest, (4) savanna to dry forest, (5) dry forest to subtropical/tropical mesic forest. As with the transitions between forest types, we can examine the processes involved and the possible time scales required for these transitions to occur.

The rates of transition from steppe to temperate dry forest with increasing precipitation will be directly limited by seed dispersal and establishment rates of woody species. The current forest–temperate grassland boundaries are dynamic, with boundaries often influenced by the interaction of climate, fire, and herbivory. Even with the absence of competition from existent woody cover, given the limits of forest migration, this transition will occur over a time scale longer than the rates of climate change discussed.

The transition from tundra to boreal forest represents a large land area that is circumpolar in distribution. The boundary between boreal forest and tundra is actually a gradient ranging from closed canopy forest to open boreal woodlands to a mosaic of open forest patches within a tundra landscape and to the eventual absence of trees (Sirois 1992). Numerous climatic features have been related to this transition, including minimum temperatures and rates of potential evapotranspiration (Sirois 1992). It also appears likely that the thermal tolerances defining the northern limits for woody plants differ with respect to photosynthesis and growth as compared to flowering and germination, with the latter having a greater temperature sensitivity.

Although the northern distributional limit of boreal tree species is concurrent with certain thermal indices, work on the black spruce treeline suggests that photosynthetic activity is more limited by light intensity than by temperature during the growing season (Vowinckel *et al.* 1974). In addition, the northern boundary of the Canadian boreal zone coincides with given values of several radiation parameters (Hare and Ritchie 1972). This possible link between a northern limit to boreal tree species and values for solar radiation has major implications on the potential for northern expansion of the boreal zone into the area now occupied by tundra. However, localized northern expansion of the forest–tundra transition (i.e., seedling establishment) has been observed in both North America (Griggs 1937; Scott *et al.* 1987; Morin and Payette 1984; Payette and Filion 1985) and Eurasia (Tikhomirov 1961, 1963; Sonesson and Hoogesteger 1983) over the past century, coinciding with a documented 0.5°C warming in the region (Jones *et al.* 1982, 1986). Rates of expansion are limited by seed dispersal and the influence of fire on successional patterns (Sirois 1992).

The transition from polar desert to boreal forest is limited by the absence of mineral soil in these regions. The transition would be dependent on pedological processes, which occur on a time scale of centuries to millennia.

The savanna–dry forest transition occurs in the subtropical/tropical zone and is a function of increased precipitation. This may be the fastest occurring of the transitions within this class, because the majority of woodland/dry forest species are represented in the savanna biome. In Africa, the savannas form a gradient of increasing canopy cover, productivity, and standing biomass from open savanna to closed woodland (Tinley 1982). This transition could well occur within the time frame of the scenarios.

In contrast to the savanna–dry forest transition, the dry forest–mesic forest transition in the subtropical/tropical region represents a major taxonomic shift. The dry forest/woodlands are characterized by a tree overstory and shrub/grass understory. The tree species characterizing the overstory are unable to regenerate in the understory (see, e.g., Smith and Goodman 1987), and therefore the transition to a multilayer mesic forest is dependent on the dispersal and establishment of mesic forest species. As in the case with other transitions dependent on the dispersal and establishment of long-lived woody species, the temporal dynamics of this shift will lag far behind the changes in climate as presented by the scenarios.

3 TRANSITIONS FROM FOREST TO NONFOREST BIOMES

The land area involved in the transition from the current distribution of forest eco-climatic zones to nonforest zones (e.g., grassland, desert) is shown in Table 5. This class of transition is directly related to increasing aridity, resulting from increased potential evapotranspirative demand (related to increased temperatures) relative to changes in precipitation. The forest dieback resulting from increased aridity could occur over the time frame of the climate change scenarios. This class of transitions is perhaps the most critical of the three discussed, in that the dieback of forest ecosystems in these regions would represent major impacts to all forest sectors (i.e., timber and fuelwood production, biodiversity, and carbon storage) over a relatively short time frame (i.e., decades to century), allowing for minimal adaptation.

Table 5. *Total land area shifting eco-climatic zone from mesic forest to a nonforest biome. Values are expressed in* $km^2 \times 10^4$

	Scenario			
	GFDL	GISS	OSU	UKMO
Boreal	683.0	185.5	270.0	423.4
Cool temperate	441.9	253.3	203.8	507.4
Warm temperate	53.8	56.2	19.0	44.9
Subtropical	299.5	367.7	69.5	299.5
Tropical	2.7	14.3	0.6	2.7

Table 6. *Percent of (a) 2,164 nature reserves listed by the World Conservation Monitoring Center and (b) 243 UNESCO Man and the Biosphere (MAB) reserves that undergo a shift in eco-climatic zone based on the analyses presented in Figure 12*

(a) WCMC Listed Reserves

OSU	GISS	GFDL	UKMO
55.3%	63.3%	66.4%	80.5%

(b) MAB Reserves

OSU	GISS	GFDL	UKMO
48.5%	59.2%	62.9%	82.7%

Potential impacts on nature conservation areas

One important feature of global assessments of potential changes in eco-climatic zones concerns the future protection of nature reserves (Leemans and Halpin 1992). To this end, a database of the distribution of over 2,100 nature reserves listed by the World Conservation Monitoring Centre (WCMC) was used to examine the potential impacts of vegetation change on the global nature reserve system (Table 6a). In addition, potential impacts on the subset of nature reserves defining the UNESCO Man and the Biosphere (MAB) reserve system (IUCN 1986) under the four climate change scenarios are shown in Table 6b.

The four scenarios analyzed clearly demonstrate that the potential shifts in eco-climatic zones expected under the climate change scenarios tested could have a significant impact on the performance of both individual reserves and also on global reserve systems in general (Halpin and Smith, in prep.). The predicted shifts in eco-climatic zones are not interpreted here as specific changes in vegetation type or species composition, but rather are interpreted to mean that the climate in the region of the reserve would change to that associated with an ecosystem type other than that currently represented by the reserve. For example, climate types associated with forest ecosystems may change to climates now associated with savanna ecosystems. For species of fauna and flora whose distributions are limited to the reserve system, these changes could result in local extinctions.

Boreal forest

The general circulation models used in the construction of the climate change scenarios for this project all showed the greatest degree of warming at the northern latitudes. As discussed previously, this warming in the boreal zone would have a major impact on the distribution and abundance of boreal forest at a global scale. Given that the time required for the expansion of forest into the tundra/polar desert regions is much greater than the time scale predicted for the changes in climate, it is important over the time frame of decades to a century to focus on the potential impacts on the existing boreal forest region.

Predicted changes in the land area currently in boreal forest cover are shown in Figure 4. The changes have been divided into three categories: (1) change from a forest eco-climatic zone to a zone associated with a drier vegetation type (i.e., grassland or shrubland), (2) change in eco-climatic zone from boreal to cool temperate forest, and (3) stable – areas that remain suitable for boreal forest. This latter shift is a function of reduced plant moisture availability. The vegetation dynamics and time scale associated with these transitions have been discussed earlier (see 'Global analyses'). The time required for the transition from boreal to temperate forest will be much longer than that associated with the predicted changes in climate. In contrast, the forest decline associated with the shift from forest to drier vegetation will occur on a time scale corresponding to the change in climate, with increasing mortality resulting from increasing aridity. This transition represents a significant decline in boreal forest cover, ranging from 132×10^6 (GISS) to 661×10^6 (GFDL) ha. This reduction would have a major impact on forest resources since the declines include areas of major forest production in the southern regions of the boreal zone (e.g. Western Canada and Siberia).

The shift in the North American boreal forest zone under the climate change scenarios shown in Figure 2 (Plate 9) involves an array of transitions between eco-climatic zones and can be used to locate sites of potential importance for the application of stand-level models. A forest gap model (Bonan 1989) was applied to a series of sites in the North American boreal zone (Figure 5) to explore the implied transitions based on the Holdridge analysis.

The model simulates plant demographics following the approach outlined above for gap models (see 'General description of models') and includes biophysical models for solar radiation, permafrost, decomposition, and soil hydrology/evapotranspiration (Bonan 1989). Sites were selected based on (1) representation of differing eco-climatic transitions based on the Holdridge analyses, and (2) availability of data for model validation under current climate. Site descriptions, parameterization, and validation of the model for the locations shown in Figure 4 are discussed in Bonan (1990a, 1990b).

For each site, stand structure and composition were simulated under current climate conditions, defined by monthly statistics (mean and standard deviation) for precipitation, temperature, and cloudiness. The resulting stand description after 500 years of simulation was used to initialize simulations for control (continued current climate statistics) and changed climate conditions based on the four GCMs discussed earlier. Transient climate change scenarios were constructed from the equilibrium scenarios using a linear interpolation and assuming a 70 year period for $2 \times CO_2$ equilibrium to be achieved (Smith *et al.* 1992a,b). Only a local seed pool was considered in light of the relatively short duration of the simulations relative to the time scale of observed patterns of species migration (Davis 1984, 1989).

The simulation of a site on a south-facing slope in central Alaska is presented in Figure 6a. The site is currently dominated by white spruce (*Picea glauca*), birch (*Betula papyrifera*), and aspen (*Populus tremuloides*). Under the warmer, drier conditions predicted under all scenarios, the stand would decline as a function of moisture stress, and the site would eventually be dominated by herbaceous ground cover. The differences in the timing of the decline among the scenarios reflect the differences in predicted warming. This result is in agreement with the transition from boreal forest to cool temperate steppe (grassland) predicted for the area using the Holdridge Life Zone System (Figure 2).

In contrast to the predicted decline in forest cover on the south-facing slope, simulation of a stand in the same location on a north-facing slope (Figure 6b) shows an increase in biomass of the dominant species, black spruce (*Picea mariana*). This difference is due to the link between soil moisture and solar radiation. The lower input of solar radiation to the north-facing slope (in comparison to the south-facing) would reduce the moisture stress associated with the regional increase in temperature.

Simulation of a site in the boreal forest–tundra transition

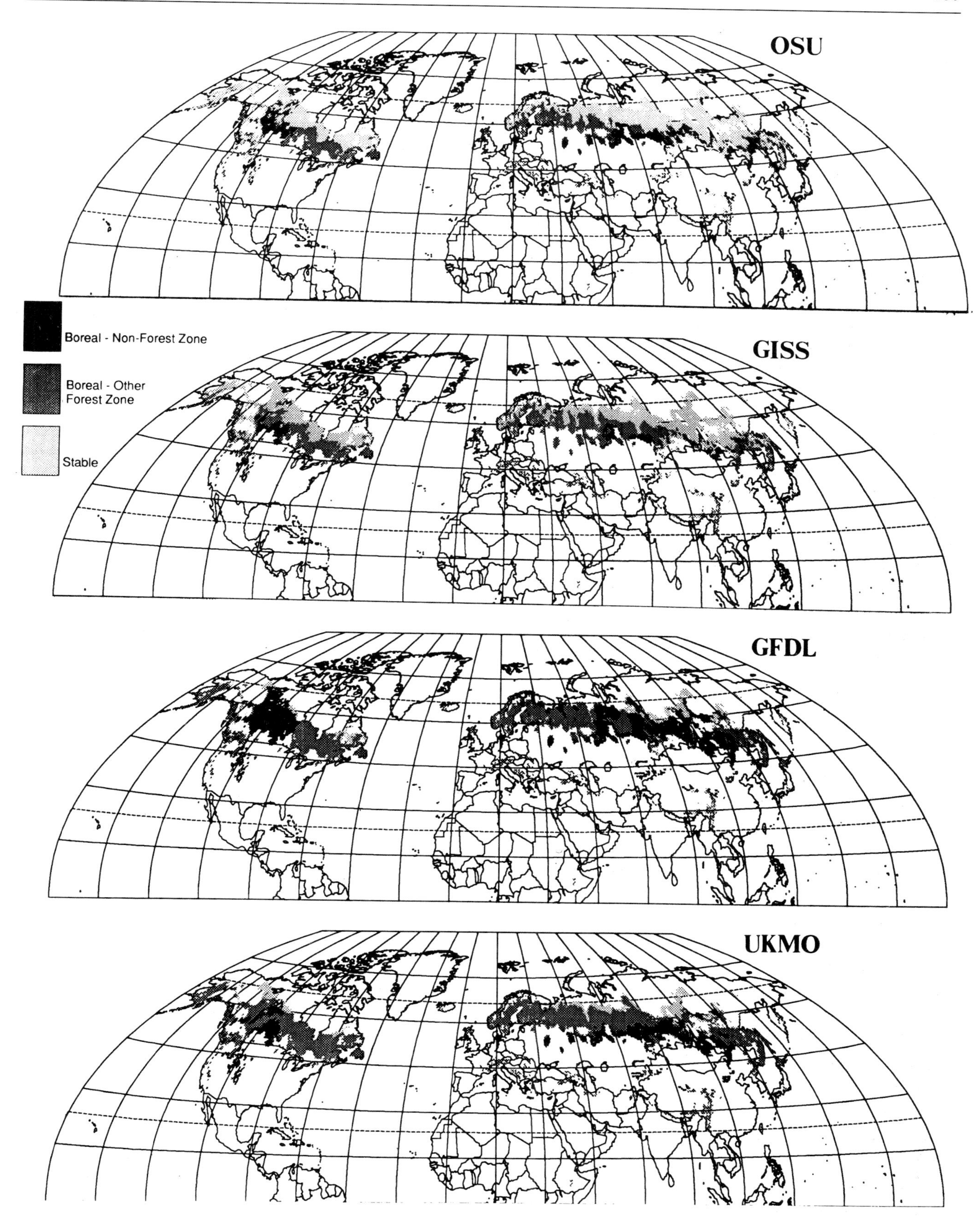

Figure 4 Changes in the current distribution of eco-climatic zones (Holdridge 1967) within the boreal forest region under four climate change scenarios. Categories of change include (1) shift from boreal forest to nonforest cover (i.e., grassland or shrubs), (2) shift to climate associated with warmer forest types (i.e., cool temperate forest), and (3) areas that remain in the boreal forest eco-climatic zone. Map also includes areas of conifer forest.

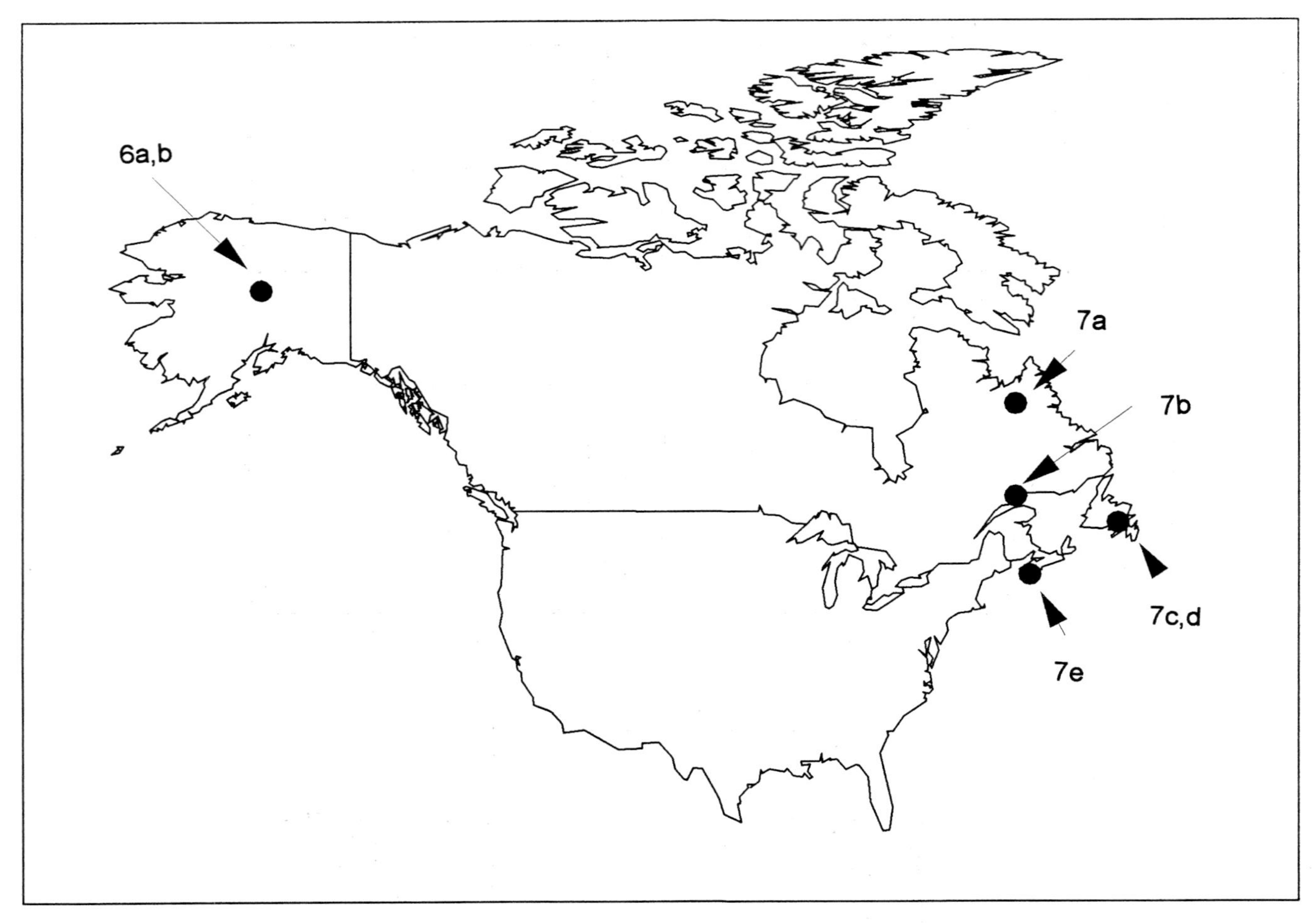

Figure 5 Location of sites for boreal forest simulations using forest gap model.

zone of northern Quebec, Canada, is shown in Figure 7a. The site is currently an open-canopy woodland of black spruce with relatively low productivity and standing biomass. With the predicted warming for this region under all scenarios, the stand would increase in biomass to a closed-canopy forest. This transition would be associated with a decline in the present ground cover of moss and lichen. The occurrence of fire would facilitate this transition because the establishment of seedlings on the site is currently hindered by ground cover. This site corresponds to the northern expansion of the boreal forest into the tundra–forest boundary in the Holdridge analyses.

In contrast to the northern Quebec site, the predicted dynamics of a black spruce stand in central Quebec (Figure 7b) vary among the scenarios. The OSU and GISS scenarios show little change from the pattern predicted under current climate (i.e., the control). These results reflect the lower warming predicted for the region by these two GCMs. In contrast, both the GFDL and UKMO scenarios predict a complete dieback of the stand, with the rates of decline directly related to the degree of warming. These declines are not a function of moisture stress but result from temperatures rising above those currently associated with the southern limit to the distribution of black spruce. The area is predicted to change to a cool temperate forest under the Holdridge Life Zone Classification, and the establishment of species characteristic of that forest type would be dependent on a range of processes operating at the landscape level (e.g., rates of dispersal and establishment). This result points out a limitation in our current understanding of the distribution of forest zones at a continental scale.

The predicted decline in the boreal forest at its southern boundary under increased temperatures has an implied assumption that the southern distributional limits of the component species are physiologically determined (e.g., the temperature exceeds the tolerance for germination, establishment, or growth). In contrast, if competition between boreal and northern hardwood (i.e., cool temperate forest) species is a determining factor in the southern distributional limit of boreal forest, then the predicted decline may not occur. In fact, the increased temperatures may result in increased productivity of these stands (Bonan and Sirois 1992), and the transition to cool temperate forest would involve the eventual invasion of northern hardwood species

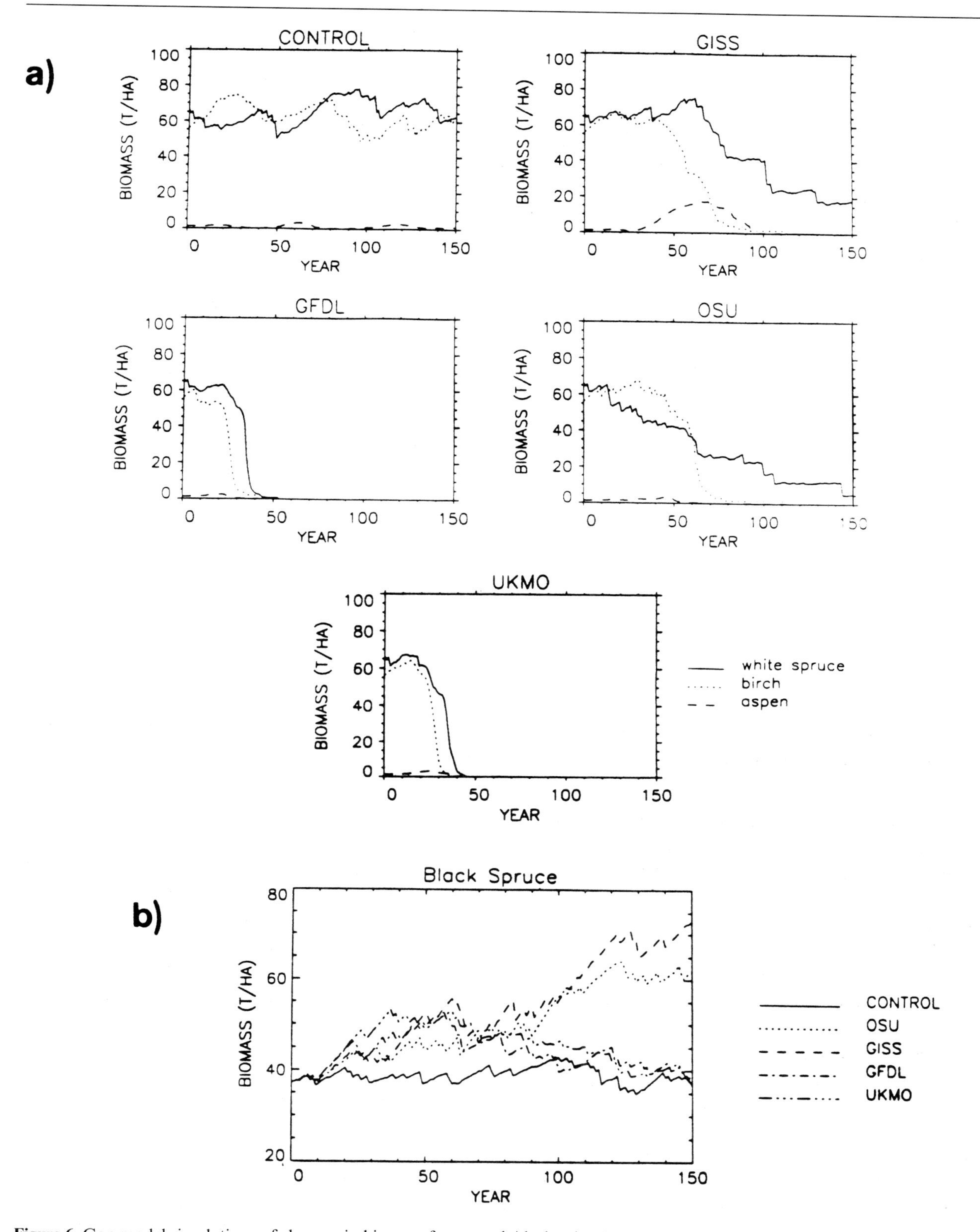

Figure 6 Gap model simulations of changes in biomass for central Alaska site: (a) south-facing slope, and (b) north-facing slope.

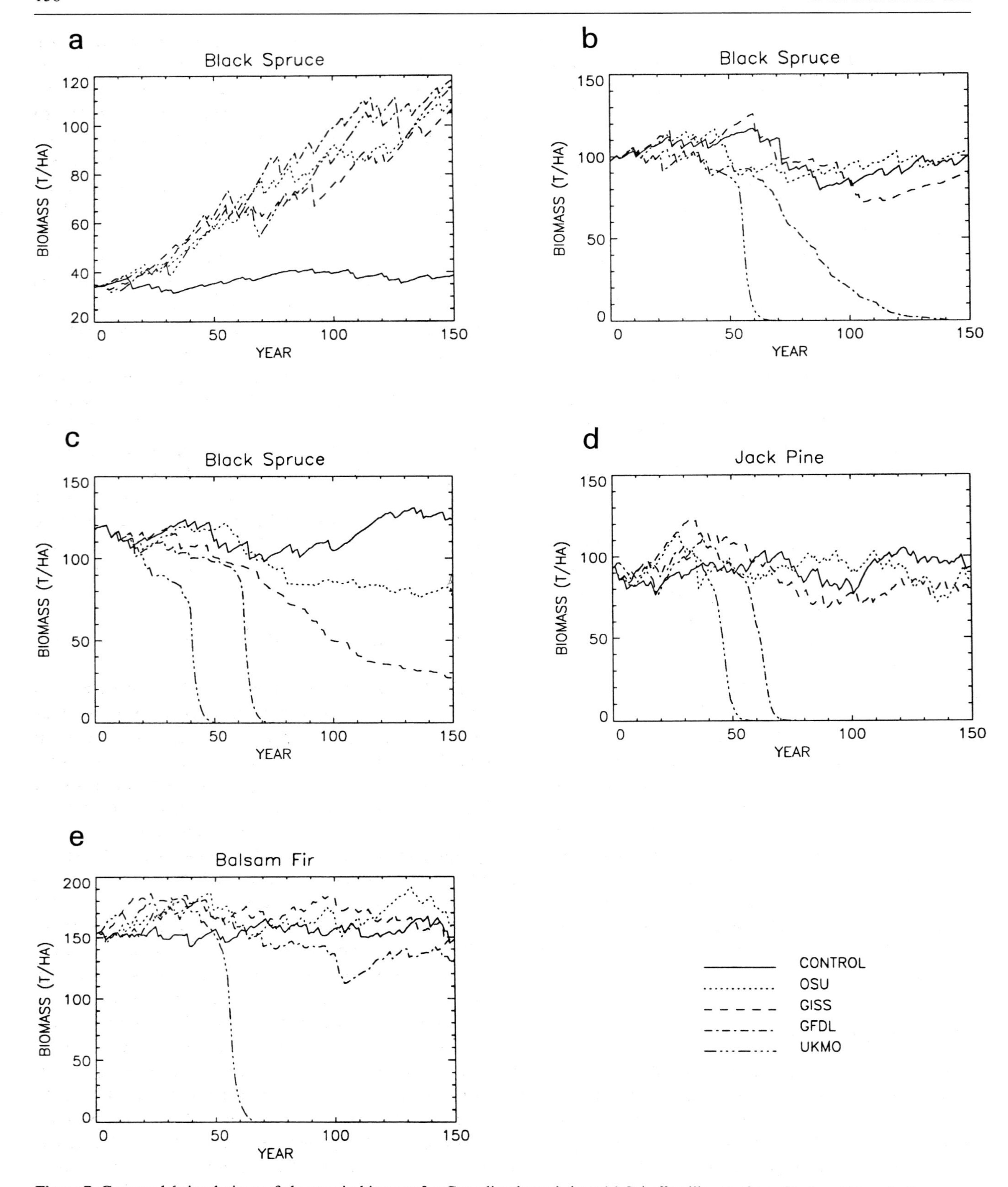

Figure 7 Gap model simulations of changes in biomass for Canadian boreal sites: (a) Schefferville, northern Quebec, (b) central Quebec, (c) Newfoundland, (d) Newfoundland, and (e) New Brunswick.

into established boreal forest. The rate of this transition would be related to the ability of the cool temperate forest species to invade established forest stands, which would be influenced by rates of disturbances such as fire (Overpeck *et al.* 1990).

Similar results to those observed for central Quebec are predicted for black spruce in Newfoundland (Figure 7c). Stand biomass would decline under all four scenarios with the rates of decline a function of the degree of warming. As with the central Quebec site, the area is predicted to change to a cool temperate forest and the decline in black spruce would be a function of the temperatures rising above those currently associated with its southern limit.

Simulation of a balsam fir (*Abies balsamea*) stand on sandy soils in the same area showed slightly different results (Figure 7d). Only the scenario with the largest temperature increase (UKMO) resulted in a significant decline in stand biomass. The difference between the two forest types (i.e., black spruce and balsam fir) for this location is a function of the differences between the two species in their growth response to temperature.

Similar results to those observed for balsam fir were predicted for Jack pine (*Pinus banksiana*) in New Brunswick (Figure 7e). The two warmest scenarios for the region (UKMO and GFDL) showed a decline in stand biomass, while the OSU and GISS scenarios differed little from current climate conditions (i.e., control) in this regard.

In general, there was a qualitative agreement between the regional patterns predicted by the bioclimatic models and the results of the gap model simulations for the representative sites. However, the results from the gap model simulations highlight the uncertainties associated with defining the factors limiting the southern distribution of the boreal forest zone as discussed earlier. There is a need for research to determine the relative importance of (1) competition with northern hardwood species, (2) the interaction between edaphic factors and patterns of nutrient cycling, and (3) the physiological limitations imposed by the thermal tolerances of establishment, growth, and reproduction of boreal tree species.

Tropical forest

All four climate change scenarios investigated showed an expansion of the current tropical forest eco-climatic zone (see Figure 2). However, forest expansion must be viewed in terms of both the biological constraints imposed by limited rates of species dispersal (see discussion in 'Global analyses') and the current and future constraints on tropical forest distribution imposed by human land-use activities. Clearing for agricultural and urban expansion has reduced the distribution of tropical rainforest to only a fraction of its previous range. The pressures of conflicting land-use practices will severely limit the movement of forest into areas currently used for agricultural purposes. Since the increase in precipitation associated with the forest zone expansion could also translate into increased agricultural production in what are currently semi-arid/arid regions, it seems unlikely that forest will be allowed to encroach upon these areas. We therefore focused the analysis on the area currently occupied by tropical forest.

The potential changes in existing tropical mesic forest cover (Olson *et al.* 1983) under the four climate change scenarios is presented in Figure 8. Unlike the boreal case, only two categories are necessary to describe the shifts in eco-climatic zone: (1) stable – remains suitable for tropical mesic forest, and (2) decline – areas no longer able to support mesic forest. The decline is a direct function of increased aridity. These expected decreases in current tropical mesic forest cover under the four climate change scenarios range from 2.7×10^6 (GFDL) to 16.9×10^6 ha (UKMO).

The patterns of change in calculated PET ratio match the direction of eco-climatic shift predicted by the Holdridge model (i.e., wetter vs. drier). This comparison supports the predicted patterns of forest decline presented in Figure 7.

REGIONAL ANALYSES

While the analyses that have been discussed succeed in identifying gross trends in potential vegetation change, they generally fail to depict potential impacts of climatic change at spatial scales appropriate to the needs of local forest planning and conservation management. Two regional-scale impact assessments were conducted to assess the potential impacts of climate change for both humid tropical forest systems and dry tropical forest/savanna systems. These two different types of climatic regions exhibit significantly different sets of ecological and natural resource management problems that influence tropical and subtropical countries.

The humid tropical forest pilot project focuses on Central American forest ecosystems with a primary concentration on forest production and nature reserve impact assessments. A case study of Costa Rica is presented to illustrate these issues. The Costa Rica case study was conducted with collaboration and generous assistance of numerous in-country participants including members of the Tropical Science Center (TSC), the National Meteorological Institute (IMN), the National Parks Service (SPN), the Forest Service (DGF), Fundacion Neotropical (FN), the Organization for Tropical Studies (OTS), and the regional office of the World Conservation Union (IUCN-ORCA).

The dry tropics pilot project focuses on southern African forest/savanna ecosystems with a primary concentration on

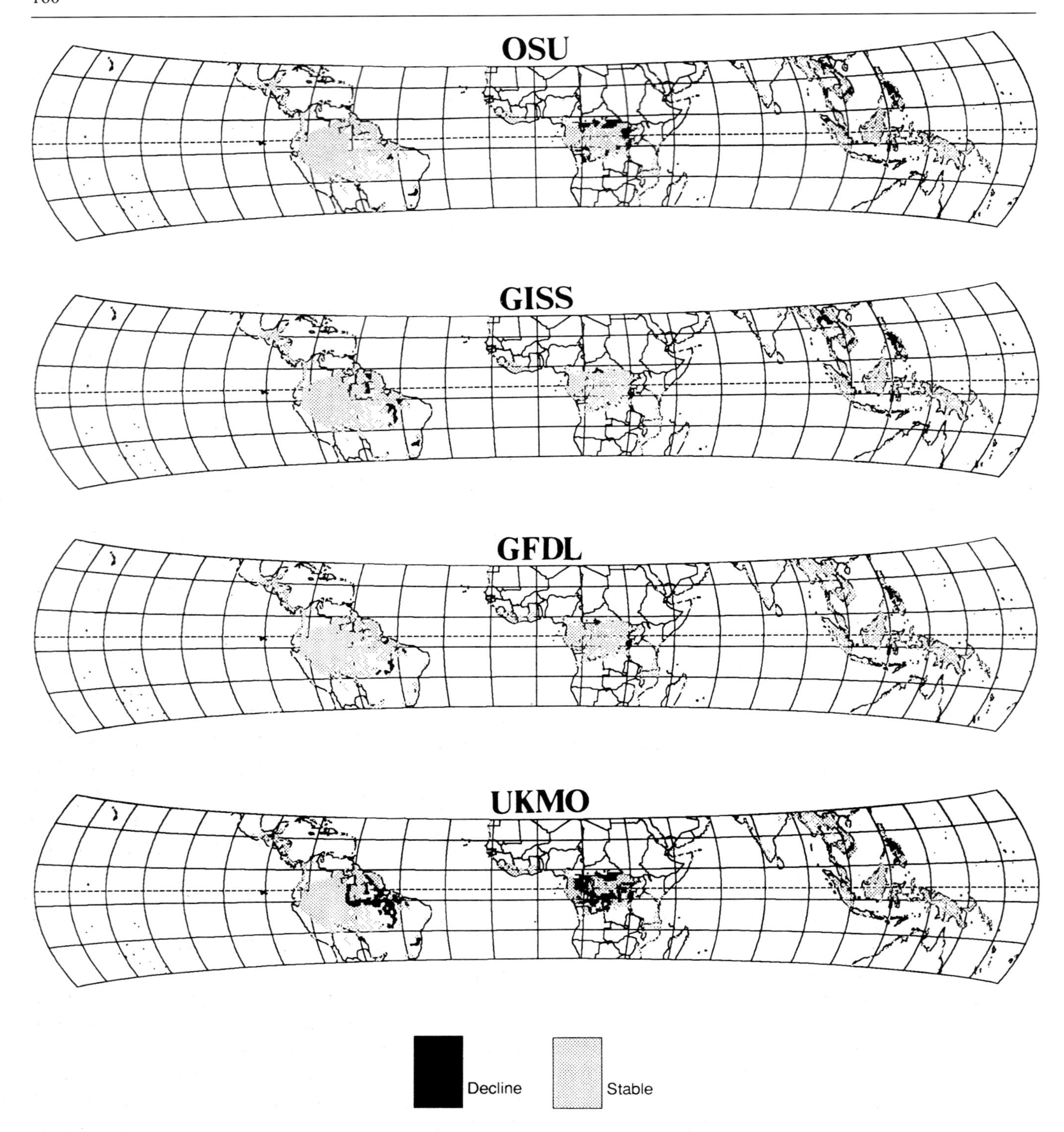

Figure 8 Changes in the current distribution of tropical mesic forest under four climate change scenarios. Areas of decline defined by shift to drier eco-climatic zone (Holdridge 1967).

fuelwood provision and natural habitat protection. A regional overview of potential impacts for the ten country Southern African Development Cooperative Conference (SADCC) and a case study of Zimbabwe are used to illustrate issues for this region. This case study also relied on generous collaboration with SADCC representatives and members of the meteorological and forestry departments of Zimbabwe.

Humid tropics: Central American pilot project

SITE SELECTION

Costa Rica was selected as an ideal tropical case study site because it exhibits a relatively complete microcosm of tropical climate regimes, ecosystem types, land use pressures, and natural resource conditions representative of a wide range of humid tropical regions in the developing world. Many general findings of this case study can be reasonably extrapolated to similar climatic regions throughout the humid tropics. Costa Rica offers an extensive range of distinct eco-climatic zones and patterns of landscape diversity due to its exceptional topographic relief (0–3800 meters) and position on the thermal equator (8° to 10° North latitude) (Holdridge 1967; Gomez and Herrera 1986; Hartshorn 1983; Walter 1985). The forest systems of Costa Rica contain noteworthy examples of a wide range of forest types from the relatively rare tropical dry forest sites of Guanacaste in the northern Pacific region, to lowland and montane tropical wet forests in the Atlantic coastal regions.

Because Costa Rica's expansive nature reserve system represents sites spanning numerous climatic and orographic gradients, the country presents a unique situation for the analysis of potential climatic impacts on ecosystem and biodiversity protection in the humid tropics (Halpin and Secrett 1994). Costa Rica contains some of the finest examples of tropical montane nature reserves and some of the few existing protected area corridor systems in the tropics (Mackintosh *et al.* 1989). Finally, Costa Rica is one of the few locations in the tropics that has a well-developed and long-standing history of climate–vegetation analysis and field-validated eco-climatic mapping relating climate to natural vegetation, productive forestry, and agricultural land use (Holdridge 1967; Holdridge *et al.* 1971; Sawyer and Lindsey 1971; Gomez and Herrera 1986).

Eco-climatic mapping for use in forestry and biodiversity management, land capability analysis and natural resource accounting serves as a vital component in existing natural resource assessment techniques in the country (Holdridge and Tosi 1972; Hartshorn *et al.* 1982; Tosi *et al.* 1988; Sader and Joyce 1988; TSC/WRI 1991; Cruz and Reppetto 1992). Because land capability and natural resource accounting techniques used in Costa Rica already incorporate climatic features into the land planning system, analysis of climatic changes on natural resources is more readily incorporated into long range national assessments. This basis of past analysis and ground verification of eco-climatic classifications coupled with ongoing forest land use debates using climatically derived land capability accounting systems makes Costa Rica an exceptionally rich site for preliminary climate change assessment.

FOREST BACKGROUND

With average precipitation exceeding 3,000 millimeters and annual temperatures ranging from 27.8°C at sea level on the Pacific Coast to 4.5°C at the highest mountain summits, Costa Rica's original vegetation represented exceptional examples of humid broadleaf tropical forest types. Of the 12 general eco-climatic zones found in the country, only the high-altitude, subalpine paramo regions (0.2 percent of the country) are naturally without full forest cover (Hartshorn *et al.* 1982). Over 4,000 species of forest trees and shrubs are estimated to exist in the region, and an exceptionally wide variety of endemic flora and fauna species are found in the forested areas (e.g., 848 species of birds, 216 species of reptiles, 205 species of mammals, and 160 species of amphibians) (Hartshorn 1983).

The extent of the remaining forest area in the country has been severely diminished in the last century through land clearing for agricultural, pasture and plantation use (Sader and Joyce 1988; Tosi *et al.* 1992). A recent World Bank/United Nations Development Program study (WB/UNDP 1990) estimates that current deforestation is due mainly to agricultural expansion (70 percent), commercial logging (20 percent), and fuelwood collection (5 percent). Historic mapping and recent satellite remote sensing of forest cover have been employed to document the decline of the forest cover in recent years. Figure 9 depicts the loss of primary forest cover in Costa Rica from 1940 to 1987 (Fundacion Neotropica 1988).

Of special interest to climatic change analysis is the fact that past trends in deforestation have been directly associated with the eco-climatic characteristics of the forest sites, with eco-climate zones suitable for agricultural uses being preferentially cleared (Tosi 1969; Sader and Joyce 1988; TSC/WRI 1991). The highest rates of recent deforestation have occurred in the tropical dry forest and pre-montane moist forest eco-climate zones while the lowest rates of deforestation have occurred in the pre-montane rain forest, lower-montane rain forest and montane rain forest sites (Sader and Joyce 1988). High rainfall, low potential evapotranspiration to precipitation ratios, and inaccessible terrain have acted together to discourage widespread deforestation in the upper montane eco-climatic regions. It is estimated that by the year 2060, all remaining primary forest coverage

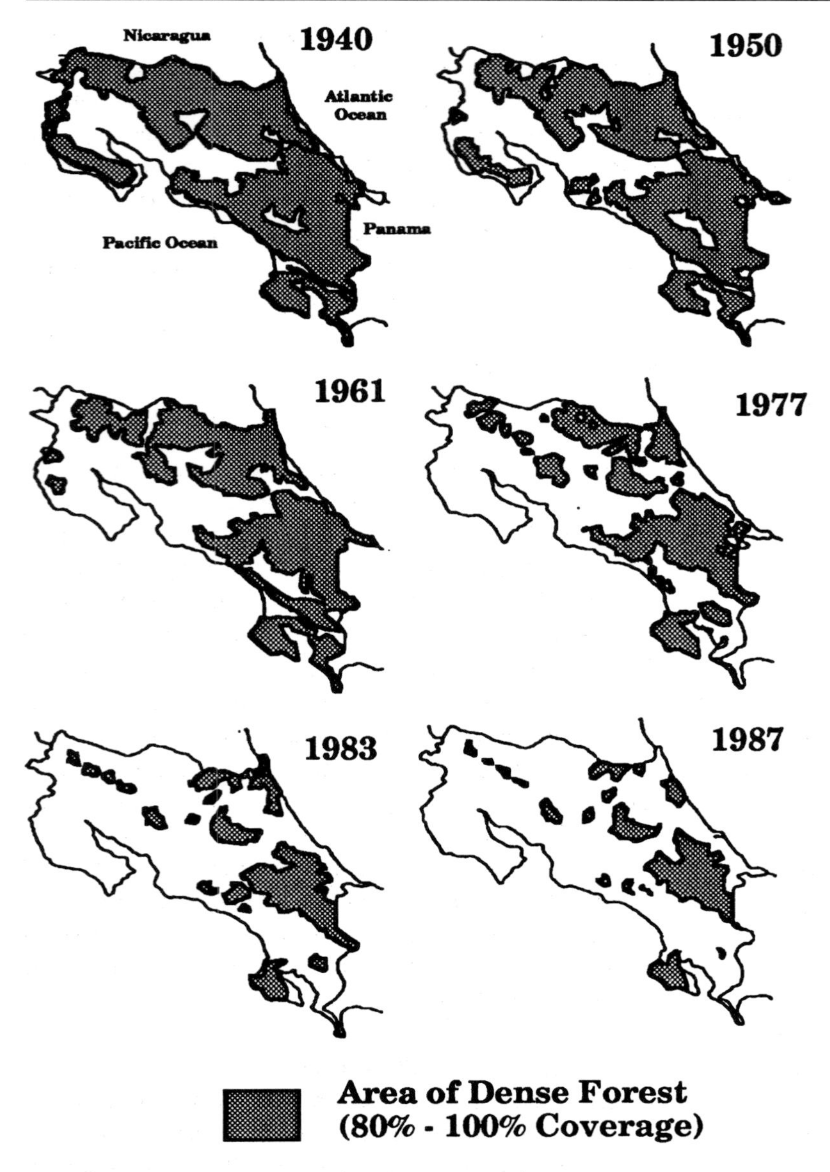

Figure 9 Deforestation in Costa Rica 1940–1987 (adapted from FN 1988).

will be contained in the reserve system and most forests outside of strict reserves will be made up of managed secondary growth stands (Tosi *et al.* 1992).

Types of forest land uses in Costa Rica have been broadly defined for this study in the categories of intensive forestry, extensive forestry, and protection forestry management (Tosi *et al.* 1988; 1992). Intensive forestry uses are defined as highly managed production forestry and short rotation plantation forestry management methods. Extensive forestry uses are defined as natural forest management and secondary forest regrowth uses exhibiting limited management intrusions and harvesting disruption. Protection forest uses are classified in a hierarchy of strict biological reserves, wildland areas, forest reserves, and indigenous Indian reserves.

The potential effects of climate change were considered for four general forest management sectors in this pilot project. These sectors were (1) sustainable forest production, (2) natural area protection, (3) fuelwood production, and (4) forest watershed protection. Brief descriptions of the current status of these forest sectors and the general susceptibility of these sectors to climatic disruption are presented below.

FOREST PRODUCTION

Total forest production capacity in Costa Rica has remained fairly constant in recent years, with the 1988 output at 3,961,000 m^3. Of this production volume, 35 percent was used for industrial purposes (sawnwood, panels, pulp, and paper) and 65 percent was used for local purposes (fuelwood, construction posts, fencing) (IDB 1987). Over 90 percent of

Table 7. *Forest resources supplies and shortfalls 1984–2010 in 1000m³ (source DGF 1988)*

Year	Natural forests	Existing plantations	Total supply	Shortfall
1984	2,705	15	2,720	0
1990	2,753	101	2,854	0
1995	2,743	267	3,010	0
2000	2,358	386	2,744	422
2005	2,028	223	2,251	1,120
2010	1,745	0	1,745	1,885

paper products used in the country are imported, leading to the recognition that Costa Rica is a net importer of wood products. The country had a $39.7 million trade deficit in this sector in 1985 (IDB 1987).

While there are approximately 2,000 species of trees in Costa Rica, only 70 are commercially used in timber production, and of these only a small percentage make up the bulk of the trade (Secrett 1991; Hartshorn *et al.* 1982; WB/UNDP 1990). This species selection coupled with high transportation costs, aging equipment, unsustainable forest extraction methods, and a lack of management incentives has led to generally inefficient forestry practices with approximately 46 percent of commercial timber volume lost during extraction (Flores Rodas 1985; TSC/WRI 1991).

Under changing climatic conditions, the ability to meet forest product demands, increase forest use efficiency, and change forestry practices to more sustainable methods may be made more difficult by increased environmental and infrastructure stresses. Table 7 depicts estimated domestic supplies and shortfalls of forest products from 1984 to 2010 under current climate conditions.

NATURAL AREA PROTECTION

The loss of the majority of forested area due to land clearing has dramatically increased the need for the immediate protection of forest areas through classification of remaining forest lands into a variety of protected management classes. The current national protected area strategy involves a mix of various types of protected areas in eight general conservation (mega-park) regions which are intended to provide both sustainable production of natural resources and the protection of forest biodiversity (PAFT-CR 1990). The regional grouping of various reserve areas in national planning is intended to promote the future development of multiple use forest resource management schemes and contiguous buffer zones around sensitive biological reserve core areas.

While Costa Rica's protected forest area system is considered to be one of the best in the world, problems in site management and enforcement of existing protection mandates exist (PAFT-CR 1990). In general, climatic change could significantly exacerbate existing management problems by (1) directly altering the ecological conditions and species composition of reserve areas; (2) lowering the natural resilience of ecosystems to disturbance; (3) increasing external land-use pressures on reserve areas; and (4) lowering the national capability to support protected area infrastructure and enforcement. These problems would be common to all countries in the region.

FUELWOOD/ENERGY USE

In 1987, 68 percent of Costa Rica's energy uses was generated within the country. Of this amount, 44 percent was attributed to fuelwood and charcoal use, 14 percent to hydroelectric power, and 10 percent to crop residues burning (Calvo 1990). Of the fuelwood use, 80 percent was consumed by households and 20 percent by industries. While domestic demand for fuelwoods is expected to be met in the future for the country as a whole, skewed geographic concentrations of population and fuelwood demand create problems in local production and distribution of this forest resource. Regional impacts of climatic change on forest production and general land use patterns could act to significantly increase existing fuelwood resource problems. Areas such as the Central Valley, which supports a high population, high industrial fuelwood use and low forest area would be especially sensitive to direct climatic changes and secondary land use shifts on sustained fuelwood provision.

WATERSHED PROTECTION

Forest vegetation serves critical roles in the provision of sustained water resources and the protection of watersheds from bank erosion, stream siltation, soil loss, and biodiversity loss. Costa Rica has been divided into 38 distinct watershed areas, and the percent of remaining forest cover and area of each watershed under protection has been recently assessed (FN/CI 1988). The northern and central watersheds of the Pacific slopes are under the most intensive land use and contain the least natural vegetation with 12 watersheds having no lands under protection. These watersheds are potentially more at risk to significant erosion losses. Changes in forest structure and function due to regional climatic changes could alter the current evapotranspiration and rainfall-runoff regimes of critical watersheds in the region, potentially exacerbating soil erosion and water management systems in critical watersheds.

APPROACH AND METHODS

Like many countries in the humid tropics, Costa Rica has in recent years faced rapid declines in the extent of natural forest area and an increase in land-use pressures on remaining forest resources. The potential impacts of climate change on forest production and protection in the humid tropics therefore must be interpreted within the context of continuing changes in forest extent and land-use capability (Halpin *et al.* 1991a). This pilot project assessed potential climate change impacts on existing forest management areas, nature reserves, and unmanaged natural areas and also identified possible future changes in potential land-use capability for the country as a whole.

Changes in potential land use capability can be used to infer both direct impacts of climate change by suggesting alteration of the productivity of a site under the current land use regime, but also offers useful insights into potential secondary changes in preferential land use which may be brought about by changing climatic conditions. Direct impacts of climatic change on forests are interpreted as changes in expected productive capability and composition of existing forest areas. Direct climatic impacts were assumed to be produced by increases in physiological heat stress, changes in seasonal water balance, and changes in the frequency and magnitude of climatic disturbances. Changes in heat stress (Holdridge *et al.* 1971) and water balance related to global GCM scenarios are assessed directly in this study. While reliable assessments concerning changes in regional disturbance regimes for climate change conditions were not available for Costa Rica, recent studies for the Caribbean region have predicted increases in both hurricane frequency and magnitude and subsequent forest damage due to increased sea surface temperatures (Maul 1989; Granger 1989; O'Brien *et al.* 1992).

Direct climatic impacts on forests are generally defined in terms of changes in mean annual growth increments (MAI) for production forest species, potential ecological stresses on protected forest areas, changes in seasonality of climatic factors, and changes in hydrologic functions for forested catchments.

Indirect or secondary impacts of climate change can be defined as changes in potential land use as a response to changing climate. For example, if the climate regime of lands currently used for cattle pasture is expected to become better suited for secondary forest regrowth, economic pressures to convert the area back to forestry uses may also increase. Conversely, a potential negative response in terms of forest use could occur if agricultural lands adjacent to forest stands lose per area productivity (land capability) due to climatic change, forcing an expansion of agricultural zones into forest areas in order to offset potential agricultural losses.

In this analysis, spatial distributions of existing eco-climatic zones using the Holdridge Life Zone system (Figure 10a, Plate 10a) were modified to represent the outcomes of potential climatic changes on a regional scale. Direct effects on production forests and nature reserves were assessed through geographic analysis of changes in eco-climatic zones overlaid onto land management and land cover type data layers. Secondary impacts measured in terms of potential changes in land capability were defined through expert interpretation on a sub-regional basis by in-country collaborators (Halpin *et al.* 1991a; Tosi *et al.* 1992).

The regional results of the four GCM scenarios used in this analysis fell into two distinct groups for the Costa Rican area. Both the GFDL and OSU models produced regional estimates of approximately a 2.5°C increase in temperature and a 10 percent increase in precipitation. Both the UKMO and GISS models, on the other hand, produced regional change values of approximately 3.6°C and 10 percent increase in precipitation. These paired results defined two working climate change scenarios for Costa Rica (Kelly 1991a).

RESULTS

The potential impacts of the moderate change scenario (+2.5°C and 10 percent precipitation increase) and the warmer change scenario (+3.6°C and 10 percent precipitation increase) on the countrywide distribution of eco-climatic (i.e., life) zones are depicted in Figures 10b–c (Plates 10b–c). In general, the two regional scenarios produced significant changes in the spatial arrangement and regional composition of eco-climatic zones. When this analysis was conducted at a high spatial resolution by the Tropical Science Center, 43 percent of the land area changed eco-climatic zones under the 2.5°C scenario, and 60 percent under the 3.6°C scenario was estimated to change (Tosi *et al.* 1992).

That more area experienced changes in eco-climatic type with the higher resolution assessment is consistent with expectations that increases in spatial resolution increase the differentiation of distinct eco-climatic zones, therefore identifying more areas of potential change. The types of eco-climatic change identified in the mapping of life zone shifts indicated three generalized trends under both scenarios. First, changes in the areal arrangement of eco-climatic zones suggested a strong trend toward the displacement of montane and subalpine zones by warmer pre-montane and lower-montane climate types. Second, changes in the calculated annual biotemperature indicated an increase in the period and magnitude of heat stress for many low elevation areas. Third, the entire climatic gradient of the region was seen to experience a change toward more tropical climatic types across all elevations.

Country-wide assessment of the changes in eco-climatic zones under the two scenarios represent similar directional trends in terms of the types of expected change, but represent different thresholds of transition and differences in area of change. A potential loss of unique eco-climatic zones along

Figure 11 Climatic subregions of Costa Rica.

altitudinal gradients is observed under both scenarios. Under the 2.5°C scenario, the principal changes would occur in the warm-dry and cool-wet climate zones, with tropical dry forest, montane rain forest, and sub-alpine paramo climatic zones being the most affected areas. The sub-alpine paramo climate zone disappears completely under this scenario (Halpin *et al.* 1991a; Tosi *et al.* 1992). Under the 3.6°C scenario, lowland tropical moist and wet forest eco-climatic zones would decrease in spatial area while large areas of increase would be expected for pre-montane forest climate types due to changes in calculated biotemperature for the region. Calculation of expected biotemperatures for the region demonstrate that both scenarios would result in significant alteration of annual growth periods for lowland forest areas (0–900 m) with an expected increase in calculated heat stress zones (Holdridge *et al.* 1971) moving up the altitudinal gradient. The expected effects of heat stress on vegetation within these zones are a general reduction in annual productivity in the short term and the potential for changes in vegetation structure, biomass, and species composition over a longer time period (Holdridge *et al.* 1971; Tosi *et al.* 1992).

SUB-REGIONAL ANALYSIS

Five general climatic sub-regions were used in the analysis to calculate specific climate change impacts. Figure 11 depicts these geographically and climatically distinct regions. A summary of the general sub-regional impacts defined by the Tropical Science Center (Tosi *et al.* 1992) is presented below.

Central Valley

In the Central Valley (800–1200 m) both the 2.5°C and 3.6°C scenarios would produce heat stress effects only on vegetation in the lower elevation extremes; however, eco-climatic zones would be expected to change over much of the region. Changes from pre-montane climate types to tropical basal zones predominate in the region. Vegetation requiring cooler temperatures, such as arabica coffee and ornamental flowers grown for export, may be forced to higher elevation zones. This potential change in land use could act to displace dairy and truck crops in the lower montane regions. Attempts to offset potential agricultural losses by encroachment into the higher montane wet forest sites could act to endanger much of the last remaining high elevation forest areas in this region which serve to maintain the nation's most reliable sources of hydroelectric power. Because this region is the most heavily populated and urbanized area of the country, limited opportunities exist for redistributing crop and forest land area under climate change scenarios. Anticipatory development policies may be best directed toward retaining the highest level of non-urbanized areas as possible into the future in this region.

Pacific South

In the Pacific South region, forest areas would be affected by an increase in heat stress conditions expressed as the difference between ambient air temperatures and calculated biotemperatures (t/tbio) for the region. The changes in biotemperature and increased rainfall would move climate types in this region from tropical pre-montane forest to shorter growing period montane forest climates. Actual evapotranspiration is expected to decrease, the number of excessively wet months is expected to increase, and runoff is expected to significantly increase in this region. This condition is expected to lower per hectare forest production for this area. Forestry would continue to be the most suitable land use under these climates, but production would be somewhat diminished.

Atlantic

In the Atlantic sub-region, heat stress and biotemperature changes from the current Caribbean climate patterns are not expected to be severe. Forest runoff is expected to increase significantly under both scenarios due to increased rainfall, but the seasonal relationship of wet months is expected to be unchanged. Excessive water in the lowland areas of this region is expected to remain a hindrance to agricultural land uses. Forest production is expected to remain the most viable land use for this region under both scenarios.

Pacific North

The dry Pacific North sub-region is expected to be negatively impacted by changes in heat stress and soil moisture deficit under both scenarios. Areas below 500 meters elevation are expected to become a belt of extreme heat stress for cultivated plants and cattle grazing land uses. Native forests in this region have already been reduced to small fragments contained in nature reserves. Relics of these dry forests should be considered as germ plasm for reintroduction of species into reclamation areas which may have similar climates in the future.

Northern Subslope

The Northern Subslope represents the most continental climate zone represented by tropical moist forest climate types. While a large percent of the area is now under agricultural use, areas of poor soil, insufficient drainage, and less favorable topography are left for forest uses. Under the 2.5°C scenario, increases in heat stress and forest runoff are expected to decrease the suitability of much of the area for agricultural pursuits and increase the land use suitability for forest production uses. Under the 3.6°C scenario, heat stress, increasing runoff, and an increase from 5 to 7 wet months per year produce conditions expected to significantly reduce forest productivity and hamper forest operations, especially commercial logging in the region.

GEOGRAPHIC IMPACTS ON FOREST RESERVES AND PROTECTED AREAS

In order to draw more direct conclusions concerning the risks posed by climate change for production forestry sites and existing nature reserves, potential changes in eco-climatic zones must be assessed in terms of actual land-use distributions.

In this study, current areas of natural vegetation coverage, production forest sites, and protected areas were assessed for changes in eco-climatic zonation. Mapping of nonagricultural areas containing 60 percent to 100 percent natural vegetation coverage was used to define the potential spatial impacts of changes in eco-climatic zones on remaining forested vegetation within the country (Figure 12). Under the 2.5°C scenario, 31.8 percent of the remaining natural vegetation areas experienced changing eco-climatic conditions, whereas the 3.6°C scenario produced a 48.2 percent change in eco-climatic zones. These nonagricultural areas represent the most likely areas for potential forest management land-use adaptation – such as the development of forest buffer zones, corridor systems, and new forest reserve areas – under changing climatic conditions.

Production forest areas and forest reserves were found to exhibit a 19.5 percent change in eco-climatic zones under the 2.5°C scenario and a 44.5 percent change under the 3.6°C scenario. In a more general assessment of production forest potential, the Tropical Science Center reported that major changes in the tropical dry, subalpine, and montane zones, although dramatic in terms of conservation areas, will have only minor effects on forest production due to a lack of forestry land uses in those zones at present. However, general increases in the area of tropical moist forest climate types are expected to raise the potential opportunity for forest growth in areas now managed for pasture and agricultural uses.

The nature reserve system of Costa Rica spans a wide range of topographic and climatic habitat types. For the purpose of this study, digital maps of 33 National Parks and biological reserves were analyzed with respect to potential changes in eco-climatic zones. Under the 2.5°C scenario 27 percent of the area contained within the reserve system experienced a change in eco-climatic zone, as compared with 53 percent for the 3.6°C scenario. The changes in eco-climatic

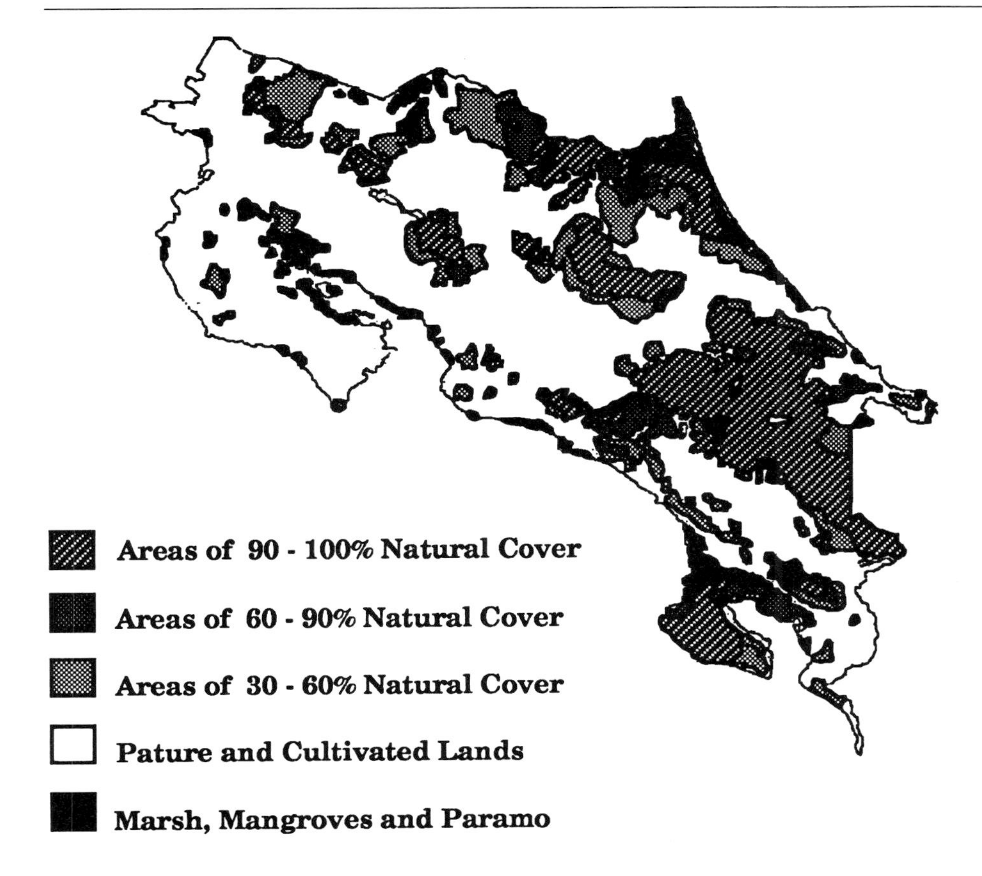

Figure 12 Natural vegetation cover classes.

zones expected to occur within the major reserve systems represent areas that experience subtle differences in temperature and moisture as well as areas that experience dramatic changes in the potential composition of eco-climatically defined habitats. For example, the la Amistad reserve complex, located along the mountainous region of southern Costa Rica, would experience a dramatic change in both types and number of unique climate zones represented in the reserve boundaries. Under the 3.6°C change scenario the eco-climatic zone associated with present pre-montane forests would significantly move up the slopes, displacing unique montane and subalpine climate types and transition zones, reducing the number of unique eco-climatic zones within the reserve area from nine to six (Figure 13). This change demonstrates an expected loss of the climate types now associated with the dwarf cloud forest and subalpine vegetation communities currently occupying the higher-altitude areas of the reserve system and the potential for significant dislocation of forest vegetation and associated faunal species within reserve areas.

While analysis of changes in the eco-climatic zones associated with present vegetation distributions offers a reasonable method for assessing general trends in the climatic features which affect floral and faunal habitats, this approach does not attempt to identify the particular responses of individual populations and species groups to specific climate changes. Individual species will obviously respond to climatic change through independent processes of acclimatization, migration, or adaptation (Peters and Darling 1985; Hunter *et al.* 1988; Graham 1988; Davis 1989). In this analysis, a change from an eco-climatic zone presently associated with a specific vegetation type to another is interpreted as a sign of potential ecological stress, not an explicit prediction of the future vegetation cover. The time frame of the change predicted by the GCMs for the regional change scenarios (e.g. 2060–2080) does not allow for an assessment of potential vegetation in terms of expected equilibria with future climate (Halpin *et al.* 1991a). The expected response times of canopy dominant tropical forest trees is in general far longer than the expected rate of change in climatic zonation.

Although the ability of individual species to tolerate rapid changes in climate is not generally known for the region, it can be assumed that these rapid changes in eco-climate would have a significant effect on individual species in terms of net primary production, regeneration, competitive ability, and dispersal. These potential changes in individual species responses to changes in climate zone could contribute to

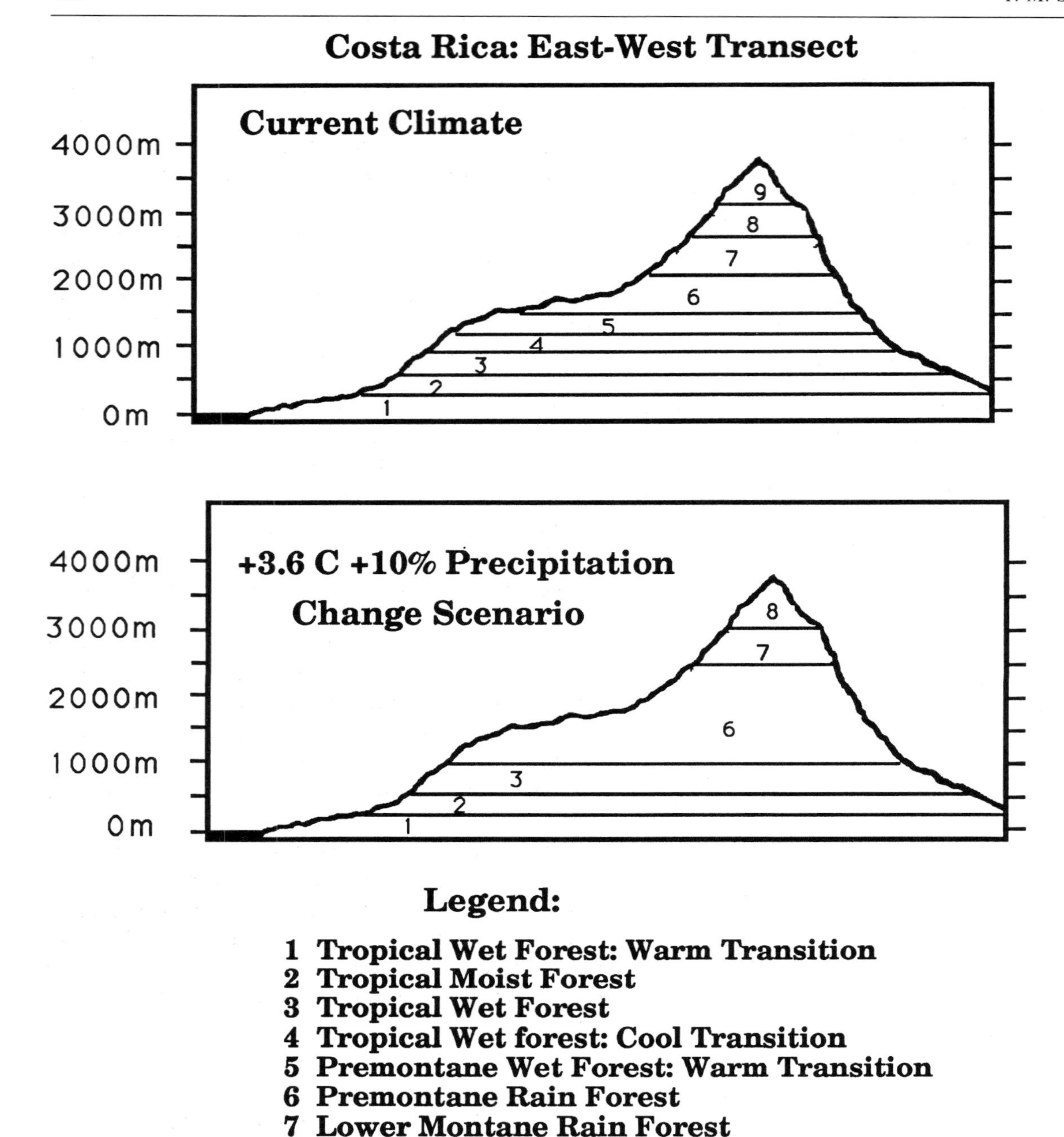

Figure 13 Distribution of eco-climatic zones along altitudinal gradients under current climate conditions and the 3.6°C scenario in Costa Rica (from Halpin and Smith 1992).

local extinctions of particular forest species, the creation of novel species associations, and significant changes in the abundance of highly specific habitat types often associated with much of the fauna of the neotropical region (Halpin and Secrett 1994).

General implications assessed by the Global Systems Analysis Program and the Tropical Science Center are summarized below:

- The economies of the forestry sector (both timber and fuelwood) and of the national system of protected areas will be particularly sensitive to any future climate changes due to their total dependencies on the natural environment.
- Global climatic change would significantly modify eco-climatic zones in the region. Land-use and forestry capability classifications would be extensively affected by the change assumed in the scenarios with significant heat stress occurring in low elevation forest areas.

- Three life zone types in the hot-dry and cold-wet extremes disappear from Costa Rica under the scenarios evaluated, indicating the possibility of significant impacts on the maintenance of habitats and species diversity in those regions.
- Loss of distinct eco-climatic zones along altitudinal gradients in the nature reserve system demonstrate a potential for a reduction in the landscape level diversity of critical habitats maintained within the reserve system.
- Agricultural production and cattle ranching land uses could be forced to higher elevations due to potential heat stress. This could increase land use pressures on the few natural forest areas remaining at these elevations.
- Future wood production is expected to decline on a per hectare basis due to the direct impacts of increased heat stress and water balance changes in lowland forest sites. Only if increases in forest area are realized will national forest productivity be maintained at current levels.
- In terms of optimal use of land capability, secondary growth (extensive forestry) practices would become, by default, the most appropriate land use in the lowland areas due to reductions in land capability for agricultural and pasture uses for these sites.
- Less productive forests could be potentially expanded to cover a wider area under the scenarios considered and potential increases in total production could be possible if secondary forest expansion is allowed to occur. This optimal use of land capability types would only occur if lands currently in agricultural use are allowed to revert back to forest regeneration.
- Increasing population pressures, critical food production requirements, expanding banana plantations, and continued forest clearing for cattle ranching are obvious obstacles to the return of lowland non-forest areas to extensive forestry practices at present.

ADAPTIVE MANAGEMENT RESPONSES FOR COSTA RICA

The regional scale analysis presented in this pilot project offers an initial assessment of potential change in eco-climatic zones at a spatial scale, which allows for the consideration of possible management responses to climatic change. Two general classes of anticipatory management responses can be developed. These are responses to direct climatic impacts on existing forest areas and responses to potential secondary changes in land use.

In general, management actions in response to an uncertain range of direct impacts on existing forest areas should include the maintenance of forest management options and natural ecosystem resilience to change (Halpin *et al.* 1991a; Secrett 1992). Direct impacts of climate change such as direct physiological stress, changes in species composition, changes in vegetation structure, changes in forest hydrology, and changes in disturbance regimes will be best met by healthy forest systems (Secrett 1992). In general, the maintenance of natural ecosystem resilience to climatic change could involve the application of management initiatives that promote natural vegetation regeneration and migration. Examples of this type of strategy would be programs that encourage the maintenance of viable populations of species, reduced protected-area fragmentation, low-impact timber harvesting techniques, and the maintenance of migration corridors spanning different habitat and altitudinal gradients. Sound, sustainable forestry practices and forest protection efforts, which would be beneficial under any climatic condition, would be especially useful in anticipation of changing environmental conditions. The possibility of future climatic change strengthens the need to invoke sustainable forest management practices, which have already been outlined in the Tropical Forest Action Plan, the Conservation Strategy for Costa Rica's Sustainable Development, and numerous existing forest policy mandates (Tosi *et al.* 1992).

Anticipatory responses to secondary changes in land use and land capability will involve the maintenance of land-use options for the future. A wider range of future management options can be maintained through conservative approaches to natural area land use that promote flexibility in the type of use over time. For example, the maintenance of multiple-use buffer zones around reserve areas increases the range of possible future land-use interventions if climatic change warrants alterations in the shape, size, or location of reserve areas in the future.

A general increase in the time frame used in assessing land-use programs and forest-use zoning would aid in the maintenance of land-use options. An increased use of geographic analysis optimizing eco-climatic land capability with expected forestry, biodiversity protection, and agricultural activities could increase efficient natural resource use and reduce risks to climatic impacts.

DRY TROPICS: SOUTHERN AFRICAN PILOT PROJECT

The Southern African Savanna/Woodlands Pilot Project included the assessment of potential impacts at two spatial scales and levels of analysis. An assessment of the potential impacts of climate change on vegetation structure, woody biomass, and nature reserves was conducted for the entire sub-equatorial region, with a special emphasis on the SADCC (Southern African Development Cooperative Council) member nations of Angola, Botswana, Lesotho, Malawi, Mozambique, Namibia, Swaziland, Tanzania, Zambia, and Zimbabwe. Country-level case studies were conducted to analyze the potential impacts of climate change on fuelwood provision, potential changes in surface erosion due to vegetation loss, as well as protected-area impacts in

Zimbabwe and fuelwood loss in Malawi. Due to similarities in the case studies, only the case study analysis of Zimbabwe will be presented here.

Regional forest and vegetation issues

Dry tropical forests and savanna regions exhibit a widely different set of ecological and management issues compared with the moist tropical sites discussed previously. Significant differences in vegetation structure and composition are related not only to annual inputs of radiation and precipitation, but also to highly variable seasonal patterns of water availability, fire disturbance regimes, herbivory, and human intervention. Production forestry is primarily limited to polewood production, industrial charcoal, and domestic fuelwood use in the region (Millington and Townsend 1989; ETC 1991; Munslow *et al.* 1988). Fuelwood resources, vegetation structure and soil loss changes, and the preservation of protected savanna–woodland habitats, are the three general management issues of concern in this analysis. All three issues are directly related to potential changes in vegetation biomass across the landscape.

Fuelwood is an extremely important renewable natural resource in Southern Africa, accounting for approximately 79 percent of energy use in the 10-nation SADCC region. With the human population in the region expected to exceed 100 million by the year 2000, fuelwood resources are expected to be in extremely high demand and short supply across the region (Munslow *et al.* 1988; Leach and Mearns 1988).

At a continental scale, potential changes in vegetation structure can be related to gross patterns of standing biomass and critical protected-area sites. In this regional overview, potential eco-climatic changes were compared to remotely sensed estimates of vegetation biomass to determine potential regional trends in biomass gain or loss. Potential changes were also assessed for nature reserve systems in order to assess potential regional impacts on biodiversity protection. Specific issues of fuelwood impacts, vegetation cover changes, and habitat protection were assessed in the country case study.

REGIONAL IMPACTS

Changes in the distribution of eco-climatic zones for sub-equatorial Africa under the four scenarios shown in Figures 2b–e (Plates 9b–e) are presented in Table 8. Eco-climatic (i.e., life) zones have been further aggregated to represent five structural classes of vegetation that categorize the region: (1) mesic forest, (2) dry forest/woodland, (3) savanna, (4) thorn shrub/grassland, and (5) desert (Figure 14, Plate 11).

Mesic forest cover would decline under the GFDL, GISS, and UKMO scenarios but would increase by 46 percent in the OSU scenario. The decrease in mesic forest represents an overall drying in the region with a subsequent transition to dry forest/woodland and savanna. In contrast, the OSU scenario predicts a significant increase in precipitation relative to potential evapotranspiration. These patterns of change in potential vegetation as predicted by the Holdridge model are in agreement with the simulated patterns of PET ratio for the region based on the Priestley–Taylor model (Priestley and Taylor 1972).

Table 8. *Changes in areal cover of five physiognomic vegetation classes for sub-equatorial Africa under the four climate change scenarios. Vegetation classes are based on aggregated Holdridge Life Zones.* Values for the scenarios represent losses or gains from current cover. All values are expressed in* $km^2 \times 10^3$

	Current	GFDL	GISS	OSU	UKMO
Mesic forest	310.5	−60.0	−57.4	142.2	−197.1
Dry forest/ woodland	385.6	23.9	26.1	−81.3	196.9
Savanna	170.8	32.0	55.2	−44.5	10.7
Shrub/grassland	51.9	−2.0	−13.5	−12.8	−12.1
Desert	38.9	5.9	−10.4	−3.6	1.6

Notes:
* Desert: warm temperate desert, subtropical desert
Shrub/grassland: subtropical desert bush, tropical desert, tropical desert bush
Savanna: subtropical thorn steppe, tropical thorn steppe, tropical very dry forest
Dry forest: warm temperate dry forest, subtropical dry forest, tropical dry forest
Mesic forest: moist, wet and rain forest warm temperate, subtropical and tropical

Although the Holdridge Life Zone Classification provides a good overall prediction of current vegetation patterns at a global scale, its basis on annual values of precipitation and evapotranspiration is weakest in semi-arid/arid areas with highly seasonal rainfall (Smith 1992). The region under study falls into this category, so as an alternative to the mapping of changes in eco-climatic zones based on the Holdridge Life Zone Classification, a bioclimatic model based on monthly patterns of evapotranspiration was developed.

Current vegetation patterns for the region were defined by developing a land cover classification for sub-equatorial Africa based on the Olson *et al.* (1983) land cover database (0.5° × 0.5° latitude/longitude resolution). The natural vegetation types for the region as defined by Olson were aggregated to provide five classes equivalent to those used in the

Holdridge analysis (as presented in Table 8): (1) mesic forest, (2) woodland/dry forest, (3) savanna, (4) shrub/grassland, and (5) desert/semi-desert. These classes are related to the absolute amount and seasonality of rainfall. A classification relating these five land cover classes to patterns of actual (AET) and potential (PET) evapotranspiration has been developed using discriminant function analysis (Smith 1992).

The distribution of the five structural classes under current climate conditions and the four climate change scenarios is shown in Figures 14a–e (Plates 11a–e). In general, the predicted patterns of change in land cover class agree with the shifts in eco-climatic zone outlined in Table 8. In both cases the shifts are a direct result of changes in potential evapotranspiration (as influenced by temperature) relative to changes in precipitation. As with the shifts in eco-climatic zone based on the Holdridge model, the changes in land cover class are dependent on the vegetation being able to track the changes in patterns of moisture availability over the time scale associated with the scenarios (i.e., 70–100 years).

REGIONAL BIOMASS IMPACTS

Recent estimates of vegetation biomass defined by phenology, productivity, and land cover derived from satellite remote sensing (AVHRR-NDVI) have been used as a basis for estimating the potential impacts of climatic change on existing biomass coverage (ETC 1991). Eight generalized classes of woody vegetation biomass have been developed for the Southern African region. Biomass use and population stresses on existing vegetation stands in southern Africa are strongly tied to the location of rural population concentrations and relative biomass distributions. Existing pressures on trees and shrubs are especially heavy in an arc of low woody biomass extending from southwest Angola, through Namibia, southern Zambia, Botswana, southern Zimbabwe, and South Africa (ETC 1991).

Three of the four climate change scenarios examined (UKMO, GISS, GFDL) show a negative impact on potential biomass production in the general regions of heavy fuelwood use due to significant decreases in effective moisture (AET/PET ratios) across all existing vegetation types. Only the OSU scenario maintains or increases potential moisture classes in the regions of intensive fuelwood use. Losses in available moisture can lead to decreases in mean annual growth increment in the short term and changes in vegetation structure over longer time scales for the region.

REGIONAL NATURE RESERVE IMPACTS

A wide distribution of regional nature reserves in the southern African region was tested to determine the general magnitude of impacts for the region as a whole. Country-scale case-studies were then used to illustrate the potential impacts at a more detailed level of analysis. The SADCC region member countries maintain a large number of highly significant protected areas, which include the Serengeti reserve complex and numerous other world renowned sites of high biodiversity and endemic species. Eighty-seven major protected areas representing sites from all ten countries of the SADCC region were located within a geographic information system to examine the potential effects of projected climate change on regional nature protection. Each site was assigned a current eco-climatic zone and then was compared to changes in climate induced from the GCM scenarios being analyzed. Significant numbers of the reserve areas shifted eco-climatic zones under all scenarios tested. Of the 87 reserves analyzed, 67.8 percent changed eco-climatic zone under the UKMO scenario, 72.4 percent under the OSU scenario, 50.5 percent under the GISS scenario, and 42.9 percent under the GFDL scenario. These changes occurred on significant expanses of important habitat throughout the region. Under the four scenarios tested, 36 million hectares (UKMO), 29 million hectares (OSU), 27.4 million hectares (GISS), and 18.9 million hectares (GFDL) of protected areas were impacted by eco-climatic changes.

Protected area managers will be concerned with changes in eco-climatic conditions which may alter the balance of habitat structure in their region. Changes that promote increases in woody biomass growth may be seen as a benefit for some species groups but could be gained at the expense of many grazing mammals that rely on more open grassland habitats. This distinction could be especially important in areas where the natural maintenance of vegetation structure by large browsers such as elephants and giraffes have been altered due to rapid declines in natural populations. The balance between herbivory and biomass maintenance could be significantly altered under rapidly changing climatic conditions.

It is important to identify fundamental changes toward more arid climates, which would bring about the most dramatic changes in habitat types within the reserve systems. This more conservative approach uses changes in the PET ratio from the Holdridge Life Zone Classification as well as a second method using the calculated relationships of actual evapotranspiration versus potential evapotranspiration for each site.

Changes in aridity derived from the ratio of actual to potential evapotranspiration (AET/PET; see 'Models of evapotranspiration') for each site are presented in Table 9. These more conservative estimates show that 17 percent of the protected areas would change to more arid habitat types under the UKMO scenario and 18.3 percent would do so under the GFDL scenario. All change recorded for the sites under the OSU and GISS models for the region show either

Table 9. *Percent of 87 Southern African nature reserves expected to experience available moisture changes (AET/PET ratio change)*

	UKMO	OSU	GISS	GFDL
All change	48	70	27.5	29.8
Change to drier climate	17	0	0	18.3
Change to wetter climate	31	70	27.5	11.4

no change in aridity index (AET/PET ratio) or increased moisture availability.

Zimbabwe fuelwood, protected area, and vegetation cover case study

Zimbabwe represents an excellent example of the convergence of arid, semi-arid, and moist climate types and a diverse spectrum of natural vegetation related to these moisture patterns (White 1983; Tinley 1982). Annual rainfall in Zimbabwe ranges from less than 300 mm in the southern portions of the country to over 2,000 mm in the highest reaches of the eastern mountain ranges. Rainfall is highly seasonal with a peaked distribution occurring between October and March and a variable mid-season dry period often occurring in January.

Zimbabwe also presents distinct differences in standing forest biomass related to land use and population density across the country. Distribution of lands into highly populated communal areas (CAs) and low-population large-scale commercial farms (LSCFs) allows for the investigation of diverse climatic change impacts on different land use sectors of a single country. In addition, Zimbabwe contains an extensive nature reserve system representing numerous sensitive vegetation habitats and faunal groups.

FOREST AND VEGETATION BACKGROUND

The natural forest and woody savanna regions of Zimbabwe contain representative sites of numerous Zambezian floral groups (White 1983). These woodland and savanna areas are distributed across both moisture and soil gradients within the country with higher density woodlands generally associated with sites exhibiting the highest effective soil moisture composition (Walter 1985; Walker 1991). Forest and woodland resources in Zimbabwe fall into two general categories: (1) indigenous natural woodlands used primarily for rural fuelwood, charcoal, and fodder production; and (2) exotic species plantations used for polewood, sawnwood, and industrial charcoal production. Remaining indigenous forests are found primarily in established nature reserve areas (2.7 million ha) or in fragmented patches within farming regions (900,000 ha). Indigenous forests currently contribute only 5 percent of the 700,000 m^3 of wood fiber used in construction, furniture, packaging, and paper products in Zimbabwe. The balance (95 percent) is derived from exotic pine and eucalyptus plantations in the region (Katerere 1991). Approximately 125,000 hectares (0.4 percent) of the country is in exotic tree plantations, mostly in the eastern highlands. State forestry offices manage 35 percent of the plantations, while the remainder are private lands.

The location of remaining stands of indigenous woodlands is extremely important with regard to population pressures on the sustainable use of wood fiber resources in the region. The area of woodlands within Communal Areas (CA), Large Scale Commercial Farms, and Protected Areas demonstrates a highly skewed distribution of wood resources by population. This distribution indicates that 44 percent of the total woodland in the country is located in the Large Scale Commercial Farms, where 14.5 percent of the total population resides, while only 20 percent of woodland resources are located in the Communal Areas, where 60 percent of the population lives (Katerere 1991). This unequal distribution of woody resources across the landscape may pose potentially significant differences in climatic change impacts and response for the different land use areas.

FUELWOOD IMPACTS

Vegetation biomass and fuelwood resources present a difficult problem for quantitative analysis due to the fact that only a small proportion of fuelwood is directly produced and traded in commodity markets (Leach and Mearns 1988; Munslow *et al.* 1988; Katerere 1991). Fuelwood has been neglected in much natural resource analysis because it has been treated as a readily available free good, directly consumed by rural populations. Closer analysis reveals that wood-based fuels are also heavily consumed for household energy uses by urban populations in the region. For example, up to 67 percent of household energy use in dense urban populations in Zimbabwe is attributed to fuelwood consumption (Munslow *et al.* 1988).

Changing patterns of biomass and fuelwood consumption may also have significant secondary and tertiary effects on landscape level resource management in the region. A SADCC regional report noted that increasing shortages of fuelwood resources for rural and urban use have resulted in the fast depletion of woody vegetative cover, which is essential for environmental protection, soil and water conservation, and the maintenance of agricultural productivity (SADCC 1986). Potential losses in fuelwood provision due to climatic change could act to exacerbate these land management problems.

Analysis of the possible effects of climate change on vegetation biomass and fuelwood provision for the region was based on projected spatial changes in the environmental potential for maintaining current levels of productivity. This eco-climatic mapping was then overlaid onto the spatial distributions of land-use management categories, remotely sensed vegetation mapping of woody biomass, human population density, and fuelwood stress areas.

The country case studies were centered on the principle that significant changes in eco-climatic type or available plant moisture (P-PET) on areas of high fuelwood use could alter the potential for sustainable fuelwood production. Areas of potential fuelwood stress were determined for this analysis to be regions that exhibit low vegetation biomass and high population or high fuelwood consumption patterns (Munslow *et al.* 1988; Millington and Townsend 1989). As an initial assessment of potential impacts of climate change, changes in eco-climatic zones and available plant moisture were overlaid onto mapping of current land use, agroecological zones, fuelwood stress areas, and maps of population concentrations.

High-resolution climatic datasets on long-term annual temperature, precipitation, effective rainfall, duration of the rainy season, soils, topography, and radiation were assembled in a geographic information system and compared to existing taxonomic vegetation mapping and remotely sensed biomass estimates for Zimbabwe. This analysis was conducted at a spatial resolution of $0.01° \times 0.01°$ latitude/longitude, or approximately 1.1 km grid cell size. The analysis identified general trends in vegetation structure across eco-climatic zones primarily in relationship to moisture gradients found in the region.

Eco-climatic (Holdridge Life Zone) maps were generated for Zimbabwe, as they were for Central Africa, and were used to estimate the spatial distribution of ecologically significant changes in climate under the four scenarios tested. Unlike the humid tropics case study undertaken for Central America, the regional outcomes of the general circulation models for Zimbabwe showed distinct variability among models.

When climate change values are converted into regional eco-climatic zone shifts, from 69 percent to 100 percent of Zimbabwe experiences a shift in eco-climatic life zone or transition zone (see 'Approach and methodology' for Central American study). The spatial areas impacted by changes in eco-climatic zone under each scenario are shown in Figure 15.

These high estimates of potential eco-climatic change must be tempered by the fact that they are identifying any change in eco-climatic zone without regard to the direction of change (i.e., wetter or drier). Because vegetative biomass in semi-arid regions is dependent on changes in moisture available for plant growth (Tinley 1982) a more conservative approach is to define only changes to more arid climate zones. In addition, changes in the distribution of eco-climatic zones must be defined in terms of the actual vegetation biomass and land-use category if one is to derive a more useful assessment of the impacts in terms of fuelwood resources.

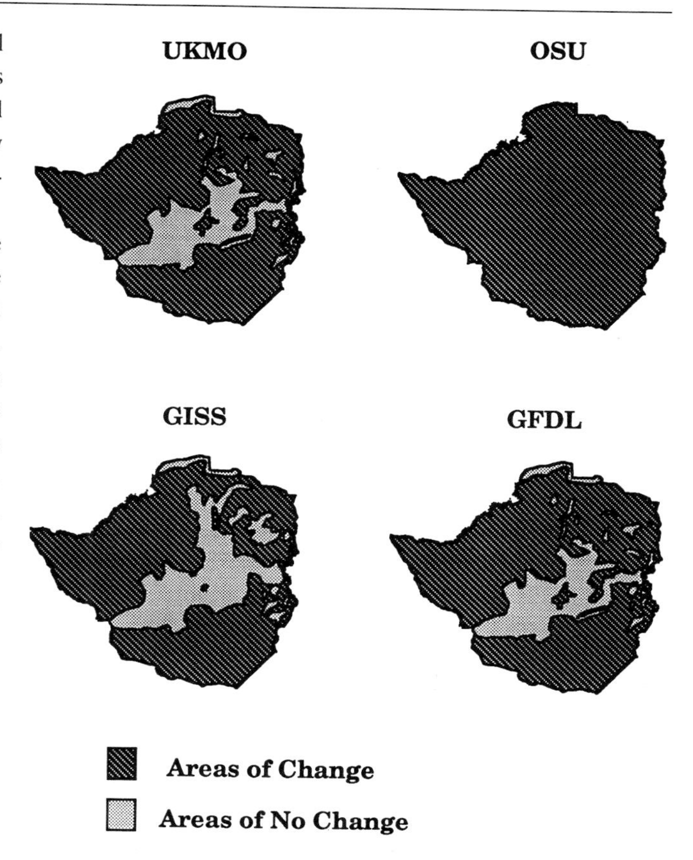

Figure 15 Areas experiencing changes in eco-climatic zone in Zimbabwe under the UKMO, OSU, GISS, and GFDL climate change scenarios.

Areas of different land management types in Zimbabwe that would change to significantly more arid conditions under the four scenarios are shown in Figure 16. As can be noted in the histograms, the OSU model, which produced 100 percent change in terms of eco-climatic zones, estimated no change to more arid types for this region. The other GCM models showed more conservative, but still significant, estimates of climate change impacts when viewed in this more highly circumscribed analysis. In particular, the UKMO and GISS models are seen to change the communal areas, forest reserves, protected areas and safari areas to significantly more arid environments. These impacts could be especially severe in the communal areas, where 75 percent of rural wood demand is located.

Another assessment was to compare the potential changes in plant moisture stress to estimates of vegetation biomass derived from AVHRR/NDVI remote sensing for the region. Milington and Townsend (1989) produced mapped estimates of structural vegetation classes associated with standing stocks of woody biomass and mean annual growth

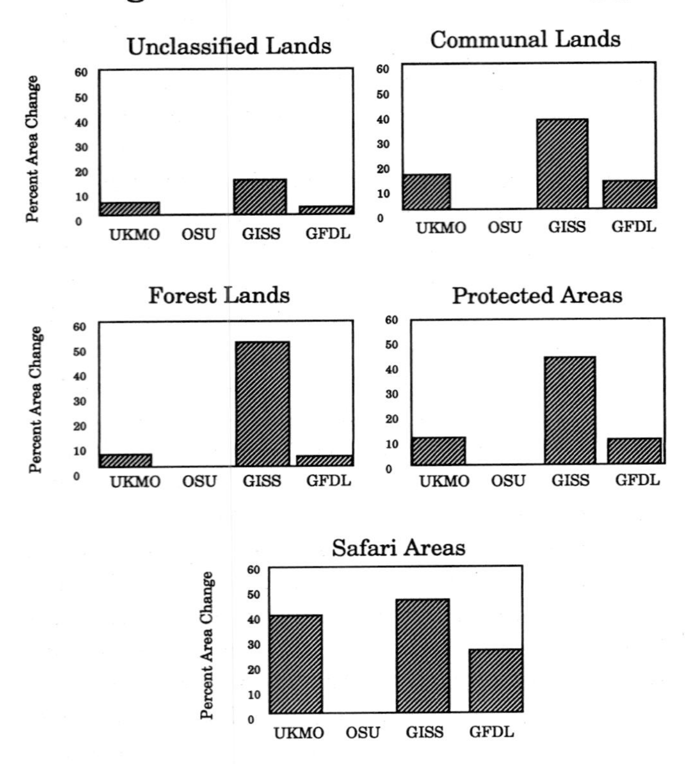

Figure 16 Percentage eco-climatic zone changes by land-use type for Zimbabwe.

increment. Mapping from the study was compared to changes in eco-climatic zones and plant moisture derived in the present study to produce estimates of change in each of the major vegetation biomass types for the region. The potential impacts on each of these biomass types in terms of the percent of each class that changed to a more arid climate zone are shown in Table 10. This type of analysis is useful because it associates directional changes in climate to specific estimates of woody standing stock of biomass. The standing stock values for woody biomass at risk to changes in effective moisture under each scenario for Zimbabwe are given in Table 11.

The final level of impact analysis required the development of population density mapping in conjunction with current biomass supply levels and rural land use. Building on work developed by Munslow *et al.* (1988), we classified communal land areas in Zimbabwe as high, medium, or low biomass supply areas based on an analysis of the underlying vegetation structure determined by remote sensing. We added high-resolution human population mapping into the analysis in order to indicate general trends of fuelwood demand in the high, medium, and low biomass districts. The final product was an identification of current fuelwood stress areas and an assessment of changes in those areas to more arid, and potentially less sustainable, biomass producing sites.

Table 10. *Percent of vegetation biomass classes which change to more arid climate types in Zimbabwe*

	UKMO	OSU	GISS	GFDL
Dense savanna woodland	1.82	0	33.56	1.41
Open savanna woodland	4.21	0	22.31	2.49
Seasonal savanna woodland	7.17	0	17.83	5.87
Dry savanna woodland	32.27	0	40.58	17.67
Mopane woodland	38.7	0	47.85	28.07
Commercial arable	3.71	0	13.10	2.12
Dry bushy savanna	0.04	0	13.69	0.04
Wooded grassland	13.83	0	20.17	11.80
Plantation	0	0	27.54	0

Table 11. *Volume of standing stock in Zimbabwe at risk due to increases in aridity under the four climate change scenarios. Standing stock in millions of metric tons*

	Current	UKMO	OSU	GISS	GFDL
Dense savanna woodland	134.2	2.44	0	45.03	1.89
Open savanna woodland	839.0	35.32	0	187.18	20.89
Seasonal savanna woodland	6.2	0.44	0	1.10	0.36
Dry savanna woodland	45.1	14.55	0	18.30	7.96
Mopane woodland	253.5	98.10	0	121.29	71.15
Commercial arable	214.1	7.94	0	28.04	4.53
Dry bushy savanna	1.7	0.006	0	0.23	0.006
Wooded grassland	7.6	1.05	0	1.53	0.89

Tables 12 and 13 depict the aggregate changes in the three general classes of rural biomass land use areas and describe the present populations supported in those areas. This assessment depicts only climate change impacts on rural fuelwood regions; it does not attempt to trace potential changes in aridity to impacts in fuelwood consumption for urban populations.

PROTECTED-AREAS IMPACTS

A country case study of the impacts of climate change on protected area habitats was conducted for Zimbabwe. This analysis identified a very high percentage of the nation's protected-area system experiencing changes in eco-climatic zones. Approximately 80 percent to 100 percent of the reserve areas in Zimbabwe would experience significant changes in eco-climatic zones. These changes would be distributed across a wide variety of reserve areas and habitat types. When the more conservative approach of examining

Table 12. *Changes to more arid climates in low, medium, and high biomass rural communal areas in Zimbabwe*

	UKMO	OSU	GISS	GFDL
High biomass	4.74	0	17.87	3.34
Medium biomass	35.55	0	50.42	24.93
Low biomass	0.01	0	16.47	0.01
High/low biomass areas	0	0	30.65	0

Table 13. *Population in high, medium, and low biomass rural communal land areas of Zimbabwe*

High biomass areas	1,094,500
Medium biomass areas	842,000
Low biomass areas	666,000
High/low biomass areas	422,500

only increases in aridity was used, 12 percent (UKMO), 0 percent (OSU), 36 percent (GISS), and 7 percent (GFDL) of the reserve areas were found to change to significantly drier habitat types.

POTENTIAL VEGETATION COVER-SOIL EROSION IMPACTS

Changes in potential vegetation structure in response to changing climate can entail significant time lags as competition and disturbance regimes alter the species composition of a site. Rapid changes in surface hydrology and denudation rates may occur in semi-arid regions before vegetation coverage regains a structural equilibrium with climatic conditions. Semi-arid regions are especially vulnerable to increased surface erosion due to changes in expected runoff (Walker 1991). In general, relatively small increases in precipitation and subsequent runoff may significantly increase erosion rates in the short term for semi-arid savanna woodlands. A nonlinear relationship between runoff and erosion has been documented for semi-arid regions experiencing between 0 and 50 mm of runoff per year (Walker 1991). Surface erosion and land degradation have been found to be especially prevalent in the high-biomass-use communal lands of Zimbabwe (Whitlow and Cambell 1989; Katerere 1991).

An assessment of erosional surfaces, soil texture, topographic relief, vegetation coverage, and current runoff was conducted for six major watershed regions in Zimbabwe under the relatively dry UKMO and relatively wet OSU climate change scenarios (Halpin 1992). Regionally developed empirical rainfall-runoff relationships were employed to derive changes under the scenarios considered (Dept. of Meteorological Services 1981). Under this general assessment, runoff could significantly increase (138 percent OSU, 58 percent UKMO) in the short term, increasing surface erosion before vegetation coverage would be expected to increase density. The southwestern region of Zimbabwe was identified as the most susceptible region to increased surface erosion, due to high levels of potential runoff, mountainous terrain, potential vegetation biomass loss, and easily transported soil surface material (Halpin 1992). The combination of short-term erosional impacts plus the expected decreases in vegetation biomass was seen to exacerbate problems in sustainable woodland management and protection under changing climatic conditions under all scenarios tested.

In conclusion, the country case studies of potential fuelwood impacts indicated that dramatic changes in eco-climatic zones would occur under all scenarios. Variability in the predicted changes in aridity between the models produced mixed results between the scenarios. Under conservative estimates, however, three scenarios (UKMO, GFDL, and GISS) showed potential changes to lower biomass vegetation types (i.e., lower woody production). When population, land-use category, and current vegetation stocks were considered, significant areas of fuelwood stress were found to lose capacity for sustainable production in the region. Reduced woody production could engender a cascade of secondary land management and economic impacts on sustainable woodland management in the already overburdened communal area regions of the country.

POLICY RESPONSES

Due to pervasive problems in providing basic human needs in the highly populated communal areas, short-term issues concerning daily survival generally displace long-term natural resource planning initiatives in the region. Policies that promote long-term sustainability of woody vegetation resources and resilience to climatic impacts in the region will have to be incorporated into management plans that demonstrate clear benefits to the population in the short term (Katerere 1991; Secrett 1992; Halpin 1992).

If infrastructure support is maintained in the future, adaptation of existing plantation areas to changing climatic conditions would not be without problem, but those areas would be significantly more amenable to intervention than the natural woodland areas. Changes in species composition and in planting and harvesting practices to match climatic conditions could be envisioned due to the short rotation periods of most tree crops used in the region. Maintenance of natural woody vegetation stands experiencing climatic stress would require more extensive controls over allowable sustainable yields as a means of reducing human impacts on the forest stands. Reduced intervention would promote the maintenance of site conditions and soil properties required for sustainable forest uses.

Protected-area management will require the retention of the highest land and buffer area possible to allow for the potential migration of fauna and the regeneration of stressed ecosystems in the short term and the natural redistribution of vegetation habitat types in the long term. Intermittent water supplies for unique vegetation and fauna may be at risk of reduction from climate change. Therefore, forest management should also consider protection of runoff sources.

CONCLUSIONS

The results of the Global Forest Project show the sensitivity of terrestrial vegetation to climate change. The changes in the global climate patterns represented by the GCM models would have a major impact on the spatial extent, productivity, and ecological composition of boreal and tropical forest zones. Potential reductions in forest cover in the high latitudes could have a significant negative impact on forest production in the boreal zone. Changes in the spatial patterns of effective moisture coupled with ongoing forest fragmentation could lead to further reductions in the distribution of mesic forests in the tropics. Although in some cases these declines could be potentially offset by the expansion of current forested areas, the time required for the migration and re-establishment of forest species would be much greater than the rate of forest decline from increased temperatures and changes in rainfall patterns.

The ability to select suitable species for plantation forestry may enable the forestry and fuelwood sectors to offset potential declines in production of native forests; however, the impacts on naturally maintained forests and nature conservation could be severe. The ability of forest managers and natural resource policy makers to respond to the risks posed to forest ecosystems by climate change is directly limited by the generally low level of management intervention possible in extensive natural forest areas in both the boreal and tropical regions. The ability of managers to respond to the risks posed to protected areas is additionally limited by species isolation due to landscape fragmentation, species losses, and introduced species. The combined forces of rapid climate change and habitat isolation could act to greatly exacerbate rates of species extinction in forest reserves.

The potential effects of climate change on forest systems can be divided into direct impacts on forest productivity and composition and indirect impacts involving secondary changes in land-use responses to changing environmental conditions. Both of these types of impacts must be considered in any assessment of forest impacts.

As indicated in this study, the assessment of specific impacts of climate change on forest systems must be conducted at a regional scale of analysis. The interacting effects of shifts in forest composition, land fragmentation, and potential changes in land capability can only be observed at such a scale. Also, inherent forest management problems vary significantly from region to region as shown in the analysis of natural forest production in Central America and fuelwood biomass in southern Africa.

The humid tropical forest system case study conducted in Central America demonstrated that tropical forest production, protection, and watershed management are at significant risk to both primary impacts of climate change due to increased heat stress and secondary impacts due to changes in land capability. Only if the area of secondary forest regrowth is allowed to increase would production be maintained under the scenarios investigated. Of special concern in this region is the high risk for increasing losses of remaining high-biodiversity rainforest stands to less diverse forest systems or other land-use types under climate change conditions.

The dry tropical forest system case study conducted in southern Africa indicated that woody biomass provision was at special risk to changes in effective moisture classes (AET/PET) and potential increases in surface erosion. Areas of high rural population and high biomass use were shown to be at increased risk to land degradation and unsustainable woodland management under changing conditions. A highly significant number of large nature reserves, containing populations of important wildlife species, were also observed to be at considerable risk to changes in habitat type under the scenarios tested.

Regional natural resource policy adaptations and forest management responses to climate change were found to rely primarily on the promotion of natural forest resilience to changing conditions. This 'no-regrets' approach to anticipatory climate change response can generally be divided into actions that promote the continued spatial integrity of forest systems and policies that promote the sustainable use of the existing forest resource base. Both of these approaches suggest activities that should be consistent with the long-term needs of tropical countries with or without the occurrence of climatic change.

Regional-scale analysis of potential forest impacts and possible anticipatory responses also promote two important features that potentially allow for the separation of local analysis and decision making from global assessments. Detailed assessments of the sensitivity of forest ecosystems to incremental changes in climate can be undertaken at regional or local scales independent from the controversy and complexity surrounding global models. This separation of forest sensitivity analysis from the long-term, centralized development of global circulation models may allow for a greater participation of regional experts in the advanced assessment of potential impacts in their regions. Advance knowledge concerning possible direct and indirect forest impacts to a wide range of climate conditions will allow regional planners the greatest opportunity to invoke sensible, anticipatory responses in the most efficient and timely manner possible if climatic conditions change in the future.

REFERENCES

Arris, L. L. and P. S. Eagleson. 1989. A Physiological Explanation for Vegetation Ecozones in Eastern North America. Ralph M. Parsons Lab. Rep. 323, 24797, Dep. Civ. Eng., Mass. Inst. of Technol., Cambridge, MA.

Bonan, G. B. 1989. A computer model of the solar radiation, soil moisture, and soil thermal regimes in boreal forests. *Ecological Modeling* 45:275–306.

Bonan, G. B. 1990a. Carbon and nitrogen cycling in North American boreal forests. I. Litter quality and soil thermal effects in interior Alaska. *Biogeochemistry* 10:1–28.

Bonan, G. B. 1990b. Carbon and nitrogen cycling in North American boreal forests. II. Biogeographic patterns. *Canadian Journal of Forest Research* 20:1077–88.

Bonan, G. B., and B. P. Hayden. 1990. Using a forest stand simulation model to examine the ecological and climatic significance of the late-Quaternary pine-spruce pollen zone in eastern Virginia, USA. *Quaternary Research* 33:204–18.

Bonan, G. B., H. H. Shugart, and D. L. Urban. 1990. The sensitivity of some high-latitude boreal forests to climatic parameters. *Climatic Change* 16:9–29.

Bonan, G. B., and L. Sirois. 1992. Air temperature, tree growth and the northern and southern range limits to *Picea mariana*. *Journal of Vegetation Science* 3:495-506.

Box, E. O. 1978. Ecoclimatic determination of terrestrial vegetation physiognomy. Ph.D. diss., University of North Carolina, Chapel Hill, NC.

Box, E. O. 1981. *Macroclimate and Plant Forms: An Introduction to Predictive Modeling in Phytogeography*. The Hague: Junk.

Calvo, J. C. 1990. The Costa Rican national conservation strategy for sustainable development: exploring the possibilities. *Environmental Conservation* 17(4):355–8

Cruz, W., and R. Reppetto. 1992. *Resource Accounting in Costa Rica*. Washington, DC: World Resources Institute.

Davis, M. B. 1976. Pleistocene biogeography of temperate deciduous forests. *Geoscience and Man* 13:13–26.

Davis, M. B. 1984. Climatic instability, time lags and community disequilibrium. In *Community Ecology*, eds. J. Diamond and T. J. Case, 269–84. New York: Harper and Row.

Davis, M. B. 1989. Lags in vegetation response to greenhouse warming. *Climatic Change* 15:75–82.

Department of Meteorological Services (DMS). 1981. *Climate Handbook of Zimbabwe*. Harare, Zimbabwe: DMS.

Emanuel, W. R., H. H. Shugart, and M. P. Stevenson. 1985. Climatic change and the broad-scale distribution of terrestrial ecosystem complexes. *Climatic Change* 7:29–43.

Erwin, T. L. 1988. The tropical forest canopy: the heart of biotic diversity. In *Biodiversity*, eds. E. O. Wilson and Frances M. Peter, 123–9. Washington, DC: National Academy Press.

ETC. 1991. Biomass Assessment in Africa. Project Report.

Flores Rodas, J. G. 1985. *Diagnostico del sector industrial forestal*. EUNED-UNED/PEA, DGF/MAG, FPN. San Jose, Costa Rica: Tinker.

Fundacion Neotropica (FN). 1988. *Desarrollo Socioeconomico y el Ambiente Natural de Costa Rica*. San Jose, Costa Rica: FN.

Fundacion Neotropica and Conservation International (FN/CI). 1988. *Costa Rica: Assessment of the Conservation of Biological Resources*. Washington, DC: CI.

Gomez, L. D., and W. Herrera. 1986. *Vegatacion y Clima de Costa Rica*, Vol 1. San Jose, Costa Rica: EUNED.

Graham, R. W. 1988. The role of climate change in the design of biological reserves: the paleoecological perspective for conservation biology. *Conservation Biology* 2(4):391–4.

Granger, O. E. 1989. Implications for Caribbean Societies of climate change, sea-level rise and shifts in storm patterns. In *Proceedings of the Second North American Conference on Preparing for Climatic Change: A Cooperative Approach*, ed. J. Topping Jr. Washington, DC: The Climate Institute.

Griggs, R. F. 1937. Timberlines as indicators of climatic trends. *Science* 85:251–5.

Grisebach, A. 1838. Ueber den Einfluss des Climas auf die Begranzung der naturlichen Floren. *Linnaea* 12:159–200.

Halpin, P. N. 1992. *Climatic Sensitivity of Vegetation Biomass and Denudation Rates in Southern Africa*.

Halpin, P. N., P. M. Kelly, C. M. Secrett and T. M. Smith. 1991a. *Climate Change and Central American Forest Systems: Costa Rica Pilot Project* (project background report).

Halpin, P. N., and C. M. Secrett. 1994. Potential impacts of climate change on forest production in the humid tropics: a case study of Costa Rica. In *Impacts of Climate Change on Ecosystems and Species (ICCES); Vol. 2 Terrestrial Ecosystems*. IUCN, Gland, Switzerland.

Halpin, P. N., and T. M. Smith. 1992. Potential impacts of climate change on forest protection in the humid tropics: A case study in Costa Rica. *Proceedings of the Symposium on Impacts of Climate Change on Ecosystems and Species*. Amersfoot, Netherlands.

Halpin, P. N., and T. M. Smith. In prep. The potential impacts of climatic change on nature reserve systems at global and regional scales.

Hansen, J., I. Fung, A. Lacis, J. Lerner, R. Ruedy, D. Rind, G. Russel, and P. Stone. 1984. Climate sensitivity: analysis of feedback mechanisms. In *Climate Processes and Climate Sensitivity*, eds. J. Hansen and R. Thompson. Geophysical Monogr. 29. Washington, DC: American Geophysical Union.

Hare, F. K., and J. C. Ritchie. 1972. The boreal bioclimate. *Geographical Review* 62:333–65.

Hartshorn, G. S. 1983. Plants. In *Costa Rica Natural History*, ed. D. Janzen. Chicago: University of Chicago Press.

Hartshorn, G. S. *et al.* 1982. *Costa Rica Country Environmental Profile: a filed study*. Tropical Science Center/USAID, San Jose, Costa Rica.

Holdridge, L. R. 1949. Determination of world plant formations from simple climate data. *Science* 105(2727):367–8.

Holdridge, L. R. 1959. Simple method for determining potential evapotranspiration from temperature data. *Science* 130:572.

Holdridge, L. R. 1967. *Life Zone Ecology*. San Jose, Costa Rica: Tropical Science Center.

Holdridge, L. R., W. C. Grenke, W. H. Hatheway, T. Liang, and J. A. Tosi, Jr. 1971. *Forest Environments in Tropical Zones: A Pilot Study*. Oxford: Pergamon Press.

Holdridge, L. R., and J. A. Tosi, Jr. 1972. *The World Life Zone Classification System and Forestry Research*. Seventh World Forestry Congress. 7CFM/C:V2G. Rome: FAO.

Hunter, M. L., G. L. Jacobson, Jr., and T. Webb III. 1988. Paleoecology and the coarse filter approach to maintaining biological diversity. *Conservation Biology* 2(4):375–85.

IDB. 1987. Forest Development Pilot Project (CR-0070). Project Report PR-1583-A. Washington, DC.

IUCN. 1986. MAB information system: Biosphere Reserves Compilation 4, 1986. IUCN Conservation Monitoring Center.

Jones, P. D., T. M. L. Wigley, and P. M. Kelly. 1982. Variations in surface air temperatures: part 1, Northern Hemisphere, 1881–1980. *Monthly Weather Review* 110:59–70.

Jones, P. D., T. M. L. Wigley, and P. B. Wright. 1986. Global temperature variation between 1861 and 1984. *Nature* 322:430–4.

Katerere, Y. 1991. Woodland Management and Utilization in Zimbabwe. Report prepared for the International Institute for Environment and Development, London.

Kelly, P. M. 1991a. *Regional Climate Change Scenarios for Costa Rica*. Climatic Research Unit, Univ. of East Anglia. (unpublished report).

Koppen, W. 1900. Versuch einer Klassification der Klimate, vorzugsweise nach ihren Beziehungen zur pflanzenwelt. *Geogr. Z.* 6:593–611.

Koppen, W. 1918. Klassification der Klimate nach Temperatur, Niedenschlag und Jahreslauf. *Petermanns Geogr. Mitt.* 64:193–203.

Koppen, W. 1936. Das Geographische System der Klimate. In *Handbuch der Klimatologie*, vol. 1, part C, eds. W. Koppen and R. Geiger. Berlin: Gebr Borntraeger.

Leach, G., and R. Mearns. 1988. *Beyond the Woodfuel Crisis: People,*

Land and Trees in Africa. London: Earthscan.
Leemans, R., and W. P. Cramer. 1990. The IIASA database for mean monthly values of temperature, precipitation and cloudiness on a global terrestrial grid. WP-90-41. Laxenburg, Austria: International Institute for Applied Systems Analysis.
Leemans, R., and P. N. Halpin. 1992. Biodiversity and global change. In *Biodiversity Status of the Earth's Living Resources*, ed. J. McComb. Cambridge, UK: World Conservation Monitoring Center.
Lugo, A.E. 1988. Estimating reductions in the diversity of tropical forest species. In *Biodiversity*, eds. E.O. Wilson and Frances M. Peter, 58–70. Washington, DC: National Academy Press.
Mackintosh, G., J. Fitzgerald, and D. Kloepfer. 1989. *Preserving Communities and Corridors. Defenders of Wildlife*. Washington, DC: G. W. Press.
Manabe, S., and R. J. Stouffer. 1980. Sensitivity of a global climate to an increase in CO_2 concentration in the atmosphere. *Journal of Geophysical Research* 85:5529–54.
Maul, G. A. 1989. Implications of climatic change for the wider Caribbean region: an overview. In *Proceedings of the Second North American Conference on Preparing for Climatic Change: A Cooperative Approach*, ed. J. Topping Jr., Washington, DC: The Climate Institute.
Millington, A., and J. Townsend. 1989. *Biomass Assessment: Woody Biomass in the SADCC Region*. London: Earthscan Publications Ltd.
Monteith, J. L. 1973. *Principles of Environmental Physics*. London: E. Arnold.
Morin, A., and S. Payette. 1984. Expansion récente du mélèze à la limite des forêts (Québec nordique). *Canadian Journal of Botany* 62:1404–08.
Munslow, B., Y. Katerere, A. Ferf, and P. O'Keefe. 1988. *The Fuelwood Trap: A Study of the SADCC Region*. London: Earthscan Publications Ltd.
O'Brien, S., B. P Hayden, and H. H. Shugart. 1992. Global climatic change, hurricanes and a tropical forest. *Climatic Change* 22:175–90.
Olson, J.S., J.A. Watts, and L. J. Allison. 1983. Carbon in live vegetation of major world ecosystems. ESD Pub. No. 1997. Oak Ridge, TN: Oak Ridge National Laboratory.
Overpeck, J. T., D. Rind, and R. Goldberg. 1990. Climate-induced changes in forest disturbance and vegetation. *Nature* 343:51–3.
PAFT-CR. 1990. Plan de Accion Forestal para Costa Rica. Symposium Document. San Jose, Costa Rica: PAFT-CR.
Pastor, J., and D. J. Mlandenoff. 1992. The southern boreal–northern hardwood forest border. In *A Systems Analysis of the Global Boreal Forest*, eds. H. H. Shugart, R. Leemans and G. B. Bonans. Cambridge University Press, New York.
Pastor, J., and W. M. Post. 1985. Development of a linked forest productivity–soil process model. ORNL/TM-9519. Oak Ridge, TN: Oak Ridge National Laboratory.
Pastor, J., and W. M. Post. 1986. Influences of climate, soil moisture, and succession on forest carbon and nitrogen cycles. *Biogeochemistry* 2:3–27.
Pastor, J., and W. M. Post. 1988. Response of northern forests to CO_2-induced climate change. *Nature* 334:55–8.
Payette, S., and L. Filion. 1985. White spruce expansion at the treeline and recent climatic change. *Canadian Journal of Forest Research* 15:241–51.
Penman, H. L. 1948. Natural evaporation from open water, bare soil and grass. *Proceedings of the Royal Society of London, Series A* 193:120–45.
Peteet, D. 1987. Late Quaternary vegetation and climatic history of the montane and lowland tropics. In *Climate–Vegetation Interactions*, eds. C. Rosenzweig and R. Dickinson, 72–6. Boulder, CO: University Corporation for Atmospheric Research.
Peters, R. L., and J. D. Darling. 1985. The greenhouse effect and nature reserves. *Bioscience* 35(11):707–17.
Prentice, K.C. 1990. Bioclimatic distribution of vegetation for GCM studies. *Journal of Geophysical Research* 95:11811–30.
Prentice, K. C., and I. Y. Fung. 1990. Bioclimatic simulations test the sensitivity of terrestrial carbon storage to perturbed climates. *Nature* 346:48–51.
Priestley, C. H. B., and R. J. Taylor. 1972. On the assessment of surface heat flux and evaporation using large-scale parameters. *Monthly Weather Review* 100:81–92.
SADCC. 1986. Food and Agricultural Report, 1986.
Sader, S. A., and A. T. Joyce. 1988. Deforestation rates and trends in Costa Rica: 1940 to 1983. *Biotropica* 20(1):11–19.
Sawyer, J., and A. Lindsey. 1971. *Vegetation of the Life Zones in Costa Rica*. Indiana Academy of Sci. Monogr. No. 2.
Scott, P. A., R. I. C. Hansell, and D. C. Fayle. 1987. Establishment of white spruce populations and responses to climatic change at the treeline, Churchill, Manitoba, Canada. *Arctic and Alpine Research* 19:45–51.
Secrett, C. M. 1991. Costa Rica Forest Resources Overview. International Institute for Environment and Development Project Paper, London.
Secrett, C. M. 1992. Adapting to Climate Change: A Strategy for the Tropical Forest Sector. International Institute for Environment and Development Project Paper, London.
Sellers, P. 1987. Canopy reflectance, photosynthesis and transpiration. II. The role of biophysics in the linearity of their interdependence. *Remote Sensing of Environment* 21:143–83.
Shugart, H. H. 1984. *A Theory of Forest Dynamics*. New York: Springer-Verlag.
Shugart, H. H., and D. C. West. 1980. Forest succession models. *BioScience* 30:308–13.
Shukla, J., and Y. Mintz. 1982. Influence of land-surface evapotranspiration on the earth's climate. *Science* 215:1498–501.
Sirois, L. 1992. A phyto-ecological investigation of the Mount Albert serpentine plateau. In *The Ecology of Areas with Serpentinized Rocks. A World View*, 115–33. Netherlands: Kluwer Academic Publishers.
Smith, J. B. 1992. Incorporation of seasonality into a bioclimatic model to predict vegetation distribution in Africa. Ph.D. diss., University of Virginia.
Smith, T. M., and P. S. Goodman. 1986. The effects of competition on the structure and dynamics of Acacia savannas in southern Africa. *Journal of Ecology* 74:1031–44.
Smith, T. M., and P. S. Goodman. 1987. Successional dynamics in an *Acacia nilotica – Euclea divinarum* savanna in southern Africa. *Journal of Ecology* 75:603–10.
Smith, T. M., H. H. Shugart, and P. N. Halpin. 1990. Global forests. In *Progress Reports on International Studies of Climate Change Impacts*. Washington, DC: USEPA.
Smith, T. M., R. Leemans, and H. H. Shugart. 1992a. Sensitivity of terrestrial carbon storage to CO_2-induced climate change: comparison of four scenarios based on general circulation models. *Climatic Change* 21:367–84.
Smith, T. M., H. H. Shugart, G. B. Bonan, and J. B. Smith. 1992b. Modeling the potential response of vegetation to global climate change. *Advances in Ecological Research* 22:93–113.
Solomon, A. M. 1986. Transient responses of forests to CO_2-induced climate change: Simulation modeling experiments in eastern North America. *Oecologia* 68:567–9.
Solomon, A. M., H. R. Delcourt, D. C. West, and T. J. Blasings. 1980. Testing a simulation model for reconstruction of prehistoric forest-stand dynamics. *Quaternary Research* 14:275–93.
Solomon, A. M., D. C. West, and J. A. Solomon. 1981. Simulating the role of climate change and species immigration in forest succession. In *Forest Succession*, eds. D. C. West, H. H. Shugart, and D. B. Botkin, 154–77. New York: Springer-Verlag.
Solomon, A. M., and H. H. Shugart. 1984. Integrating forest-stand simulations with paleoecological records to examine long-term forest dynamics. In *State and Change of Forest Ecosystems: Indicators in Current Research*, ed. G. I. Agren, 333–57. Report Number 13. Upsalla, Sweden: Swedish University of Agricultural Science.
Solomon, A. M., M. L. Tharp, D. C. West, G. E. Taylor, J. M. Webb, and J. L. Trimble. 1984. Response of unmanaged forests to CO_2-induced climate change: available information, initial tests and data requirements. Tech. Report TR009. Washington, DC: U.S. DOE Carbon Dioxide Research Division.
Solomon, A. M., and T. Webb III. 1985. Computer-aided reconstruction of late-quaternary landscape dynamics. *Ann. Rev. Ecol. Syst.* 16:63–84.
Sonesson, M., and J. Hoogesteger. 1983. Recent tree-line dynamics (Betula pubescens Ehrh. ssp. tortuosa (Ledeb.) Nyman) in northern Sweden. *Nordicana* 47:47–54.
Thornthwaite, C. W. 1931. The climates of North America according to a new classification. *Geographical Review* 21:633–55.
Thornthwaite, C. W. 1933. The climates of the earth. *Geographical Review* 23:433–40.
Thornthwaite, C. W. 1948. An approach toward a rational classification

of climate. *Geographical Review* 38:55–89.
Tikhomirov, B. A. 1961. The changes in biogeographical boundaries in the north of USSR as related with climatic fluctuations and activity of man. *Botanisk Tidsskrift* 5:284–92.
Tikhomirov, B. A. 1963. Principal stages of vegetation development in northern USSR as related to climatic fluctuations and the activity of man. *Canadian Geographer* 7:55–71.
Tinley, K. L. 1982. The influence of soil moisture balance on ecosystem patterns in southern Africa. In *Ecology of Tropical Savannas*, eds. B. J. Huntley and B. H. Walker, 175–92. Heidelberg: Springer.
Tosi, J. A. Jr. 1969. *Mapa Ecologico Republica de Costa Rica*. San Jose, Costa Rica: Tropical Science Center.
Tosi, J. A., *et al.* 1988. Manual para la determinacion de la capacidad de uso de las tierras de Costa Rica. San Jose, Costa Rica: Tropical Science Center.
Tosi, J. A., V. Watson, and J. Echeverria. 1992. *Potential Impacts of Climate Change on the Productive Capacity of Costa Rican Forests: A Case Study*. San Jose, Costa Rica: Tropical Science Center.
Troll, C., and K. H. Paffen. 1964. Karte Der Jahreszeitenklimate der Erde. *Erkund. Arch. Wiss Geogr.* 18:5–28.
Tropical Science Center and World Resources Institute (TSC/WRI). 1991. *Costa Rica Natural Resources Accounting Study*. San Jose, Costa Rica: TSC/WRI.
Urban, D. L., and H. H. Shugart. 1989. Forest response to climate change: a simulation study for Southeastern forests. In *The Potential Effects of Global Climate Change on the United States*, eds. J. Smith and D. Tirpak, 3–1 to 3–45. EPA-230-05-89-054. Washington, DC: U.S. Environmental Protection Agency.
von Humbolt, A. 1867. *Ideen zu einer Geographie der Pflanzen nebst einem Naturgemälde der Tropenländer*. Tübingen, Germany.
Vowinckel, T., W. C. Oechel, and W. C. Boll. 1974. The effect of climate on the photosynthesis of Picea mariana at the subarctic tree line. 1. Field measurements. *Canadian Journal of Botany* 53:604–20.
Walker, B. H. 1991. Ecological consequences of atmospheric climate change. *Climatic Change* 18:301–16.
Walter, H. 1985. *Vegetation of the Earth and Ecological Systems of the Geo-Biosphere*. 3d ed. Berlin: Springer-Verlag.
Webb, T. III. 1987. The appearance and disappearance of major vegetational assemblages: long-term vegetation dynamics in eastern North America. *Vegetatio* 69:177–87.
Whitlow, R., and B. Cambell. 1989. Factors influencing erosion in Zimbabwe: a statistical analysis. *Journal of Environmental Management* 29:17–29.
World Bank/United Nations Development Program (WB/UNDP). 1990. Forestry Residues Utilization Study: Costa Rica. WB/UNDP Bilateral Aid Energy Sector Management Assistance Program, Activity Completion Reports No. 108A/90 and 108B/90. Washington, DC: WB/UNDP.

7 An Assessment of Integrated Climate Change Impacts on Egypt

KENNETH M. STRZEPEK
University of Colorado at Boulder, USA
International Institute for Applied Systems Analysis, Laxenburg, Austria

S. CHIBỌ ONYEJI
University of Colorado at Boulder, USA
International Institute for Applied Systems Analysis, Laxenburg, Austria

MAGDY SALEH
Cairo University, Egypt

DAVID N. YATES
University of Colorado at Boulder, USA
International Institute for Applied Systems Analysis, Laxenburg, Austria

INTRODUCTION

The waters of the river will dry up,
and the riverbed will be parched and dry.
The canals will stink;
the streams of Egypt will dwindle and dry up.
The reeds and rushes will wither,
also the plants along the Nile,
at the mouth of the river.
Every sown field along the Nile will become parched,
will blow away and be no more.
The fishermen will groan and lament,
all who cast hooks into the Nile;
those who throw nets on the water will pine away.
Those who work with combed flax will despair,
the weavers of fine linen will lose hope.
The workers in cloth will be dejected,
and all the wage earners will be sick at heart.

ISAIAH 19:5–10
(*Holy Bible, New International Version*, 1984)

This text, from the prophet Isaiah, describes the potential impact of a climatic fluctuation in the Nile Basin on Egypt and the impact of this fluctuation on water resources (reduction in Nile flow) and agriculture. It then describes an integrated impact assessment, not only on direct water users, such as farmers and fisherman, but also on industries dependent upon agricultural output, such as the agro-industries of flax, weaving, and the clothing industry. Finally, it looks beyond the first and second order impacts and addresses the impacts on the entire economy ('all wage earners') and upon society: '(they) ... will lose hope,' '... will be dejected,' and '... will be sick at heart.'

This integrated economic analysis of the impact of a prolonged drought on Egypt, described by the Old Testament prophet over 2700 years ago, is amazingly similar in structure and framework to the work reported on in this chapter. Surprisingly, the prophecies of Isaiah are very similar to the results of the integrated assessment for one of the GCM climate change scenarios analyzed in the book. Old Testament scholars (Alexander, 1994) suggest that Isaiah's prophecy describes a long but not permanent drought as a punishment to Egypt, which will be followed by a time of blessing that has yet to occur. The significance to us of this passage is that (1) for Egypt, with its unique water resources and riverine economy and society, climatic fluctuations can have a catastrophic impact upon the entire economy and society, and (2) the vulnerability of Egypt to climatic change relates especially to the impacts upon the Nile River.

The insight we gain from Isaiah's words is that climate

change impacts can affect all of a society, and the total impact on society must be investigated systematically. This chapter discusses a study that is just such an attempt to look not at first-order effects of climate change on individual sectors, but on the integrated impacts of all the sectors on the nation as a whole.

AN INTEGRATED IMPACT STUDY

Why do an integrated study?

In the preceding chapters, analysis was given of the potential impacts of climate change on a variety of natural and human-based systems or sectors. For each sector, the analysis focused only on the direct impacts of a 'changed climate' (temperature and precipitation changes) on the particular sector. For example, the river basin study (Chapter 3) did not address increased water demands due to increased irrigation or municipal water use as a result of a changing climate. The agriculture study (Chapter 2) did not take into account the increase or decrease in the availability of water resources as a result of climatic change. The sea-level-rise study (Chapter 4) did not address the changing land-use patterns in the vulnerable areas that may result from changes in crop yields, agricultural economics, or water resources, due to a changing climate.

While examining all of these indirect or feedback mechanisms is beyond the scope of each of the studies, the question remains: how will impacts on different, related sectors interact? For example, if agriculture in a region was less profitable due to climate change, the economic impacts of land loss due to sea-level rise may be significantly reduced. Or, if increased temperatures lead to increased water demands for irrigated agriculture, but decreased precipitation results in reduced water resource availability, then there will be an even greater decrease in agricultural production than the agricultural or water resource impact assessments would suggest.

Performing such a study requires an interdisciplinary effort that integrates information and assumptions across many disciplines and uses a consistent and systematic analytic framework. This chapter will discuss such an undertaking.

Examples of other integrated studies include: the MINK Study of climate change impacts on Missouri, Iowa, Nebraska, and Kansas (Bowes and Crosson, 1993; Crosson and Rosenberg, 1993); the Great Lakes–St. Lawrence Basin Project (Mortsch *et al.* 1993); and Southeast Asia (Parry *et al.* 1992).

Why do an integrated study of Egypt?

Previous climate change impact studies that focused on Egypt in some detail point to Egypt as being extremely vulnerable to climatic change. Broadus *et al.* (1986) and El-Raey (1991) suggest land losses of 12 to 15 percent of Egypt's current arable land for a one-meter sea-level rise. Gleick (1991), in an aggregated study of the Nile Basin, suggested that it is extremely sensitive to changes in temperature and precipitation.

The issues presented above and the implications that they posed for climate change impact assessment led the U.S. Environmental Protection Agency to initiate an integrated study. Egypt was chosen because detailed sectoral studies were being developed for it involving the Agricultural Study, the Sea-Level-Rise Study, the River Basin Study and the Human Health Study, all of which enabled an integrated study of impacts on Egypt.

In addition, the Egyptian Ministry of Public Works and Water Resources and USAID/U.S. Bureau of Reclamation were jointly extending a model of Egyptian agriculture, EASM (Egyptian Agricultural Sector Model) (Humphries, 1991). EASM provides a systematic assessment of national agricultural economic impacts with detailed modeling of agronomics, economics, land and water resources, and technology. The model could be modified to accept as inputs the results of the Agricultural, River Basin, and Sea-Level-Rise studies to provide an integrated assessment.

Despite the difficulties in undertaking an integrated study after all the sectoral studies were essentially complete, the project leadership felt there was much to be gained from the analysis.

EGYPT: A BACKGROUND SKETCH

Egypt today

GEOGRAPHY AND CLIMATE

Egypt occupies the northeastern corner of Africa (Figure 1) from 22° to 31° North latitude and 24° to 36° East longitude. It is bounded in the east by the Red Sea, in the west by Libya, in the north by the Mediterranean Sea, and in the south by Sudan. The total land area is 997,688 square kilometers that comprise five major geographical regions: the Nile Valley (upper Egypt, lower Egypt), the Nile Delta, the Eastern Desert, Sinai, and the Western Desert. These geographical areas are divided into 26 administrative units or governorates that are grouped into four regions: Urban Egypt, Lower Egypt, Upper Egypt, and Frontier Egypt.

Figure 2 is a map of Egypt which shows that agricultural activity is located in the narrow corridor of the Nile valley and Delta. The Nile Valley winds approximately 1000 kilometers from the High Aswan Dam in the south to the Nile Delta in the north. The figure shows that major cities and industries are concentrated in this corridor.

The Nile Delta is an important economic region. It makes

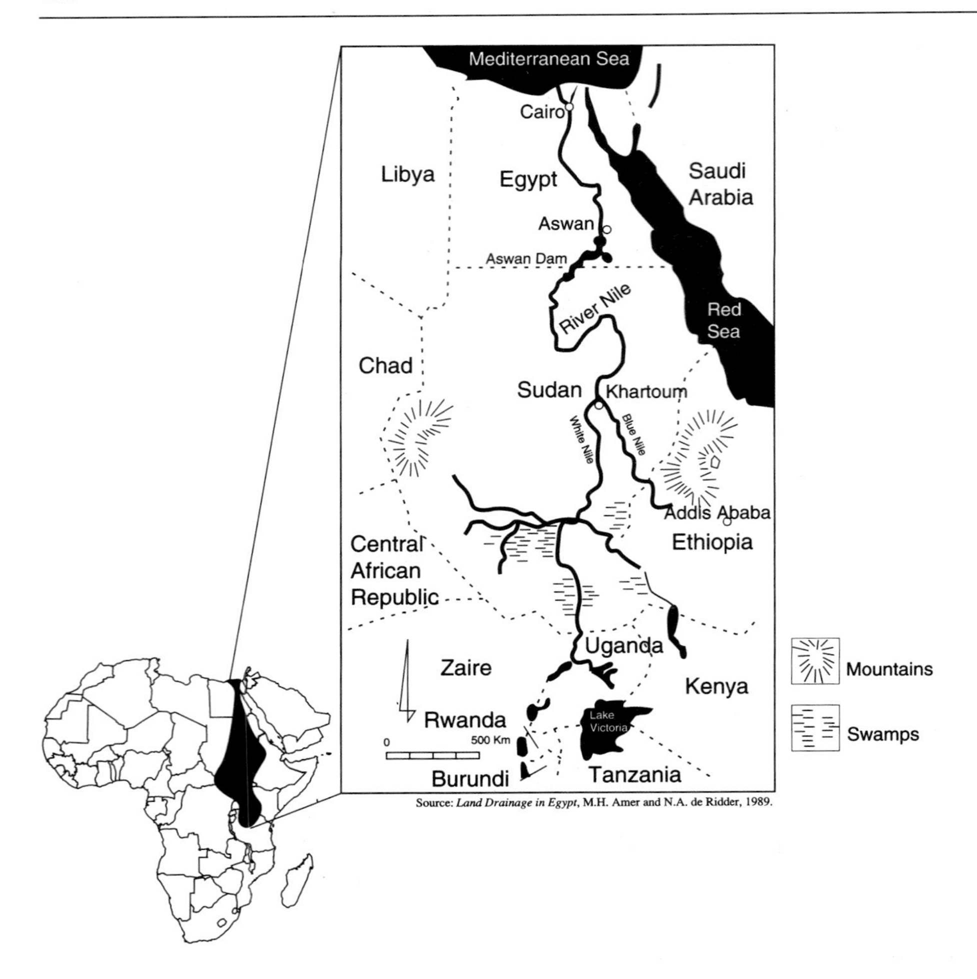

Figure 1 The Nile Basin in relation to Africa and Egypt.

up 60 percent of current agricultural land and is home to over 60 percent of the population. It is the focal point for industry and commerce, bounded by Alexandria in the west, the Suez Canal in the east, and Cairo in the south. Its coastline is currently subsiding due to the loss of Nile sediments to the High Aswan Dam. This results in the potential loss of agricultural land and reduced productivity in coastal lands due to waterlogging and salinity (Nicholls, 1991).

Figure 3a shows the annual rainfall distribution. Egypt is an arid country with only trace rainfall from the southern border to just south of Cairo, limited rainfall in the Delta, and up to 200 mm/yr in a narrow strip along the Mediterranean Coast. Figure 3b shows potential evapotranspiration estimates from the Penman equation for Egypt. When Figures 3a and 3b are compared, it is clear that Egypt is desert except for a narrow strip along the coast and the long oasis of the Nile Valley.

Table 1 presents values of three climatological variables for key points in Egypt: Central Nile Delta, Cairo, and Aswan, while Figure 4 shows the mean monthly distribution of precipitation and potential evapotranspiration for the agriculturally important Nile Delta region. Figures 3 and 4 and Table 1 describe a hot, dry region capable of agriculture year-round, but with very high potential evapotranspiration, especially in the summer months.

POPULATION

Egypt's population in 1990 was 52.4 million, of which 27.9 million lived in rural areas and 24.5 million lived in urban centers (FAO, 1993). Almost the entire population is located in the area of cultivation of the Nile Valley and Delta. This area represents only 3 percent of the total land area (Figure 2), providing an effective population density of over 5000 persons per square kilometer. Figure 5 shows population trends in Egypt from 1961 to 1992. Egypt's gross population density in 1990 of 55.4 persons per square kilometer is one of the most dense in the world (Onyeji, 1992).

Table 2 shows the distribution of population, income, and household size by region. Rural population accounts for 56 percent of the population but only 42 percent of the income. The annual rate of population growth in Egypt is currently 2.39 percent (WRI *et al.*, 1992). The rapid growth of popula-

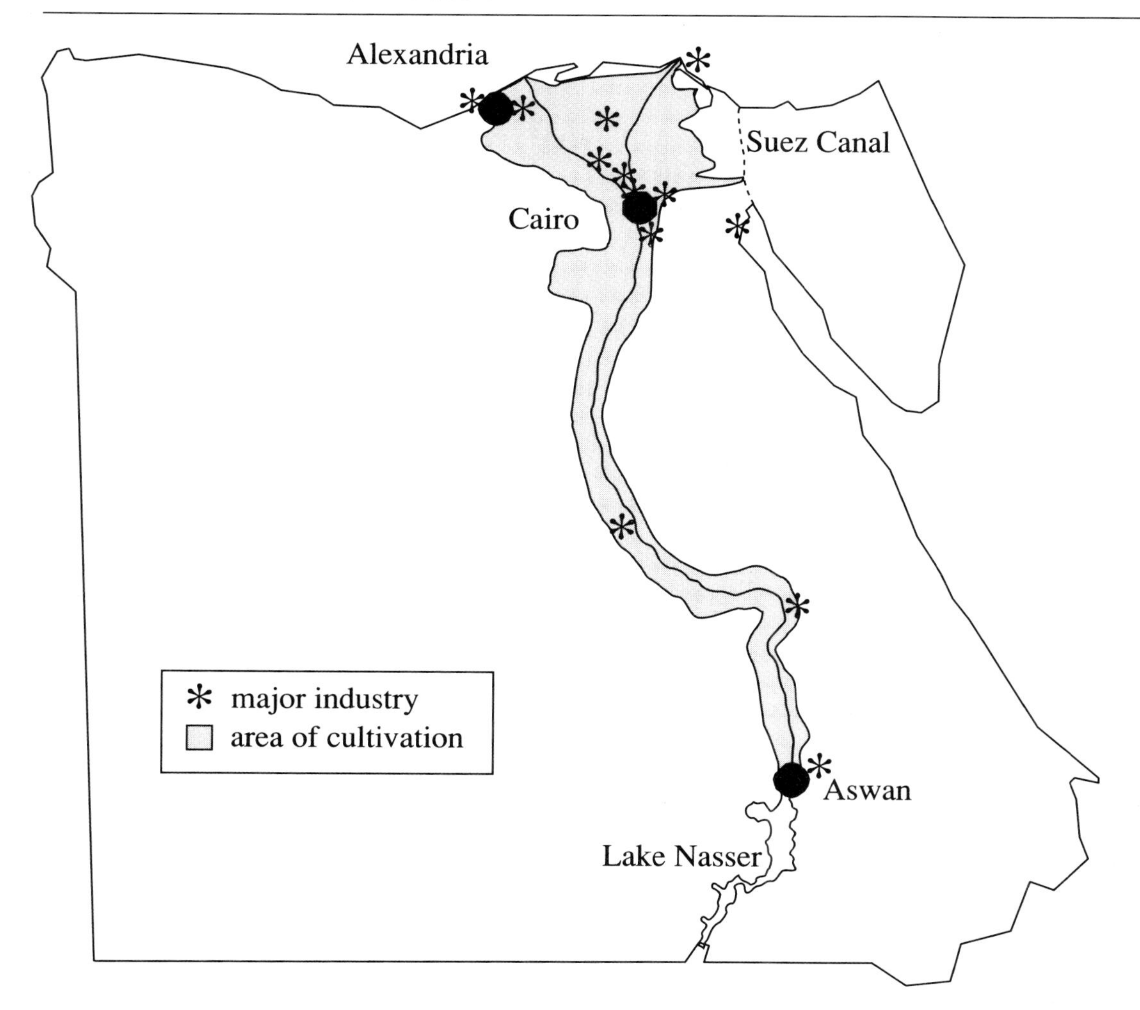

Figure 2 Map of Egypt showing agriculture, population, and economic activity confined to Nile Valley and Delta.

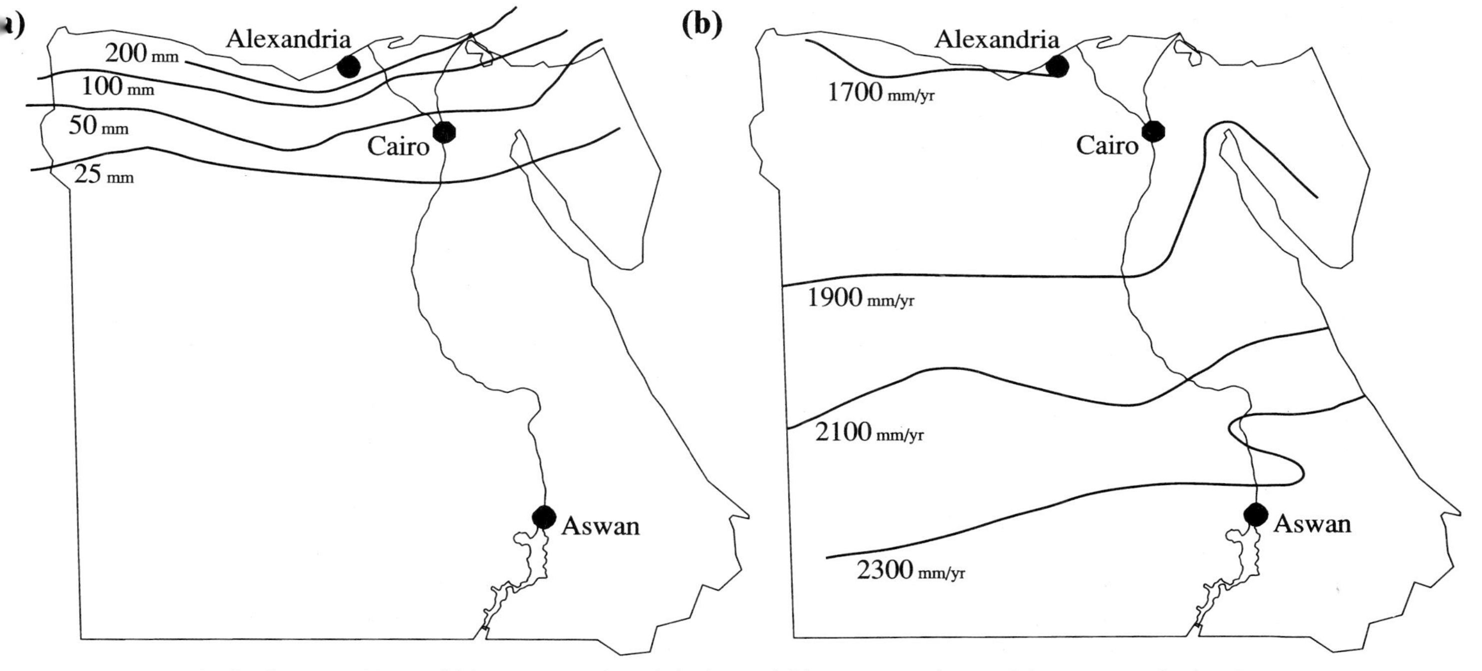

Figure 3 Distributions over Egypt of (a) mean annual precipitation and (b) mean annual potential evapotranspiration (Penman).

Table 1. *Climatological parameters for Egypt*

Location	Mean annual precipitation (mm)	Mean annual potential evapotranspiration (mm)	Mean annual temperature (°C)
Mid Delta	48	1,390	19.8
Cairo	25.5	1,780	20.8
Aswan	1	2,480	25.8

Source: Shahin, 1985.

Table 2. *Socio-economic indicators for Egypt in 1990*

Region	% Population	Per-capita income (Egyptian pounds)	Average household size
Total Egypt	100	1,540	4.9
Urban	44	2,036	4.5
Rural	56	1,151	5.1

Source: Onyeji, 1992.

tion has been accompanied by strong rural–urban migration. One result is that Cairo is estimated to have 12 to 15 million inhabitants, or approximately 25 percent of the population.

STANDARD OF LIVING

An overview of the Standard of Living Indicators for Egypt in 1960 and 1985 (Table 3) shows a country that has undergone dramatic changes. Egypt's standard of living has increased significantly in the last generation. For example, income per capita has more than doubled. How were these changes accomplished, are they sustainable, and how could a changing climate affect them? To examine these questions in detail, the following sections will describe key factors affecting Egyptian society and economy.

THE ECONOMY

From the 1880s until Nasser's taking control in the 1950s, private enterprise and relatively free trade were the main features of the economy. Development was driven by public investment in agriculture, and, from 1930, limited protection for industry and some protection for a few agricultural products. From Nasser's program of Arab Socialism until the late 1980s, public ownership of the modern sector, major trade restrictions, and import substitutions predominated (Hansen, 1991). Governmental policies of the 50s, 60s, and 70s were quite successful in improving the standard of living, as discussed above. Much of this success was due to the liberalization and open-door policies of 1974 that resulted in annual real economic growth of 9 percent and an inflation rate of 10 percent. Balance of payments improved from foreign exchange earnings that accrued from expatriate

Table 3. *Standard of living indicators for Egypt*

Indicator	1960	1985
GNP per capita (1985 $)	275	610
Daily caloric supply (per capita)	2,435	3,203
Life expectancy (at birth)	46	61
Crude death rate (per 1000 population)	19	10
Infant mortality (per 1000 live births)	109	93
Crude birth rate (per 1000 population)	44	36
Urbanization (per cent of population)	38	46
Physicians (persons per physician)	2,560	760

Source: Hansen, 1991.

worker remittances, tourism, and foreign aid (Zeineldin, 1986). However, following the subsequent collapse of oil prices and dwindling worker remittances, industrial growth began to lag behind the rapidly growing domestic demand which was fueled by the growing population and government policy. Since 1985, economic growth has declined, and unemployment and inflation have increased (Onyeji, 1992).

The growth in living standard has not been evenly distributed across the population, and the cost has been high. To achieve such progress, Egypt spent more than it earned. Average annual expenditure growth was 1 percent greater than average annual income growth. This income deficit was made up with foreign loans, leaving Egypt as one of the most indebted nations in the world (Hansen, 1991).

In May 1987, the Egyptian government and the International Monetary Fund concluded a stabilization and structural adjustment agreement purportedly to revitalize economic growth. This included a gradual move back to private enterprise and relatively free trade, particularly in agriculture. The full impact of the new policies was not in place in 1990, the base year for our analysis, but it has been observed in subsequent years (O'Mara and Hawary, 1992). For example, cotton and wheat cultivation have responded dramatically to new pricing policies (El-Din, 1993).

Table 4 presents a highly aggregated view of the Egyptian economy in 1984 and 1987. Agriculture is a major sector at 20 percent of GDP. Industry, mining, and petroleum declined from 1984 to 1987. Petroleum is not seen as a long-term resource, as new major reserves have not been found. The rapidly increasing trade and finance sector, combined with a declining manufacturing sector, indicates a basic weakness in the Egyptian economy (Hansen, 1991). Table 5 presents employment by sector for 1990. It should be noted that agriculture is the dominant employer at 39 percent, with service next at 22 percent.

The Egyptian industrial structure is dominated by basic consumer goods (textiles, shoes, food, beverages, and cigarettes) and essential intermediate goods (building materials,

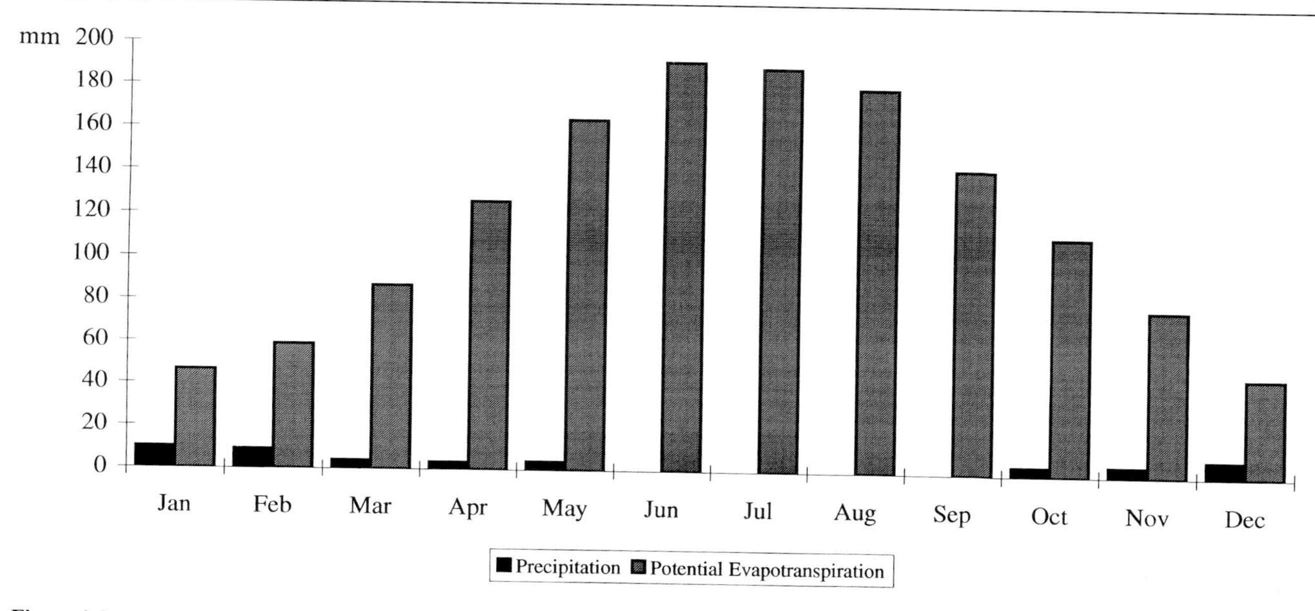

Figure 4 Mean monthly distribution of precipitation and potential evapotranspiration (Penman) for a location in the Nile Delta showing the extreme aridity of the region.

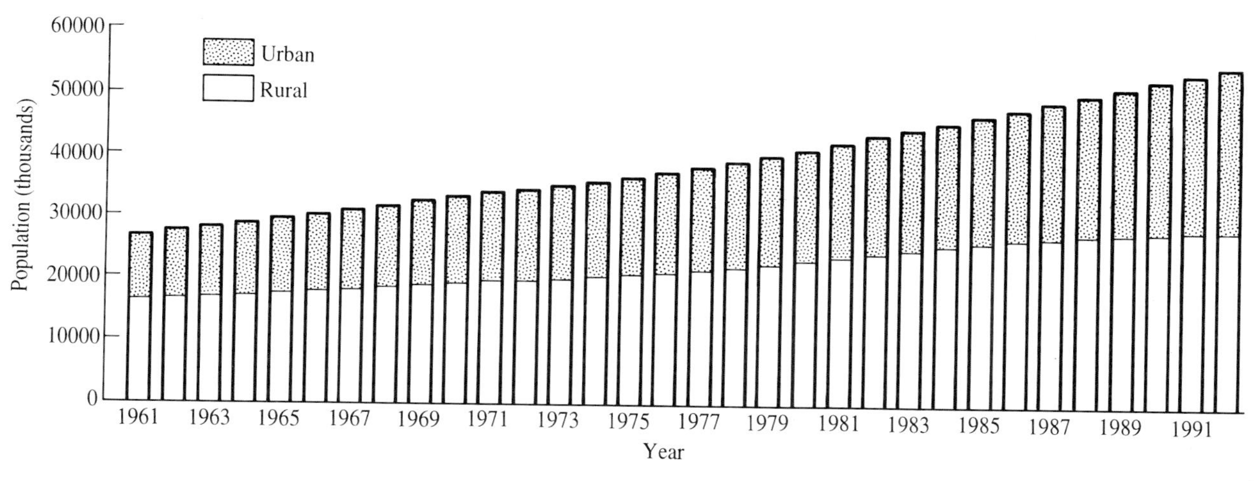

Figure 5 Egyptian population growth from 1961 to 1992 with urban/rural distribution.

fertilizer, chemicals, paper, petroleum products, and some metals) sold primarily on the domestic market. The capital goods industry (machinery, tools, implements) is small (Onyeji, 1992).

Table 6 is a list of the major exported agricultural commodities for 1984. While petroleum is the major export revenue source, agriculture was responsible for over 70 percent of the remaining exports. From the standpoint of employment, trade, and intermediate inputs to industry, agriculture is a key economic sector. The next section will take a detailed look at this sector.

AGRICULTURE SECTOR

With its very arid climate, Egypt has a unique agricultural system. Because there is effectively no rainfall, almost all of the agricultural land is irrigated. The water supply for this irrigation comes solely from the Nile River waters stored in Lake Nasser behind the High Aswan Dam. (See Chapter 3 for a detailed discussion of the effect of climate change on the Nile Basin.)

Currently, agriculture is practiced on 3 million hectares (5.892 million *feddans*[1]), or only 3 percent of the area of Egypt. It is limited to 9,300 square kilometers of the fertile lands in the narrow Nile Valley from Aswan to Cairo and the Nile Delta north of Cairo. New, less fertile lands on the fringe of the Delta and in the Sinai are being developed. Table 7 shows the distribution of agricultural land within Egypt.

[1] A *feddan* is the Egyptian unit of land measurement. It is equal to 1.04 acres, 0.42 hectares and 0.004 square kilometers.

Table 4. *Economic structure of Egypt*

	1984	1987
Total value of GDP (million Egyptian pounds)	25,961	43,685
Sector	Percent of GDP	
Agriculture	20.0	20.1
Industry, mining & petroleum	33.1	28.7
Other sectors	46.9	51.2

Source: Hansen, 1991.

Table 5. *Employment figures for Egypt in 1990*[a]

Sector	1000s	Percent
Agriculture	4,797	38.7
Mining	45	0.4
Manufacturing	1,540	12.4
Electricity, gas, water	98	0.8
Construction	900	7.3
Commerce and hotels	882	7.1
Transport and communication	671	5.4
Trade and finance	243	2.0
Services	2,702	21.8
Other	505	4.1
Total	12,386	100.0

Note:
[a] These figures underestimate female employment.
Source: Onyeji, 1992.

Table 6. *Exports in 1984 as percent of GNP*

Commodity	Percent of GNP
Cotton	1.5
Rice	0.1
Fruits and vegetables	0.5
Cotton textiles	0.7
Petroleum	16.4
Industrial goods	1.0
Total	20.2

Source: Hansen, 1991.

Table 7. *Agricultural land in Egypt 1990*

Agro-climatic zone	Area (10^3 feddans)	Percentage
Upper Egypt	1076	18
Middle Egypt	1192	20
Delta	3624	62
Total	5892	100

Source: Humphries, 1991.

Over three-fifths of the land used for agriculture is in the Delta (Humphries, 1991).

The agricultural year has three crop seasons. The winter season starts between October and December and ends between April and June. Its main crops are wheat and barley, berseem[2] and lentils, winter onions, and vegetables. The summer crops – cotton, rice, maize, sorghum, sesame, groundnuts, summer onions, and vegetables – are sown from March to June and harvested from August to November. A third growing season known as '*nili*' is a delayed summer season, when rice, sorghum, berseem, and some vegetables are grown. Since *nili* and summer cropping seasons overlap, a piece of land cannot be planted with both summer and *nili* crops in any one year. To prevent soil degradation and pest-induced losses, an elaborate crop rotation is practiced in Egypt. This leads to many small plots with multiple crops and a general three-year crop rotation to preserve soil fertility. There are significant perennial crops such as sugar-cane in Upper and Middle Egypt and citrus, grapes, bananas, mangoes, olives, and dates (USU, 1986). Table 8 is a summary of the 1990 observed cropping pattern.

In 1990, agriculture (including livestock) accounted for nearly 20 percent of the gross domestic product (GDP) and employed 39 percent of the labor force. These figures do not include agro-industries such as textiles or food processing (Onyeji, 1992). Agricultural exports accounted for approximately 20 percent of export earnings. Even with some of the most productive agricultural land in the world and a plentiful water supply, Egypt is currently importing over two-thirds of its wheat and vegetable oils and one-third of its corn. Agricultural imports have increased three-fold since 1975, resulting in annual agricultural import costs of over $3 billion (FAO, 1993). This is partly due to governmental policy, international commodity prices, and foreign food aid and population growth. While annual agricultural production has increased by 46 percent over the period from 1978 to 1990, the population has grown 28 percent, resulting in a per capita increase of only 14 percent over this period.

The net growth of agricultural lands is slowed to 20,000 *feddans* per year by the loss of highly productive lands to housing and urban construction around population centers. Highly productive agricultural land is limited, and land reclamation is slow and costly. The reclaimed lands are primarily desert, with lower productivity than the Nile Valley or Delta lands. Figure 6 plots the growth of agricultural land

[2] Berseem is an Egyptian clover used for fodder.

Table 8. *Egyptian agricultural production 1990*

Crop	Production [ton × 1000]
Barley	175
Citrus	2734
Cotton lint	219
Cotton seed oil	115
Eggs	156
Flax fiber	1138
Groundnut	24
Horse bean	528
Legumes	26
Lentils	100
Maize	4606
Meat	340
Milk	2370
Onion	728
Potato	1937
Poultry	203
Rice	2093
Sesame	32
Sorghum	610
Soybeans	176
Sugar	872
Tomato	1657
Vegetable oil	615
Vegetables	2166
Wheat	4218

Source: FAO, 1993.

from 1961 to 1991. The rate of increase of land utilization is greater than the rate of increase in production, since most growth is in reclaimed desert lands. Even with the growth of agricultural production, the demand for food outstrips these gains, again driven mainly by population growth.

Egypt's future without climate change

Based on World Bank estimates, an Egyptian population of 115 million, i.e. 2.2 times that in 1990, is forecast for 2060. This is derived from a forecast of annual population growth rates of 1.66 percent from 1990 to 2020, 0.95 percent from 2020 to 2040, and 0.58 percent from 2040 to 2060; a long-term average annual growth rate over 70 years of 1 percent. The growing population will put priority on efforts to increase agricultural production through land reclamation and technological improvement. Increasing agricultural production must be accomplished despite fixed water resources and harsh climatic and geographic features of the 97 percent of land that is undeveloped. The Central Planning Bureau of the Netherlands (1992) forecasts 2.1 to 3.6 percent annual GDP growth for Middle-Eastern economies over the period from 1990 to 2015. Egypt's future depends on the sign and magnitude of difference in these two growth rates.

There are two quite opposite views about Egypt's long-range economic development (Yates *et al.*, forthcoming). The pessimists see that Egypt has dug itself into a deep economic hole. With its growing population, large foreign debt, and continuing tendencies towards increased debt and low investments, they forecast that this 'hole' will actually get deeper. This group sees very low growth in the agricultural as well as the non-agricultural sectors and the prospect of unchecked population growth and slow changes in government policies leading to short-term consumption rather than long-term investments.

The optimists believe that, in spite of its current debt and low-technology industries, Egypt will invest in new, effective technologies in the mid- and long-term and will actually surpass some countries with aging and less productive technologies. This group sees Egypt as the South Korea of the twenty-first century. One economic modeling effort (Fischer *et al.*, 1994; Onyeji and Fischer, 1994; Chapter 2, this volume), which was based on optimistic growth parameters, projects a doubling of the 1990 per capita GDP in Egypt by the year 2060, fueled by six-fold and three-fold increases in non-agricultural and agricultural GDP, respectively. This is all accomplished while accommodating a population that has more than doubled.

While we cannot estimate the probability of these scenarios, the assumptions about the future are crucial to any analysis that attempts to assess climatic change impacts on a future economy. Given the structure of the economy, the impact may vary greatly. We look at Egypt's vulnerability to climatic change in the next section.

Egypt's changed climate

GCM results for doubled CO_2 uniformly estimate an increase of slightly over 4°C in average annual temperature for Cairo (Table 9). This increase will lead to higher potential evapotranspiration. These results suggest that Cairo and the Delta region will have a climate similar to that currently found in Aswan (see Table 1). Humphries (1991) found that there is a significant difference in crop water requirement between Upper Egypt and the Nile Delta. With only trace amounts of rainfall in the current climate, the precipitation changes estimated by the GCM are insignificant.

Almost all of Egypt's water supply comes from the Nile River, so climatic change over the entire Nile basin, as far south as Lake Victoria – some 5000 kilometers south of Aswan – will have an impact on Egypt. Table 9 presents the GCM results averaged over the 10 to 16 grid cells that cover the approximately 2,000,000 square kilometers of the Nile

Table 9. *Annual changes in temperature and precipitation for Cairo and the Nile Basin under three climate change scenarios*

	Cairo			Nile Basin		
	UKMO	GISS	GFDL	UKMO	GISS	GFDL
Change in temp. (°C) ($2 \times CO_2 - 1 \times CO_2$)	4.43	4.16	4.20	4.70	3.4	3.1
Precipitation ratio ($2 \times CO_2 / 1 \times CO_2$)	0.86	1.56	0.85	1.22	1.31	1.05

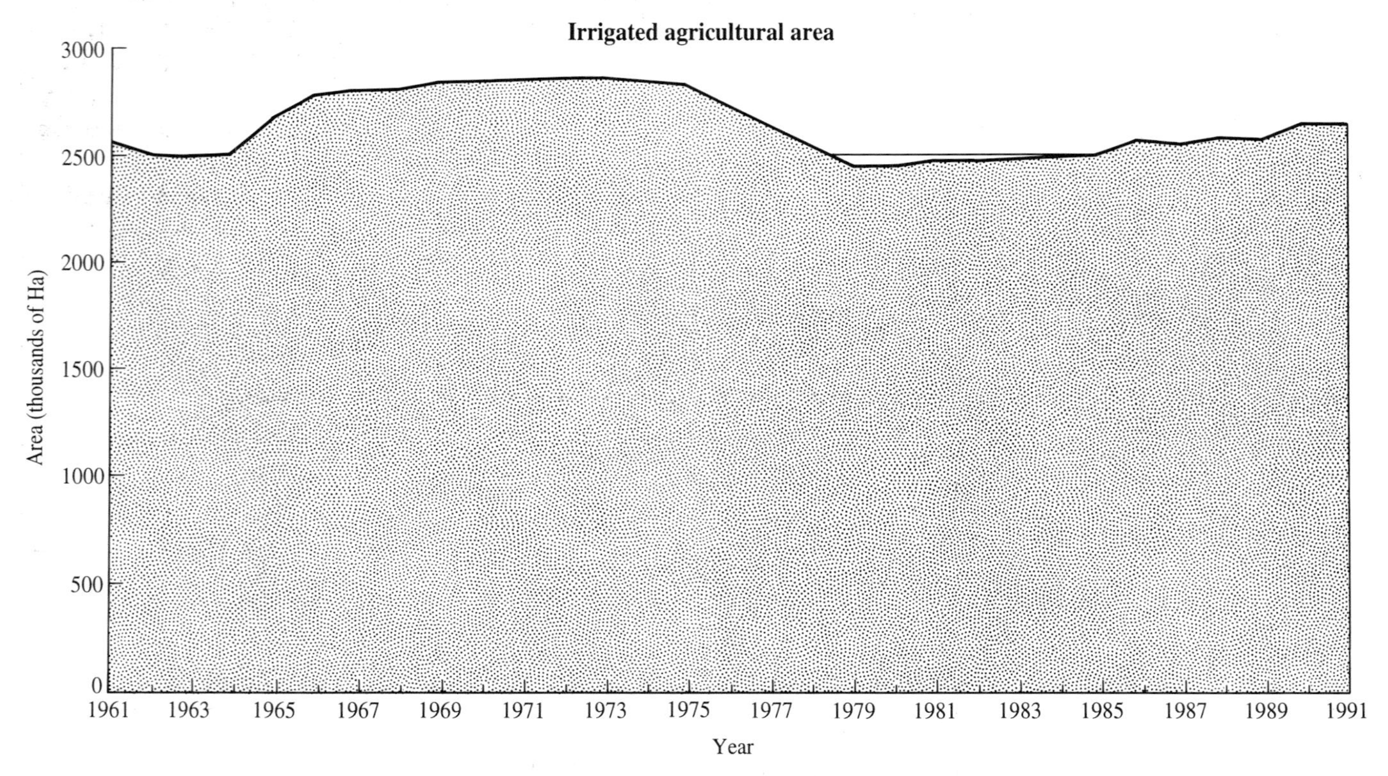

Figure 6 Egyptian cultivated land under irrigation from 1961 to 1991.

Basin upstream of the High Aswan Dam (see Figure 1). Annual average temperature increases of 3.1°C to 4.7°C and precipitation increases from 5 to 31 percent are estimated. A hydrologic model of the Nile River estimates that the GCM-based climate changes will lead to impacts on flow at Aswan varying from an increase of 30 percent to a decrease of 77 percent (Chapter 3).

Egypt's vulnerability to climate change

Egypt is very dependent on natural resources that are vulnerable to climate change. The land used for growing crops is mainly in the Nile Delta, a low-lying area vulnerable to sea-level rise. Agriculture needs water from the Nile for irrigation. Climate change will likely impact water supply. In addition, crop yields and water use will also be directly affected by climate change.

The prospect of a climate-change-induced sea-level rise should be of serious concern to Egypt, since a majority of its population, agriculture, and industrial activity is located in the low-lying Nile Delta. Egypt's vulnerability to sea-level rise is further heightened by the High Aswan Dam's effect on the Nile. Under natural conditions, sediments reaching the Mediterranean coast would accumulate and compensate for delta subsidence. However, with the High Aswan Dam in place, sediment-laden Nile waters are no longer able to reach the Mediterranean Sea, effectively starving the delta of fluvial deposition. The lack of freshwater influx has led to subsidence of the Delta, an increase in soil salinization, and a loss of cultivatable land.

Egypt has been granted an annual yield of 55 billion cubic meters of Nile water from Lake Nasser via the Nile Waters agreement with Sudan. Annual per capita water availability in 1990 was 1005 m^3 and is projected to drop to 452 m^3 in the

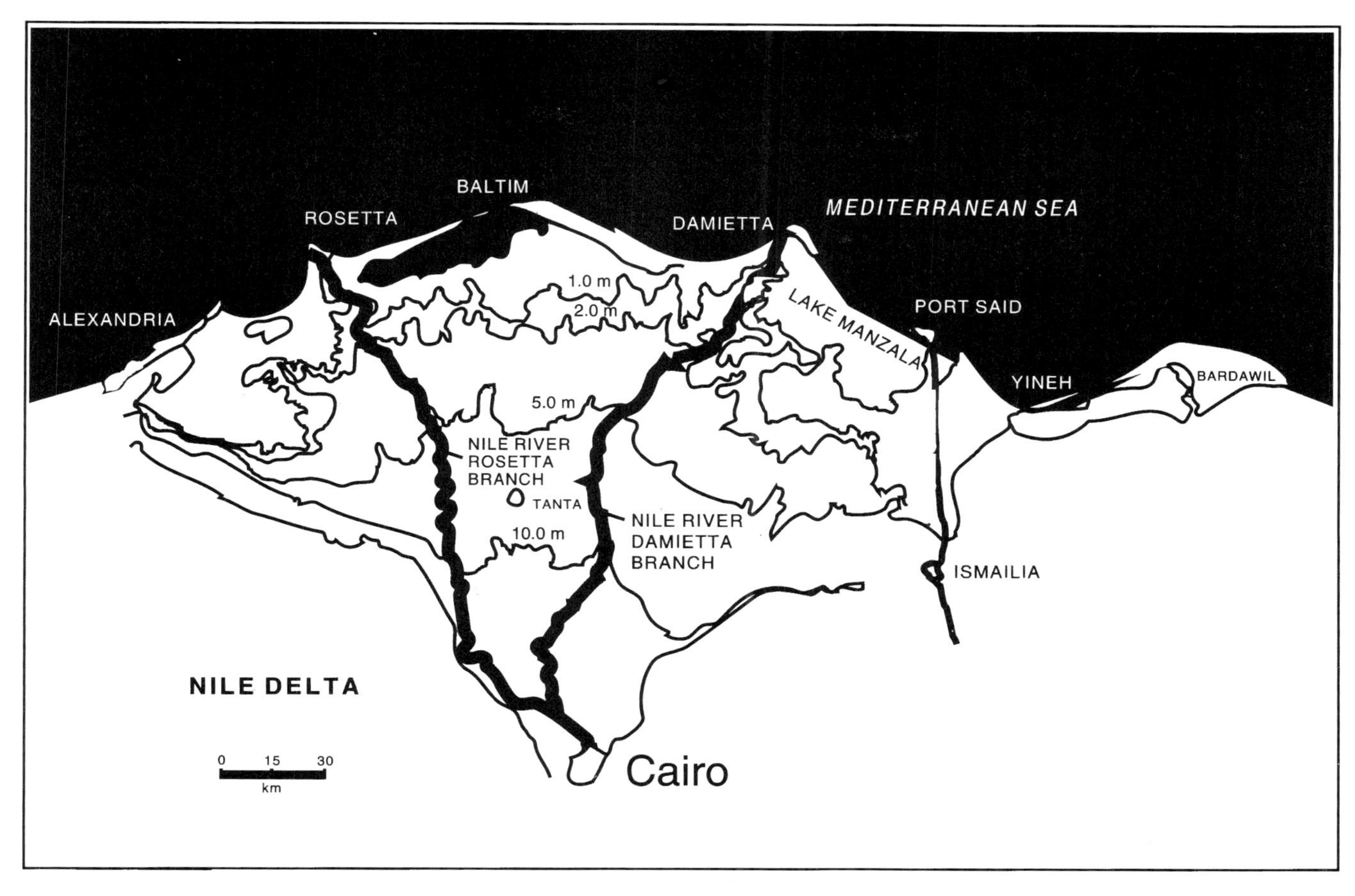

Figure 7 Topographical map of the Nile Delta (Nicholls, 1991).

year 2060 based on World Bank forecasts (Onyeji, 1992). Kulshreshtha (1993) suggests that a nation with less than 1000 cubic meters per person per year results in a water-scarcity condition. With few economically feasible alternative water supply options, Egypt must rely almost entirely on the Nile. Climate change could affect the flow of the Nile and the availability of water for Egypt.

GCM results for Egypt show substantial warming. An increase in temperature will lead to increased evapotranspiration by crops. This translates into increased crop water requirements and lower yields (Eid and Saleh, 1992).

The description of the Egyptian economy in 1990 suggests that it might be sensitive to the type of climatic change projected for Egypt. With 20 percent of GDP and 39 percent of the labor forces involved directly in agriculture, any impact will be felt throughout the economy. In addition, loss of land and infrastructure due to sea-level rise will have economic impacts, while major investments to mitigate impacts will take away from more productive investments. Because of the importance of this sector, this study focuses on the agricultural economy.

SECTORAL IMPACTS OF CLIMATE CHANGE

Sea-level rise

Analysis performed as part of the sea-level-rise study reported in Chapter 4 estimates that by 2060, the mean impact of a global sea-level rise of 1 meter by 2100 would produce a 0.37 meter sea-level rise at the Nile Delta. This effect, combined with a non-climate-induced subsidence of the Nile Delta of 0.38 meters, is estimated to result in the movement of the shoreline to the current 0.75-meter contour by 2060 (Figure 7). El-Raey (1991) and others assume that agriculture cannot take place within a 1 meter buffer zone from sea level to the 1-meter contour. Currently, irrigation takes place in this buffer zone and Egyptian government plans call for reclamation of delta lands well below the current 0.75-meter contour, including large areas of inland salt lakes and sand dunes. However, due to salinization and sea-water intrusion, agriculture below an elevation of 1 meter is very difficult and requires careful water management (Rosenzweig and Hillel, forthcoming). A 0.75-meter net rise based on agriculture in the buffer zone will result in a direct loss of 4 percent of Egyptian agricultural land in 2060. If, in

Table 10. *Climate change impacts on Nile River flow at Aswan and High Aswan Dam yield as percent of current conditions*

Scenario	Nile yield	Egypt yield
UKMO	88	87
GISS	130	118
GFDL	23	17
GISS-A	118	114

fact, irrigation is not sustainable within the 1-meter buffer, then sea-level rise in 2060 will mean the loss of land between the current 1.00-meter and 1.75-meter contours (Strzepek and Saidin, 1994).

Water resources

The GCM-based results do not agree on the direction of change of Nile River flow. Chapter 3 presents a full discussion of climate change impacts on the Nile Basin and Egyptian water resources at Aswan. Table 10 is a summary of the impacts. The UKMO scenario suggests a 12 percent decrease in the flow of the Nile at Aswan. The GISS and GISS-A[3] scenarios suggest increases in the flow of the Nile at Aswan that would become only 18 percent and 14 percent increases, respectively, in water supply for Egypt.[4] This is due to increased evaporation at Aswan and the nature of the Nile Waters Treaty with Sudan, which evenly shares any increase in flow at Aswan. Finally, the GFDL results show a catastrophic decline of 77 percent of the Nile flow at Aswan.

Agriculture

The study of the agriculture sector, described in Chapter 2, had two components: agricultural (crop yields) and economic (world food trade study). Both of these were used for the integration analysis. The crop-yield analysis also considered the direct effects of increased CO_2 on plant growth and water requirements. The agricultural impacts were studied in depth for wheat, maize, soybean, and rice. In general, crop yields decrease at a minimum of 20 percent without direct effects of CO_2 and, on average, the direct effects lessen this decrease by 5 percent. Water requirements of crops increase about 7 percent in the Delta and 15 percent in upper and middle Egypt. Tables 11 and 12 report on the results of the climate change scenarios on crop yield and water use (Eid and Saleh, 1992).

[3] The GISS-A scenario has a lesser amount of warming than UKMO, GISS, and GFDL. It was defined as the average change in the 2030s of the GISS-A transient run and was used as a scenario for doubled CO_2 conditions. Note that the IPCC's most likely warming range is 1.5 to 4.5°C for CO_2 doubling. for CO_2 doubling.

[4] There currently exists an agreement between Egypt and the Sudan on the allocation of Nile River flows. This allocation is based on the current annual mean flow at Aswan of 84 MCM, taking into account losses, evaporation, and seepage caused by storing the water in Lake Nasser. However, there are formal procedures for dealing with any increases to the Nile flow due to the Upper Nile projects which call for an equal sharing between Sudan and Egypt of any flow above the current 84. For the GISS scenario, which projects a 30 percent increase in Nile flow, both Sudan and Egypt would get 11.5 MCM of additional allowable annual withdrawals after accounting for the increased evaporation due to warming over the reservoir area.

Table 11. *Egyptian crop yield impacts under climate change. (Units are ton/ha except where noted.)*

Scenario	REF	GISS	GFDL	UKMO	GISS-A
CO_2 (ppm)	330	555	555	555	555
Wheat	4.92	3.59	3.82	1.49	4.72
Maize	10.60	8.63	8.28	8.71	8.30
Rice	6.93	6.88	6.70	6.58	—
Soybean	2.98	2.28	2.08	2.00	—

Source: Eid and Saleh, 1992.

Table 12. *Egyptian crop water use impacts under climate change. (Units are mm/yr except where noted.)*

Scenario	REF	GISS	GFDL	UKMO	GISS-A
CO_2 (ppm)	330	555	555	555	555
Wheat	557	454	499	515	476
Maize	571	480	516	569	515
Rice	1,062	1,073	1,131	1,195	—
Soybean	816	910	981	1,018	—

Source: Eid and Saleh, 1992.

The results in Tables 11 and 12 show rather dramatic impacts on Egyptian agricultural production and water demand. It is likely, though, that Egyptian farmers will respond to changing conditions by changing agricultural practices and water use.

To examine the combined impacts of local and global crop yield and economic changes, a world food trade analysis was performed. The Basic Linked System (BLS) (Fischer *et al.* 1988) was used to forecast Egyptian domestic demand and net exports for each future scenario (see Chapter 2). Table 13 shows the national domestic consumption in absolute terms for the year 2060 Base scenario. The results for the GCM scenarios are presented as percentage changes from the reference scenario.

The BLS projects relatively small changes in national demand (consumption) for most agricultural commodities. The largest reduction in any category is a 5.2 percent decrease in wheat consumption for the UKMO scenario. The UKMO

Table 13. *BLS results for Egypt under climate change*

	Domestic consumption				
	Million tons	Percentage change from base			
Commodity	2060 Base	GISS	GFDL	UKMO	GISS-A
Wheat	13.8	−1.0	−0.7	−5.2	0.6
Rice	3.5	−0.6	−0.4	−2.7	−0.2
Coarse grains	6.5	−0.8	−2.1	−2.4	0.3
Protein feed	0.2	8.4	−3.1	1.6	18.4
Other food	7.5	−0.6	0.1	−1.2	−0.2
Non food ag	0.9	−0.8	−0.4	−4.8	1.0
Bovine & ovine	2.1	−1.2	0.0	−1.3	−1.1
Dairy products	7.6	−0.6	−0.1	−1.0	−0.4
Other animal	0.1	−0.8	−0.2	−1.3	−0.5

Source: Fischer, 1993.

scenario projects the greatest overall impact on domestic consumption, while the GFDL scenario has the least impact.

While total domestic consumption is not greatly affected, the agricultural trade balance is significantly impacted with all scenarios showing, on average, a 10 to 20 percent increase in imports with a 120 to 237 percent increase in the import of rice. The net exports in non-food agricultural products (Egyptian cotton and flax) range from a 455 percent increase to a 722 percent decrease. The BLS is a collection of applied general equilibrium models, with Egypt as one of the country models. For the economy to be in equilibrium, the model projects increases in non-agricultural net exports to fund the increases of the import of agricultural commodities. In all the scenarios, the non-agricultural sector expands and provides the national income to maintain domestic food demand, and provides non-agricultural exports to make up for the negative agricultural balance of trade. These results follow the 'optimistic' economic future projections for Egypt as discussed in the section on Egypt's future.

APPROACH AND METHODOLOGY TO MODELING THE INTEGRATED CLIMATE CHANGE EFFECTS

A forward-linkage impact integration

The analytical framework under which this study was done is shown in Figure 8. The applied approach for integrated assessment takes the individual sectoral impacts on Egypt as inputs to the EASM model. There are no feedback linkages in the analysis. BLS modeled demands and domestic agricultural production do not reflect the direct impacts of the water supply, water requirements, or sea-level rise. Neglecting these impacts could lead the BLS to under- or overestimate the impacts of climate change on domestic demand and production. This lack of feedback is referred to as a forward-linkage analysis.

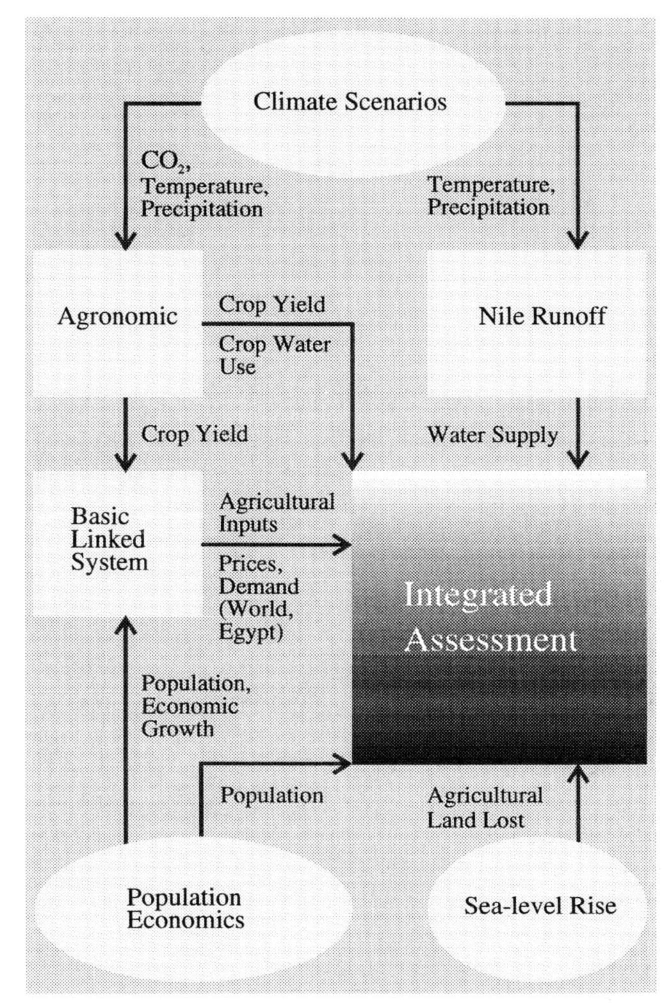

Figure 8 Schematic of integrated assessment framework showing sectoral model linkages.

The schematic representation in Figure 8 gives an outline of the major elements of the study. Climate scenarios based on an effective doubling of CO_2 came from GCM runs. These scenarios were used to model hydrologic and agronomic impacts (see Nile River flow, Chapter 3, and global and Egyptian crop yields and water use, Chapter 2, respectively). Only the crop yields were input into the IIASA Basic Linked System (BLS) for modeling world food supply and trade to the year 2060, when full CO_2 effects were assumed to be realized. Egypt's economy and its linkage to the rest of the world is specifically modeled in the BLS up to the year 2060 using World Bank forecasts of population growth. Egyptian domestic and foreign commodity demands are generated by the BLS and input, along with Nile flow, land availability (based on a 37-cm sea-level rise in 2060), crop yield, and water use, to a modified Egyptian Agricultural Sector Model (EASM-CC).

EASM-CC models the economics of the agricultural sector as well as the water, land, crop, livestock, and labor components on a sub-national scale. EASM models economic behavior by maximizing the consumer–producer surplus (CPS) over an annual cycle (Yates *et al.* forthcoming).

A consistent 2060 base scenario

A consistent 2060 Base scenario was needed in order to create a plausible scheme for Egyptian agriculture in 2060.

ADAPTATION OF COSTS, PRICES, AND DEMAND

For the 2060 Base scenario, restrictive taxes and subsidies were removed from the marketing costs. Marketing costs were assumed to be true markups between domestic prices and farm gate prices. Export marketing prices were assumed to be 20 percent greater than the domestic marketing costs. International import and export prices were taken from the BLS and are referenced as percentage changes from the 1990 Base.

To derive a demand curve for 2060, it was assumed that a parallel shift of 1990 demand would take place. For the 2060 Base scenario, it was assumed that the slope of the demand curve did not change from 1990; only the intercept of the demand curve was recomputed based on the price and demand equilibrium from the BLS.

TECHNOLOGY

Technology changes in agriculture primarily affect crop yield responses. Endogenously-calculated 2060-Base crop yields were taken from the BLS, based on exogenous technological growth rates and production functions of inputs. Crops in 2060 were assumed to require more water as yields increase and more inputs are applied. Crop water requirements were assumed to be 25 percent of the yield change based on experimental data from the southwestern USA (Hexem and Heady, 1978). Inputs such as mechanization, fertilizer, etc., were taken from the BLS as percent increases or decreases from 1990 data. Technological inputs affect the utilization of labor, capital, and fertilizer in EASM-CC.

CONSTRAINTS AND ASSUMPTIONS

Several constraints and assumptions made for 1990 conditions were removed or adapted for 2060. Changes include:

- Removing the upper limits on sugar processing and on citrus, cotton, and vegetable cultivation;
- Removing the exogenous herds of camels, donkeys, and goats;
- Utilizing only 50 percent of 1990 crop by-product fodder (assumption of specialization);
- Increasing labor availability linearly with total population;
- Adding new lands in the Upper Valley and the Nile Delta comprising 2 million *feddans* (distributed equally) with a limited cropping pattern (USU, 1986);
- Removing 124,000 *feddans* of agricultural area due to land subsidence in the Nile Delta;
- Removing the navigational requirement;
- Removing bounds on imports and exports;
- Keeping industrial water use at 1990 levels;
- Keeping domestic water use per capita at 1990 levels, resulting in a 120 percent increase of domestic water use.

The model runs

To assess the integrated impacts of climatic change, a primary set of model runs was made using EASM-CC. These runs are listed below.

B1990 (Base 1990)
Based on the 1990 observed population, land and water resources, and technology, but with a free market economy with no trade restrictions as postulated for 2060. This run was made using current climate conditions to provide a comparison between a 1990 hypothetical situation (removal of trade restrictions) and the 2060 scenario postulated by the BLS.

B2060 (Base 2060)
Based on a consistent scenario for 2060, with population, economic growth, agricultural production, and commodity demands from the BLS, land and water resources based on the current climate, and water use based on agricultural production and technology postulated for 2060.

GCMs
The four GCM scenarios discussed in the introductory chapter were run to analyze the integrated impacts on the Egyptian agricultural economic sector due to climatic change.

All of these runs were performed with the standard EASM-CC objective to maximize consumer-producer surplus. Additional runs were made to assess possible adaptations to mitigate the impacts of climate change. These adaptations are described below.

Adaptations

Adaptations were examined for land resources, water resources, and irrigation and agricultural technologies.

Land resources
In addition to runs assuming land loss to sea-level rise, each of the climate change scenarios was run assuming no land

lost due to global sea-level rise, because some form of coastal protection would be constructed. The land lost from natural subsidence remains, however.

There are hundreds of possible adaptation combinations that could be examined. It was not practical to consider them all. To provide some focus, the following adaptations were applied only to the UKMO scenario:

Water resources

There are plans (most notably the Jonglei Canal Project) to develop major water projects in the Upper Nile Basin (the Sudd Swamps) to divert runoff before it enters the swamps and evapotranspires. These projects would provide Egypt with an additional 10 MCM of water, or 20 percent of its current water supply. It is assumed that the projects will be realized and still be effective under climate change conditions.

Irrigation technology

A 5 percent increase in irrigation system efficiency was assumed.

Agricultural technology

The low-cost adaptations discussed in the agricultural chapter as adaptation Level 1 were used (see Chapter 2). These adaptations affect crop yield, water use, and commodity demand.

Adaptation scenario 1 (AD1) includes the improvement in irrigation technology, the Level 1 agricultural adaptations, and protection against sea-level rise. Adaptation scenario 2 (AD2) is the same as AD1 with an additional 20 percent increase in Egyptian water availability.

Policy adaptations

Maximizing social welfare by way of maximizing economic performance makes sense to an economist, but doing so might involve reduced agricultural output and increased agricultural imports. The 2060 Base scenario that maximizes social welfare for Egypt would have only a 10 percent food self-sufficiency value. An alternative policy, which many countries follow based on national security grounds, is to maximize food self-sufficiency. This policy could have substantial economic costs. Most governments adopt a policy somewhere between the free-market economy and the maximizing of food self-sufficiency. Since we will examine a policy of maximizing social welfare defined by the consumer–producer surplus, an alternative food self-sufficiency policy was developed. The EASM-CC objective function was changed to maximize domestic production of calories to increase food self-sufficiency while meeting the BLS estimated domestic demands.

Measures of climate change impacts

The objective function of EASM-CC is to maximize Consumer–Producer Surplus [CPS]. In addition to this integrative economic measure of social welfare, other economic and societal welfare measures are recorded for each run. They include the following:

Economic measures

- Consumer Surplus [CS]
- Producer Surplus [PS]; exports are considered as positive and imports as negative
- Agricultural Trade Balance [ATB] in Egyptian Pounds, calculated as exports minus imports
- Exports in Egyptian Pounds [EXP]
- Imports in Egyptian Pounds [IMP]
- Marginal Value of Water [MVW]
- Marginal Value of Land [MVL]

Societal welfare measures

- Calories per person per day exogenously determined by BLS and 1990 consumption patterns [CAL]
- Food Self-Sufficiency-Calories [FSS-CAL]: ratio of domestically-produced calories to total demanded calories per year

These specific measures provide insight into the nature of the economic impacts as well as providing some measure of the impact upon direct societal welfare.

THE INTEGRATED ASSESSMENT

Model Results

THE BASE SCENARIOS

While the Base 2060 scenario is not intended to be a prediction of Egypt's future, it is a self-consistent and plausible scenario considered optimistic by our definition. In this scenario the BLS Egyptian economic model suggests a 5.2-fold increase in GDP, made up of 3.1- and 5.6-fold increases in the agricultural and non-agricultural sectors, respectively. This growth is fueled by a 2.2-fold increase in population and 3.4- and 7.1-fold increases in the agricultural and non-agricultural capital, respectively (Onyeji and Fischer, 1994). The first analysis to be presented is a comparison of the Egyptian agricultural economic sector between 1990 and 2060 from the viewpoint of an integrated analysis, with the current climate.

Tables 14a–b present the set of economic measures recorded for each of the model runs. Comparing the 1990 Base (B1990) with the 2060 Base (B2060), four results stand out:

- Food consumption (calories per capita per day) decreases from 1990 to 2060. In 1990, Egyptian food consumption

Table 14a. *Economic measures of integrated impact assessment results under climate change (absolute numbers). (Units are Egyptian pounds except where noted.)*

	B 1990	B 2060	UKMO	GISS	GFDL*	GISS-A
Consumer–producer surplus	36,039	222,554	171,998	208,822	106,114	199,318
Consumer surplus	27,673	140,999	132,426	138,820	139,746	141,391
Producer surplus	8,366	81,555	39,571	70,003	−33,631	57,928
Agricultural trade balance	−926	83,280	4,681	67,783	−82,013	46,295
Food consumption (cal/day)	3,519	3,261	3,159	3,235	3,238	3,263
Food self-sufficiency (%)	53	10	9	12	4	16
Imports	4,244	43,116	69,488	44,430	102,760	34,250
Exports	3,318	126,396	74,169	112,213	20,747	80,545
Marginal value of water	0.00	0.53	1.29	0.32	6.58	0.00
Marginal value of land	0.30	0.49	0.59	1.09	0.00	2.54

Note:
* GFDL results only feasible with imports of winter, summer, and nili vegetables, summer and nili tomato, and sorghum allowed.

Table 14b. *Economic measures of integrated impact assessment results under climate change (relative numbers)*

	B1990	B2060	UKMO	GISS	GFDL*	GISS-A
Consumer–producer surplus	0.16	1	0.77	0.94	0.48	0.90
Consumer surplus	0.20	1	0.94	0.98	0.99	1.00
Producer surplus	0.10	1	0.49	0.86	−0.41	0.71
Agricultural trade balance	−0.01	1	0.06	0.81	−0.98	0.56
Food consumption	1.08	1	0.97	0.99	0.99	1.00
Food self-sufficiency	5.30	1	0.90	1.20	0.40	1.60
Imports	0.10	1	1.61	1.03	2.38	0.79
Exports	0.03	1	0.59	0.89	0.16	0.64
Marginal value of water	0.00	1	2.43	0.60	12.42	0.00
Marginal value of land	0.62	1	1.21	2.23	0.00	5.20

Note:
* GFDL results only feasible with imports of winter, summer, and nili vegetables, summer and nili tomato, and sorghum allowed.

was well above what the economy could afford, due to foreign food aid (principally wheat and grain) and heavy government subsidies funded by foreign aid and loans. Food consumption in 2060 is more in line with per capita income.

- The agricultural trade balance for 1990 is negative, and food self-sufficiency is close to sixty percent. This is the result of trade restrictions, pricing policies that affect the domestic farmer, and consumer policies.
- In 1990, water had a zero marginal value[5] while land had a positive value. This means that in 1990, land was the limiting resource to agricultural production under the observed economic conditions, while in 2060 both land and water are limiting resources.
- For the Base 2060 case, food self-sufficiency plummets to 10 percent, but agricultural trade balance is 40 percent of CPS and exports outnumber imports by 3 to 1. This shows a nation using its comparative agricultural advantage to export high valued crops and importing inexpensive grains and feeds that can be grown elsewhere more efficiently. Egypt benefits economically from free trade, but becomes quite vulnerable to international markets.

[5] Marginal value or shadow price is the change of the model objective function at the margin to the addition or subtraction of an additional unit of the resource. If a resource is in *surplus* at the optimal solution, the marginal value is zero. If the marginal value is positive it means that the resource is *limiting* and the objective will increase by the marginal value if an additional unit of the resource can be supplied or decrease if a unit is removed.

THE CLIMATE CHANGE SCENARIOS

A summary of the sectoral climate change impacts follows.

- Under all scenarios crop yields decline and crop water requirements increase;
- Agricultural land is lost to sea-level rise;
- Domestic commodity demands decrease only slightly with a slight increase under the GISS-A scenario;

Table 15. *Summary of integrated impacts measured as changes from the base scenario (percent)*

	UKMO	GISS	GFDL	GISS-A
Consumer–producer surplus	−23	−6	−52	−10
Consumer surplus	94	98	99	100
Producer surplus	49	86	−41	71
Agricultural trade balance	−94	−19	−198	−44
Food consumption	−3	−1	−1	0

- Nile water resources decline under UKMO and GFDL, and increase under GISS and GISS-A scenarios.

Chapters 2, 3, and 4 provide details of the sectoral impacts. Below, a discussion of the differences between the integrative results and sectoral results will be presented.

Tables 14a–b present the set of model results for the four climate change scenarios used for this study. Table 15 and Figure 9 provide a summary of the relative impact of climate change on four agro-economic measures. The four measures are indicators of climate change impact on the sector as a whole (consumer–producer surplus), the consumer (consumer surplus), the producer (producer surplus), and balance of payments (agricultural trade balance).

The summaries show that while consumer–producer surplus decreases from 6 to 52 percent from the 2060 Base Run, the consumer surplus ranges from a slight increase in the GISS-A scenario to a decrease of 6 percent in UKMO. Consumer surplus is driven by domestic commodity demands that come from the BLS. The BLS determines commodity demands based on prices and income. The non-agricultural sector is modeled as being unaffected by climate change. The BLS Base 2060 and climate change scenarios show the non-agricultural sector accounting for approximately 90 percent of Egyptian GDP (Onyeji and Fischer, 1994). Thus, under climate change scenarios, national income, as well as food consumption and commodity demands, remains relatively unaffected.

Agricultural production, which is modeled in great detail in EASM-CC, is most directly affected by climate change, and the results reflect this fact with decreases in producer surplus from 14 to 141 percent.[6] Decreased domestic production leads to changes in the agricultural trade balance with decreases from 19 to 198 percent. The GFDL results are extremely dramatic and lead to a negative agricultural trade balance.

The impact of climate change on the marginal values of land and water is an important measure. Tables 14a–b show these values, which are discussed below.

[6] Producer surplus includes imports as negative and exports as positive. So, if imports far exceed exports, as in the case in the GFDL scenario, then the producer surplus can become negative.

Table 16. *Sea-level-rise effects on integrated impacts under climate change*

	UKMO	GISS	GFDL	LOWEND
Percentage change in consumer–producer surplus	99	99	100	99

- UKMO: Both land and water are limiting constraints, with water at 2.43 and land at 1.21 times the Base 2060 marginal value. The 1.21 factor for land reflects the seasonal limitation on land, as the model simulates planting more winter crops when crop consumptive water use per output value is lowest.
- GISS: With an 18 percent increase in Nile water there is still a marginal value on water due to reductions in agricultural water productivity from reduced crop yields and increased water requirements. With increased water availability, land will become the limiting resource, leading to a 2.23-fold increase in marginal value.
- GFDL: With so little water, land is in surplus and marginal value is zero, but water has a very high marginal value.
- GISS-A: With a 14 percent increase in Nile water and a decrease in average water use per ton of agricultural production, the marginal value on water is zero and land is the only limiting resource.

Adaptation results

SEA-LEVEL RISE

All the above scenarios were run assuming loss of land from a 37-cm sea-level rise in 2060. Runs were made to examine the net impact to Egypt of mitigating sea-level rise. Table 16 reveals that the land loss due to 37 cm of sea-level rise has very little economic impact on the agricultural economic sector. This insensitivity results from high-valued, higher water-use crops being substituted for some lower water-use crops. Thus, water is spread less thin over fewer areas, and the economic yields are only slightly less.

INTEGRATED ADAPTATIONS: WATER, IRRIGATION, AND AGRICULTURE

To examine the effects of a series of integrated adaptations to mitigate climate change impacts, two adaptation runs were made. Each run has adaptations for land resources, water resources, and agricultural and irrigation technologies. The UKMO scenario was chosen as the climate change scenario upon which to measure the impacts of the adaptations. The only difference between AD1 and AD2 is that AD2 has an additional 20 percent increase in available water due to the upper Nile projects. Tables 17a–b present the results in absolute and relative numbers, respectively.

The model results (Table 17b) show only modest increases

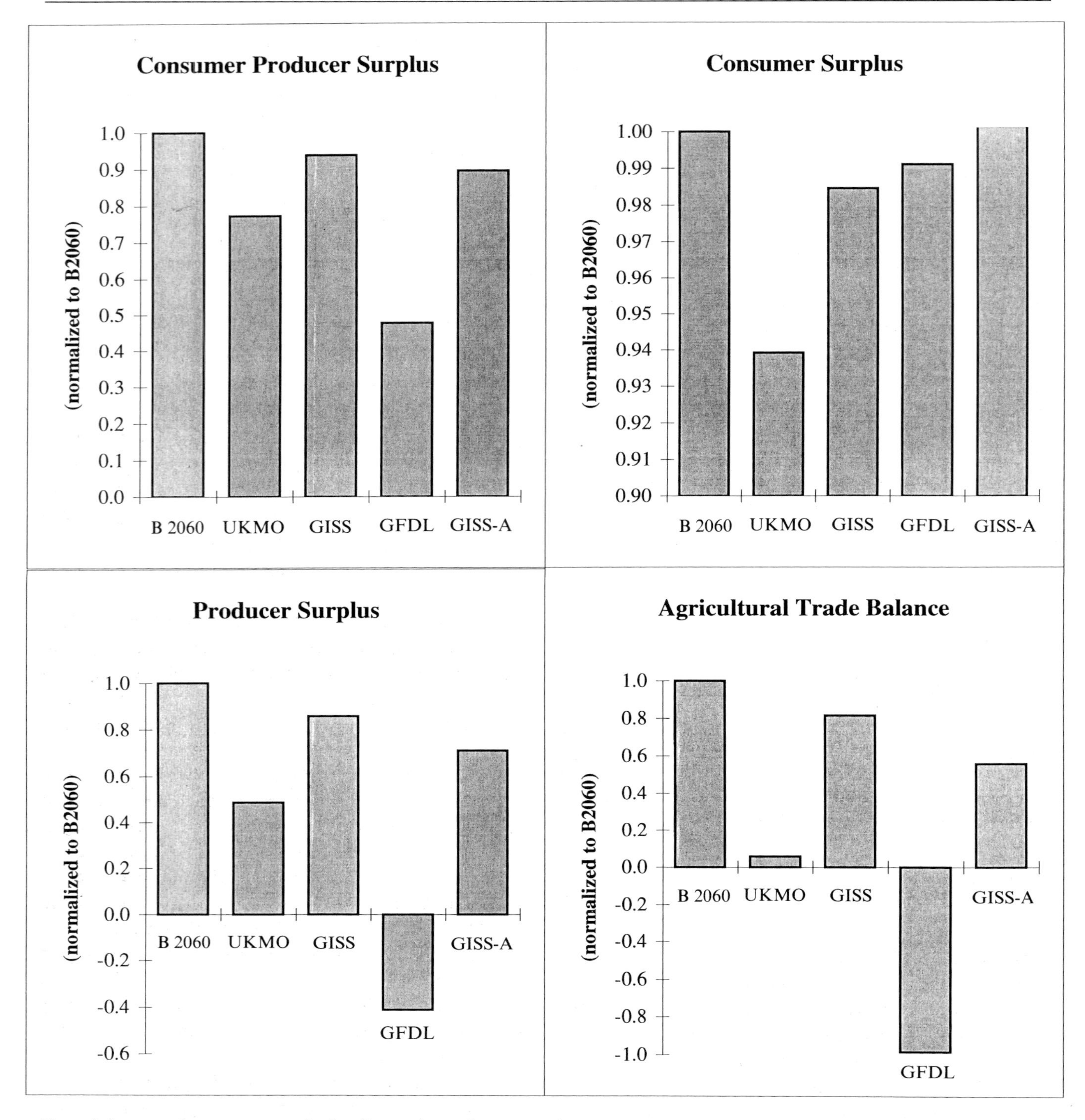

Figure 9 Integrated assessment results for climate change impacts on four economic welfare measures.

in the CPS of 7 and 8 percent for AD1 and AD2, respectively. However, there are significant increases in the ATB of 3.5-fold for AD1 and 4-fold for AD2.

The UKMO scenario has a 13 percent decrease in water resources. AD1, with improved irrigation efficiencies and decreases in crop water use, shows the marginal value of water decreasing by 75 percent, while the marginal value of land increases 3-fold even with an increase in land resources. This is because water supply has been increased. The result of AD2, however, shows that a 20 percent increase of water provides for only a minor improvement in the CPS and moderate improvement in the ATB.

Table 17a. *Integrated impacts results for adaptation to the UKMO scenario. (Units are Egyptian pounds except where noted.)*

	UKMO	AD1*	AD2*
Consumer–producer surplus	171,998	184,238	185,063
Consumer surplus	132,426	134,205	134,205
Producer surplus	39,571	50,033	50,858
Agricultural trade balance	4,681	20,863	23,550
Food consumption (cal/day)	3,159	3,184	3,184
Food self-sufficiency (%)	9	9	9
Imports	69,488	42,417	42,364
Exports	74,169	63,280	65,915
Marginal value of water	1.29	0.30	0.28
Marginal value of land	0.59	1.87	2.00

Notes:
*1 No Upper Lakes Projects
2 Upper Lakes Projects

Table 17b. *Adaptation results measured as percentage changes to the UKMO scenario results*

	AD1*	AD2*
Consumer–producer surplus	+7.1	+7.6
Consumer surplus	+1.3	+1.3
Producer surplus	+26.4	+28.5
Agricultural trade balance	+346	+403
Food consumption (cal/day)	+0.8	+0.8
Food self-sufficiency (%)	0	0
Imports	−39.0	−39.0
Exports	−14.7	−11.1
Marginal value of water	−76.7	−78.3
Marginal value of land	+217	+239

Notes:
*1 No Upper Lakes Projects
2 Upper Lakes Projects

FOOD SELF-SUFFICIENCY

All the previous results were based on maximizing economic welfare. Table 18 provides the results for a model run that maximizes food self-sufficiency under the UKMO scenario. For this model run, the effect of climate change is to reduce food self-sufficiency by 42 percent and economic welfare by 56 percent. By comparison, the UKMO scenario for maximizing economic welfare has food self-sufficiency dropping 10 percent and economic welfare dropping 23 percent. It appears that economic welfare maximizing policy is less vulnerable to climate change.

Summary

The integrated climate change impacts upon the agricultural economic sector of Egypt result in decreases of consumer–producer surplus from 6 to 52 percent from the Base 2060. Under all scenarios, the consumers fare relatively well with consumer surplus impacts ranging from a 0.3 percent increase to a 6 percent decrease from the Base 2060. The producers and the balance of trade are the parts of the agricultural sector that are affected the most by climate change.

Table 18. *UKMO climate change impacts on food self-sufficiency policy*

	Percentage change from base
Consumer–producer surplus	44
Consumer surplus	94
Producer surplus	−2244
Agricultural trade balance	−234
Food consumption (cal/day)	97
Food self-sufficiency (%)	58
Imports	234
Exports	—
Marginal value of water	—
Marginal value of land	15

Producer surplus drops 14 to 141 percent from the 2060 Base, while the agricultural trade balance decreases by 19 to 198 percent. These impacts are the results of strong agricultural commodity demands fueled by income from the non-agricultural sector, which is not directly impacted by climate change.

Dramatic declines of 141 percent and 198 percent in producer surplus and agricultural trade balance are found for the GFDL scenario. From a water resource and agro-economic perspective, the GFDL scenario is a catastrophe. Such a decline in water resources leading to a major reduction in agricultural production and massive imports will most likely result in major socio-economic impacts.

Land loss from a 0.37 meter sea-level rise by 2060 has little impact on economic performance. However, a full 1 meter or more of sea-level rise on top of continued subsidence may present more significant agro-economic impacts in addition to major social impacts.

Adaptations in the water resources, irrigation and agricultural technologies, as well as protection against sea-level rise, provide for a modest 7 to 8 percent increase in the agricultural sector performance above the no-adaptation results. The adaptation results are still 17 percent below the Base 2060 results. These extremely expensive adaptations are unable to mitigate the climate change impacts, but investments in improving irrigation efficiency appear to be a robust policy that would be beneficial regardless of whether the climate changes.

The analysis of a food self-sufficiency policy alternative

Table 19. *A comparison of integrated and sectoral impacts to climate change. (Percentage change from Base 2060 results)*

Scenario	Sea-level (land)	Food demand (kcal)	Ag. water productivity (ton/m³)	Water resources (Aswan)	Integrated (CPS)
UKMO	−4	−3	−45	−13	−23
GISS	−4	−1	−13	+18	−6
GFDL	−4	−1	−36	−78	−52
GISS-A	−4	0	+10	+14	−10

shows that the climate change impacts are more severe when compared to a 2060 food self-sufficiency base than to the impacts of maximizing agricultural sector consumer-producer surplus.

If food consumption is not significantly impacted, should Egypt worry about climate change impacts? This food consumption outlook is based on an 'optimistic' view of future non-agricultural performance as modeled as a single sector in the BLS.

As Onyeji (1992) points out, in Egypt the close link between the agricultural and non-agricultural sectors means that any change in agricultural output or its composition will, in turn, affect the level of industrial activity, which will extend to different socio-economic groups. How these climate changes affect government control and markets will undoubtedly directly affect all socio-economic groups. Thus, the primary impact of global climatic change on people is measured by how the change affects the people's entitlement to food (income), and the ability to import the food.

Lessons from an integrated assessment

Table 19 provides a summary of the impacts of climate change on Egypt as measured by the biophysical impacts of each sector together with the integrated economic welfare measure. To provide an aggregate biophysical measure of climate change impacts on Egyptian agriculture, the Agricultural Water Productivity Index [tons/m³] was developed. The index is defined as total agricultural production [tons] divided by total agricultural water use [m³]. It is calculated by taking the cropping pattern from the Base 2060 model run, multiplying the cropped acreage by its yield, and summing over all crops to provide total agricultural production. The next step is multiplying the cropped acreage by its water use and summing over all crops to provide total water use, then dividing total agricultural production by total water use. These steps are then repeated for each of the climate change scenarios, but with the crop yields and water use determined by the crop modeling described in Chapter 2.

There are two main conclusions to be drawn about the

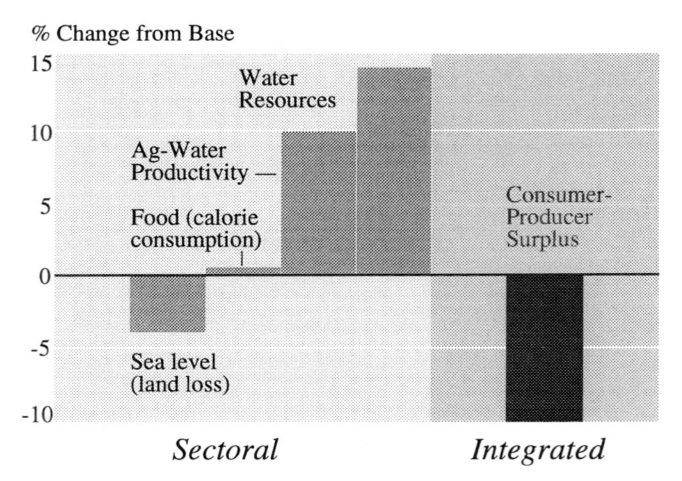

Figure 10 Comparison of sectoral versus integrated impacts for the GISS-A scenario.

integrated analysis. First, the interactions and importance of the sectoral impacts depend very much on the collective impact of all sectors. For example, under the GFDL scenario where water resources are greatly reduced, the impact of sea-level rise is zero. In the GISS scenario, where water resources increased by 18 percent, the 4 percent loss of land and the 13 percent reduction in agricultural water productivity leads to a 6 percent reduction in economic welfare. Under the UKMO scenario, sectoral reductions of 4 percent in land, 13 percent in water, and 3 percent reduction in food intake result in a 23 percent reduction of economic welfare due to the 45 percent reduction in agricultural water productivity.

Second, integrated analysis shows a substantially different picture about climate change impacts than if each individual sectoral impact had been studied. For example, in Table 19, food intake reductions are 0 to 3 percent, due to the climatic change, while the reductions in economic welfare are on the order of 6 to 50 percent. The GISS-A scenario is the best example for this point. Figure 10 presents, graphically, the last row of Table 19. For the GISS-A scenario, each sector, except sea-level rise, shows significant increases over current climate conditions.[7] However, there is a 10 percent decline in economic welfare. This result is because the rest of the world does even better under the GISS-A scenario and Egypt therefore loses some of its competitive advantage for exports, and the trade balance declines. This result could be found only via an integrated (and global) analysis. Sectoral results point to a positive impact, but when using the BLS results with global impacts and their implications for Egypt, this negative impact appears. Even the BLS alone would not have shown this result so clearly.

[7] Agricultural water productivity is greater than current climate conditions due to the CO_2 effects that reduce water requirements, so that even though yields decline, water requirements decline slightly more to give a 10 percent increase in agricultural water productivity.

While much can be learned about climate change impacts from separate bio-geophysical sectoral impacts, a much better measure of the economic and societal impacts is an integrated impacts approach.

Lessons for Egypt about vulnerability to climate change

The analysis presented in this chapter has identified a number of insights or lessons about Egypt's vulnerability to climate change. Some of the important lessons are listed below.

- Population is a dominant factor in all future scenarios with or without climate change.

GENERAL LESSONS LEARNED

The results of this analysis not only provide lessons for Egypt, but provide for more general insights that may be applicable to analyzing climate change effects on nations or regions of the world. These are as follows:

- CONSISTENT SCENARIO AND ASSUMPTIONS
  In performing integrated analysis, it is absolutely essential that all scenarios and assumptions are consistent across the sectoral analysis, or integrated results will be meaningless. The level of impact of climate change on an economy can vary greatly, depending on which measure or criterion is chosen to measure the impact. The level of economic development of a country should be taken into account when choosing a macro-economic measure.
- THE DISTRIBUTION OF IMPACTS ACROSS SOCIO-ECONOMIC LEVELS
  When examining socio-economic impacts, a single macro-economic measure may not reflect the situation when the impacts are not being felt equally across the population; and in many cases, the population, already struggling, is being burdened on the lower levels.
- THE IMPACT ON FOOD ENTITLEMENT
  In many cases the amount of food available to the population at large may not be greatly reduced, but certain sectors of the population may be significantly affected in terms of entitlement to food, either by reduced income or lack of resources for subsistence agriculture.
- ADAPTATIONS MAY NOT BE JUSTIFIED
  Some of the easier adaptations that appear justifiable from a sector standpoint (e.g. increased water supply) may turn out not to be justifiable under an integrated assessment (e.g. sea-level rise may reduce land, reducing need for water supply).
- NEED TO HAVE FEEDBACK, POPULATION/CONSUMPTION
  The economic and population growth must be analyzed with feedback of resource and economic impacts due to climate change over a dynamic time horizon.
- FLEXIBLE SYSTEMS/POLICY
  Government social and economic policies can result in a system that is relatively flexible and more easily adaptable to climatic change or produce a rigid system which is more vulnerable to climate change (e.g., food self-sufficiency policies).

Conclusions and recommendations

Egypt is highly vulnerable to the warming and changes in precipitation and river runoff that are forecast to accompany greenhouse-gas-induced climate change. It has been shown that water is an important element in the agricultural economy and must be explicitly included in any analysis of climate change impacts. Finally, this study has shown that in addressing climate change impacts on an economy or nation, the entire economy must be analyzed. The linkage of the sectors directly affected by climate change must be analyzed in concert with the other sectors of the economy in sufficient detail so that feedback can be part of the analysis.

Future research must include the development of more crop models that include CO_2 effects, and the incorporation of crop water use, irrigation, and water resources directly into the macro-economic policy models. The development of a better understanding of potential agricultural/water adaptations to climate change and sea-level-rise impacts on deltaic agricultural lands is needed. Better macro- and sectoral economic understanding and models of bio-geophysical resources, particularly those impacted by climate change, and their role in the economy need to be developed.

REFERENCES

Alexander, R. 1994. Personal communication.

Amer, M. H., and N. A. de Ridder. 1989. *Land Drainage in Egypt*. Drainage Research Institute. Cairo: Nubar Printing House.

Bowes, M. D., and P. R. Crosson. 1993. Paper 6: Consequences of Climate Change for the MINK Economy: Impacts and Responses. *Climatic Change* 24(1–2):131–58.

Broadus, J., J. Milliman, S. Edwards, D. Aubrey, and F. Gable. 1986. Rising Sea Level and Damming of Rivers: Possible Effects in Egypt and Bangladesh. In *Effects of Changes in Stratospheric Ozone and Global Climate, Volume 4: Sea Level Rise*, ed. J. Titus. Washington, DC: United Nations Environment Programme and U.S. Environmental Protection Agency.

Central Planning Bureau (CPB). 1992. *Scanning the Future, A long-term scenario study of the world economy 1990–2015*. The Hague: Sdu Publishers.

Crosson, P. R., and N. J. Rosenberg. 1993. An Overview of the MINK Study. *Climatic Change* 24(1–2):159–73.

Eid, H., and M. Saleh. 1992. Modeling of Climate Change Impacts on Egyptian Agriculture. A report prepared for the University of Colorado at Boulder.

El-Din, G. E. 1993. It's no bed of flours, say local wheat millers. *Egyptian Gazette* (Cairo), 3 October.

El-Raey, M. 1991. Responses to the Impacts of Greenhouse-Induced Sea Level Rise on Egypt. In *Changing Climate and the Coast, Volume 2: West Africa, the Americas, the Mediterranean Basin, and the Rest of Europe*, ed. J. G. Titus. Washington, DC: United Nations Environment Programme and U.S. Environmental Protection Agency.

Food and Agricultural Organization (FAO). 1993. AGROSTAT/PC. Rome: United Nations FAO.

Fischer, G. 1993. BLS Model Results. Personal communication.

Fischer, G., K. Frohberg, M. A. Keyzer, and K. S. Parikh. 1988. *Linked National Models: A Tool for International Policy Analysis*. Dordrecht: Kluwer.

Fischer, G., K. Frohberg, M. L. Parry, and C. Rosenzweig. 1994. Climate change and world food supply, demand, and trade: Who benefits, who loses? *Global Environmental Change* 4(1):7–23.

Gleick, P. H. 1991. The Vulnerability of Runoff in the Nile Basin to Climate Changes. *Environmental Professional* 13:66–73.

Hansen, B. 1991. *The Political Economy of Poverty, Equity, and Growth: Egypt and Turkey*. New York, Oxford: Oxford University Press.

Hexem, R. W., and E. O. Heady. 1978. *Water Production Functions for Irrigated Agriculture*. Ames, Iowa: Iowa State University Press.

Holy Bible, New International Version. 1984. Grand Rapids, Mich.: Zondervan Bible Publishers.

Humphries, J. 1991. EASM91, A users manual. A Planning Studies and Models Project report of the Ministry of Public Works and Water Resources, Government of Egypt, Cairo.

Kulshreshtha, S. N. 1993. *World Water Resources and Regional Vulnerability: Impact of Future Changes*. RR-93-10. Laxenburg, Austria: International Institute for Applied Systems Analysis.

Mortsch, L., G. Koshida, and D. Tavares, eds. 1993. *Adapting to the Impacts of Climate Change and Variability. Proceedings of the Great Lakes–St. Lawrence Basin Project, 9–11 February, 1993, Quebec City*. Downsview, Ontario: Environment Canada.

Nicholls, R. 1991. Land loss scenarios for the Nile Delta, Egypt, for the years 2060 and 2100. A report prepared for the United States Environmental Protection Agency, Office of Policy Analysis.

O'Mara, G., and E. Hawary. 1992. The Response of Egyptian Farmers to Cotton Policy Intervention. A report to the Ministry of Agriculture of the Arab Republic of Egypt and the U.S. Agency for International Development. (unpublished).

Onyeji, S. C. 1992. A Socioeconomic Analysis of Integrated Climate Change Impacts in Egypt. A report prepared for the United States Environmental Protection Agency, Office of Policy Analysis.

Onyeji, S. C., and G. Fischer. 1994. An Economic Analysis of Potential Impacts of Climate Change in Egypt. *Global Environmental Change* 4(4):281–99.

Parry, M. L., M. Blantran de Rozari, A. L. Chong, and S. Panich. 1992. *The Potential Socio-Economic Effects of Climate Change in Southeast Asia*. Nairobi: United Nations Environment Programme.

Rosenzweig, C., and D. Hillel. Forthcoming. *Egyptian Agriculture in the Mid-21st Century*. Collaborative Paper. Laxenburg, Austria: International Institute for Applied Systems Analysis.

Shahin, M. 1985. *Hydrology of the Nile Basin*. Amsterdam: Elsevier.

Strzepek, K., and S. Saidin. 1994. *A GIS Assessment of Sea-level Rise on the Nile Delta*. WP-94-48. Laxenburg, Austria: International Institute for Applied Systems Analysis.

Utah State University (USU). 1986. Strategies for Irrigation Development. Water Management Synthesis Report 42. Logan, Utah: USU.

World Resources Institute (WRI), United Nations Environment Programme (UNEP), and United Nations Development Programme (UNDP). 1992. *World Resources 1992–93*. New York, Oxford: Oxford University Press.

Yates, D., K. Strzepek, and G. O'Mara. Forthcoming. *A Sectoral Modeling Approach to Assessing Climate Change Impacts on the Agricultural Economy of Egypt*. Working Paper. Laxenburg, Austria: International Institute for Applied Systems Analysis.

Zeineldin, A. 1986. The Egyptian economy in 1999: An input-output study. *Economic Modelling* 3(2):140–6.

8 Adaptation Policy

JOEL B. SMITH
Hagler Bailly Consulting, Inc.
Formerly Deputy Director, Climate Change Division
U.S. Environmental Protection Agency

JEFFREY J. CARMICHAEL
Department of Economics
University of Colorado at Boulder

JAMES G. TITUS
Office of Policy, Planning and Evaluation
U.S. Environmental Protection Agency

DO WE NEED TO PREPARE FOR CLIMATE CHANGE?

This book shows that global warming could have far reaching impacts across the world. In many places ecosystems could be transformed, coastal lands inundated, river flows dropped, crop yields in many countries could be reduced, and even human health may suffer. In other places, crop yields and water supplies may increase. A consensus has existed for over a decade that some global warming is inevitable (NAS, 1979; Seidel and Keyes, 1983). Recent assessments have reaffirmed the likelihood of significant warming (Houghton *et al.* 1992), although studies have found that the rate of warming could be lower than originally thought (Wigley and Raper, 1992). Within this consensus, however, large uncertainties exist concerning the rate and magnitude of global change, particularly involving how it will be manifested at the local level, what the effects of regional climate changes will be, and how human and natural systems will adapt to those effects.

Assuming the consensus on global warming is correct, society has two fundamental options:

1. Take steps to counteract these changes, by limiting emissions of greenhouse gases; or
2. Adapt to whatever warming cannot be prevented.

Neither of these options is likely to be sufficient by itself. It is unlikely that global emissions of greenhouse gases will be reduced enough to eliminate the possibility of significant warming.[1] Thus, society must also adapt to climate change.

This chapter focuses on what can and should be done to adapt to the effects of climate change. It examines what types of resources and policies should be candidates for adaptation, what types of adaptation are possible, the issues involved in determining whether adaptation in advance of climate change is necessary, and how to design and implement appropriate policies that incorporate climate change effects.

WHAT TYPES OF RESOURCES SHOULD PREPARE FOR CLIMATE CHANGE?

Table 1 lists examples of resources where analysis of adaptation to climate change should occur. The resources listed

[1] Even if emissions limitations efforts currently being contemplated are successful, significant climate warming is still likely. In the United Nations Framework Convention on Climate Change, signed at the 1992 Earth Summit, developed countries agreed to an aim of stabilizing their emissions of greenhouse gases by the year 2000 at 1990 levels. Stabilizing emissions from developed countries will only reduce global temperatures in 2100 by a few tenths of a degree Celsius (to about 2.5°C) compared to continuing to emit greenhouse gases at current trends (Houghton *et al.*, 1992). If global emissions of greenhouse gases are stabilized and emissions of chlorofluorocarbons eliminated, temperatures could still rise by 1.5°C by 2100 (*ibid.*) To stabilize climate, carbon dioxide emissions would have to be reduced by more than 60 percent from current levels (Houghton *et al.*, 1990). Draconian measures would probably be required to obtain such reductions, particularly for carbon dioxide. Such measures are unlikely to be adopted.

Table 1. *Resources sensitive to climate change*

Resource	Vulnerability
Agriculture	Crop yields Soil erosion
Forests	Change in species composition Shifts in geographic distribution
Conservation areas	Disruption or loss of habitat Invasion of new species
Coastal areas	Inundation of coastal dry and wetlands and property
Fisheries	Change in composition of stocks Change in location
Water resources	Alteration of supply Shifts in flood and drought frequency
Human health	Changes in heat and cold stress Shifts in prevalence of infectious diseases
Energy demand and production	Increases in summer energy demand and decreases in winter demand Changes in supply

Sources: Smith and Tirpak, 1989; Tegart *et al.*, 1990; NAS 1992; Rosenberg and Crosson, 1991.

below have been shown by studies such as this to be potentially vulnerable to climate change. The impacts listed in the table are first order effects of climate change. These effects will also cause second and third order effects on society. For example, changes in crop yields will affect food prices and agricultural income, and beach erosion could reduce tourist income in affected areas.

WHAT TYPES OF ADAPTATION ARE POSSIBLE?

Assuming that global warming will occur, two types of adaptation response are possible. *Reactive adaptation* may be chosen by responding to climate change after it occurs. Alternatively, *anticipatory adaptation* may be preferred, which means taking steps in advance of climate change to minimize any potentially negative effects or to enhance the ability of society and nature to rapidly and inexpensively adapt to changes. The debate on adaptation to global climate change has focused on the issue of whether reactive adaptation is capable of offsetting the negative consequences of climate change rather than whether measures should be taken in advance of climate change to prepare for any negative effects.

There is no consensus on whether adaptation in response to climate change will completely offset adverse impacts. Some studies conclude that human and natural systems may be able to adapt to the most likely scenarios of climate change (NAS, 1992; CAST, 1992). Heat and drought resistant crops can be substituted for those whose yields are reduced. People could move to more favorable climates. Infrastructure, which is replaced on a much faster time scale than climate change, could be adapted to changes in climate (Ausubel, 1991).

Yet, these same studies also conclude that a rapid climate change or significant increases in the intensity and frequency of extreme events could make societal adaptation much more difficult. Many studies agree that a rapid climate change could cause severe dislocations in natural ecosystems, especially among plants, because rates of change will be too fast for many flora and fauna to adapt (Tegart *et al.*, 1990; Peters and Lovejoy, 1992; World Wildlife Fund, 1992). This makes a case for the necessity of anticipatory adaptation.

WHAT ISSUES ARE INVOLVED IN DETERMINING THE NEED FOR ANTICIPATORY ADAPTATION?

Determining in which circumstances a reactive adaptation policy is sufficient and when, in addition, anticipatory adaptation is necessary involves a number of concerns.

- *Uncertainty*
  The timing, magnitude, direction of climate change impacts, and their results are uncertain, as is the ability of society to adjust to climate change. This is particularly true at a regional level (Smith, 1990).
- *Irreversibility*
  The impacts of climate change on valued resources may be irreversible or may cause significant long term damage in some situations, creating a need for anticipatory adaptation.
- *Effective life of policy*
  Decisions made today may last long enough to be affected by climate change. If decisions are made assuming continuation of current climate conditions, their effectiveness may be reduced by climate change.
- *Importance of climate trend*
  Reacting to climate change may mean reacting to climate extremes, such as floods, droughts, or storms. Natural resource management policies are often changed in reaction to the last storm or drought. This policy of reaction is sensible but runs the risk of taking short term incremental approaches and not anticipating potential greater variability in the future (Glantz, 1988; Washington, 1992). A short term approach may cause needed anticipatory measures to be ignored.

- *Rate of climate change*
  Climate change may come suddenly rather than gradually (Dansgaard *et al.*, 1993). Reactive adaptation to a rapidly changing climate will be more difficult than to a gradually changing climate.

- *Health and safety*
  In some situations, climate change may threaten human life or increase risks of disease and injury. The dangers posed are difficult to quantify directly as costs and benefits, but should nevertheless come into consideration in determining whether anticipatory adaptation steps are appropriate.

DESIGNING AND IMPLEMENTING ANTICIPATORY POLICIES

The development or incorporation of anticipatory adaptation measures will be appropriate in many resource areas, despite the uncertainties involved. Given the need for the design of appropriate policies, what criteria and types of policies are useful?

- *Benefit/cost analysis*
  The highest priority adaptation measures should be those whose discounted benefits substantially exceed the discounted cost. Clearly, if this criterion is not met, the policy is not appropriate, unless other special conditions exist (such as threat of species extinction) that override the excessive costs.

- *Urgency*
  Anticipatory policies should be designed and implemented quickly, if current trends will make delayed adaptation measures less successful or likely to succeed. If a policy is much more expensive if implemented later, or if its benefits would be greatly reduced, such a policy should be a high priority for implementation now[2] (Coastal Management Subgroup, 1992).

- *Benefits independent of climate change*
  When possible, anticipatory policies that bring benefits or reduce costs even if climate change does not occur, should be designed. Examples of such policies are discussed later.

- *Flexibility*
  A policy that enables a resource to successfully adapt to a wide range of future climates increases flexibility. Success in this context involves either minimal costs from climate change (robustness) or the ability of a system to quickly recover from climate changes (resiliency). This is useful, given the uncertainties about the direction and magnitude of climate change.

[2] A policy is economically efficient if Benefit-Cost > 0. We can define urgency as a policy that is efficient and for which Cost/t-Benefit/t > 0. Thus, policies that are efficient but are not urgent should not be implemented now because the benefits of delay are greater than the costs of delay. See Titus (1990) for examples of urgent policies.

- *Use of planning*
  Policies which are contingency plans or marginal actions on predetermined policies enhance the ability of resource systems to react to climate change (Office of Technology Assessment, 1993). Such changes have low costs, but are very useful should climate change.

- *Mitigation effects*
  Policies should be designed not only to adapt to potential climate change, but also to help mitigate climate change impacts. Examples of such policies are discussed later.

- *Research*
  The incorporation of research into policy can expand the options available for adapting to climate change. Since such research can take years to develop, anticipatory action will be most useful.

- *Equity and values*
  Consideration should be given in policy design to issues of equity: who will benefit from this policy, and who will lose? Policies that will place most of the burden on the poor may not be most appropriate. Consideration should also be given to the value framework: does the policy consist of adjusting human desires (such as through water and energy conservation), or of changing the natural environment even more to meet current human desires (such as by constructing dams to assure availability of water and energy)? Policy makers should consider such aspects.

EXAMPLES OF ANTICIPATORY POLICIES

In this section, anticipatory policies are divided into three categories and examples are discussed to illustrate each, including some drawn from this book. The categories are:

1. Policies that result in mitigation and adaptation
2. Policies that are justified without consideration of climate change
3. Policies that are justified by the risks of climate change

Policies that result in mitigation and adaptation

Some policies help reduce greenhouse gas emissions and enhance the ability to adapt to its impacts. This category substantially overlaps with the next category, policies that are justified without consideration of climate change, as many of the policies have significant benefits under current climate conditions.

REDUCED POPULATION GROWTH

Reducing population growth will help reduce the demand for fossil fuels, deforestation, and other sources of greenhouse gases. It also reduces demand for food, water, and land, thus

Table 2. *Number of people at risk of hunger in 2060 (millions)*

Population	Number of people at risk			
	Climate scenario			
	No change	GISS	GFDL	UKMO
10.3 billion				
Total	641	704	750	1011
Increase due to climate change	n/a	63	109	370
8.6 Billion				
Total	395	412	445	578
Increase due to climate change	n/a	17	50	183
Percent reduction in people at risk due to lower population	38%	41%	41%	43%

Note: All GCM scenarios assume 555 ppm of CO_2.

reducing vulnerability to change in availability of these resources. The need for population control primarily exists in the developing countries.

Although it was not analyzed in this study, the rate of population growth will have a major effect on emissions of greenhouse gases. Most of the increase in global population over the next century is expected to take place in developing countries. Evidence exists that even small gains in per capita and total economic growth by developing countries will result in greenhouse gas emissions exceeding those from the developed world within a few decades (Office of Technology Assessment, 1991).

Reduced population growth rates also lead to reduced demand for food and water, and therefore to lower expectations of the number of people at risk of hunger. Table 2 (based on Chapter 2, World Food Supply) displays the effect of lower population growth assumptions on the number of people at risk of hunger.[3] This metric, which is described in Chapter 2, is an indicator of how well global food supplies can meet global food demand. It should be noted that it is not an estimate of the number of people starving, as it does not address non-market food distribution systems. This metric tends to magnify any shortfall in food production, particularly in developing countries.

The base case, using UN Mid Estimates, assumes a population level of 10.3 billion people in 2060, while the low population growth scenario, using UN Low Estimates, assumes there will be 8.6 billion people by 2060, which is about 17 percent lower than the base case. Virtually all of the people estimated to be at risk of hunger are in developing countries. Higher per capita GDP works together with lower world population to produce approximately 38 percent fewer people at risk from hunger in the year 2060 compared to the reference scenario.

When climate change is considered, the analysis shows that there is a positive synergy between climate change and population growth. The lower population growth reduces hunger by 38 percent under current climate, while reducing it by 41 to 43 percent from a base that is already much higher in the climate change scenarios. Lower population growth has even greater benefits when climate change is considered than when it is not.

REDUCE DEFORESTATION AND PROMOTE AFFORESTATION

Increasing forest cover creates more sinks for carbon dioxide, which reduces the amount of greenhouse warming taking place, therefore partly mitigating global warming. Greater forest cover also assists adaptation in a number of ways. More forests increase species population and range, which enhances the ability to adapt to climate change. Larger contiguous forest stands that cover different habitats, types of moisture content, and altitudes allow more potential for migration of both animal and tree species under climate change (see Chapter 6).

Policies that are justified without consideration of climate change

Policies in this category are designed in preparation for potential climate change, and are designed in such a way that the benefits outweigh the costs. If, however, climate change does not take place, such policies continue to be valuable from a benefit/cost perspective. Such policies are known as 'no regrets' policies because they will succeed in either circumstance, meaning that policy makers never have to regret their adoption. However, that doesn't mean that such policies are inexpensive. Policies in this area are not mutually exclusive from policies that result in mitigation and adaptation; some may incorporate aspects of both. Table 3 gives examples of no regrets anticipatory adaptation policies.

Policies in this category are not exclusively useful as anticipatory responses. Many of these policies may also be appropriate as reactive adaptation responses to climate change. For example, free trade of agricultural products or market mechanisms for water allocation can be introduced in the future as needed. However, imposition of such measures as a reaction to climate change would be politically difficult as winners and losers become identified. Impacts and costs would be incurred that could have been avoided, if these policies had been in place. The argument for instituting these as anticipatory policies is that they are economically justified now. By not using these measures, society is foregoing net benefits.

[3] The Rosenzweig *et al.* study did not examine policies to reduce population growth.

Table 3. *No regrets anticipatory adaptation policies*

Coastal zone management	
Wetland preservation and migrations	Maintains healthy wetlands which are more likely to withstand sea-level rise and have higher value than artificially created replacements. Maintains existing coastal fisheries that are difficult to relocate.
Integrated development of coastal datasets	Integrated data allows formation of comprehensive planning and identification of regions most likely to be affected by physical or social changes. Allows effects of changes to be examined beyond the local or regional scale.
Improved development of coastal models	Improved modeling allows more accurate evaluation of how coastal systems respond to climate change and also to other shocks.
Land use planning	Sensible land-use planning, such as the use of land setbacks to control shoreline development, better preserves the landscape and also minimizes the concerns of beach erosion from any cause.
Water resources	
Conservation	Reducing demand can increase excess supply, giving more safety margin for future droughts. Using efficient technologies such as drip irrigation reduces demand to some extent. Preserving some flexibility of demand is useful as less valuable uses allow reduced demand during droughts.*
Market allocation	Market-based allocation allows water to be diverted to its most efficient uses, in contrast with non-market mechanisms that can result in wasteful uses. Market allocations are able to respond more rapidly to changing supply conditions, and also tend to lower demand, conserving water.
Pollution control	Improving water quality by improving the quality of incoming emissions provides greater water quality safety margins during droughts and makes water supply systems less vulnerable to declines in quality because of climate change.
River basin planning	Comprehensive planning across a river basin can allow for imposition of cost-effective solutions to water quality and water supply problems. Planning can also help cope with population growth and changes in supply and demand from many causes, including climate change.
Drought contingency planning	Plans for short-term measures to adapt to droughts. These measures would help offset droughts of known or greater intensity and duration.
Human health	
Weather/health watch warning systems	Warning systems to notify people of heat stress conditions or other dangerous weather situations will allow people to take necessary precautions. This can reduce heat stress and other types of fatalities both now and if heat waves become more severe.
Improve public health and pest management procedures	Many diseases which will spread if climate changes are curable or controllable, and efforts in these areas will raise the quality of human life both now and if climate change occurs.
Improve surveillance systems	More and better data on the incidence and spread of diseases is necessary to better determine the future patterns of infection and disease spread. This information is helpful under any scenario.
Ecosystems	
Protect biodiversity and nature	Biodiversity protection maintains ecological diversity, and richness preserves variety in genotypes for medical and other research. A more diverse gene pool provides more candidates for successful adaptation to climate change. One possibility is to preserve endangered species outside of their natural habitat, such as in zoos.
Protect and enhance migration corridors	Such policies help maintain an ecosystem and animal and tree species diversity. Corridors and buffer zones around current reserve areas that include different altitudes and ecosystems are more likely to withstand climate change by increasing the likelihood of successful animal and tree migration.

Table 3. *No regrets anticipatory adaptation policies (cont.)*

Watershed protection	Forest cover provides watershed protection, including protection from bank erosion, siltation, and soil loss. All of these functions are extremely valuable whether climate changes occur or not.
Agriculture	
Irrigation efficiency	Many improvements are possible and efficient from a benefit/cost standpoint. Improvements allow greater flexibility to future change by reducing water consumption without reducing crop yields.
Development of new crop types	Development of more and better heat- and drought-resistant crops will help alleviate current and future world food demand by enabling production in marginal areas to expand. Improvements will be critical, as world population continues to increase, with or without climate change.

Note:
* Water resource managers prefer to maintain a cushion of excess demand that can easily be reduced during droughts. Voluntary measures or mandatory curtailments can be imposed when water supplies are reduced below normal and lifted when plentiful supplies return. Permanent water conservation measures can reduce this buffer for drought, if excess supplies created by conservation are devoted to new sources of demand such as new construction.
Sources: Smith and Tirpak, 1989; Smith *et al.*, 1991; Titus, 1990.

Detailed examples of four types of no regrets adaptation policies analyzed in the earlier chapters are given below. Liberalization of world food trade in agricultural products, improvements in human health, integration and improvements in water resource management, and improvements in information are discussed.

AGRICULTURAL TRADE LIBERALIZATION

Economic theory shows that free trade raises global net income because it lets regions or countries produce goods at their comparative advantage. Rosenzweig *et al.* (Chapter 2) tested the effect of free trade on global agriculture production with and without climate change.

If climate does not change, full trade liberalization in agriculture provides for more efficient resource use, compared to the reference scenario of partial trade liberalization. Full trade liberalization leads to a 3.2 percent increase by 2060 in agricultural GDP worldwide, and a 5.2 percent increase by 2060 in agricultural GDP in developing countries (excluding China). Full trade liberalization also causes global cereal production to increase 2.1 percent, with most of the production increases occurring in developing countries. This policy change also results in almost 20 percent fewer people at risk from hunger. On the basis of increased agricultural output and of the reduction in the number of people at risk of hunger, a policy which liberalizes trade is justified, independent of climate change.

Table 4 displays the estimates of global cereal crop production levels in 2060 with and without climate change and for partial and full trade liberalization. Partial trade liberalization is defined as reducing agricultural trade barriers to half by 2020, and full trade liberalization is defined as eliminating all agricultural trade barriers by 2020.[4]

Under each climate change scenario, global cereal production is higher with full trade liberalization than with partial liberalization. The advantage is not equally distributed, however; enhanced gains in production accrue to developed countries, but losses in production occur in developing countries. Global cereal production in 2060 under the GISS climate scenario is estimated to be 2.4 percent higher with full trade liberalization than with partial liberalization. In fact, cereal production is estimated to be greater with the GISS climate scenario and full trade liberalization than with no climate change and partial trade liberalization.

IMPROVEMENTS IN HUMAN HEALTH

Situations that are dangerous to human health already pose problems for humankind. Heat stress is a significant cause of mortality, particularly in developing countries. Diseases are another important dilemma, again particularly in developing countries. Many of these diseases are carried by pests, requiring attention to pest management. All of these human health dangers could worsen under conditions of climate change.

[4] A trade barrier increases the price of domestic agricultural product above world market prices at the raw materials level (G. Fischer, personal communication). Assuming the elimination of trade barriers means that import prices are equal to global market prices. Reducing barriers by half assumes that the differences between import prices and global prices are reduced by half.

Table 4. *Impact of climate change on global cereal production levels in 2060 with partial and full trade liberalization (millions of metric tons)*

Production	Total production under climate change scenarios			
	No climate change	GISS*	GFDL*	UKMO*
Partial Liberalization				
Developed	1449	1613	1524	1398
Developing	1836	1634	1668	1637
Total	3285	3247	3192	3035
Full Liberalization				
Developed	1472	1656	1568	1418
Developing	1884	1671	1701	1666
Total	3356	3327	3269	3084
Absolute Gain from Policy Change	70	79	76	48

Note:
* Assumes 555 ppm of CO_2.

Development of watch and warning systems is a necessary step to reducing the impacts of dangerous weather conditions. Elderly populations are particularly at risk. These conditions already exist, and development of such systems should be considered even if climate change does not occur.

Many diseases, such as onchocerciasis and malaria which are addressed in this text, are curable or at least controllable. The technology and the administrational structures already exist, which can provide more medicines, more immunization programs, and more care to individuals in the area. The need for improvement in these areas already exists, independent of climate change.

Integrated programs for pest management are already needed, as many of these pests carry diseases. Well-designed pest management not only improves human health, but also can increase the amount of habitable land, as has occurred in several west African countries through the eradication of blackflies infected with onchocerciasis. This increases the agricultural capacity of the region, and also may reduce pressure on urban areas brought about by population migration away from infected regions.

WATER RESOURCES DEVELOPMENTS

Improvements in water resource development are justified, independent of climate change. Integrated river basin planning, conjunctive uses of water resources, market allocation of water, and hydropower energy needs are all very worthwhile to society, both under climate change scenarios and without climate change.

The river basin is the natural spatial scale for management. Integrated management seeks to combine water planning across an entire basin with economic and social development concerns. For example, the Mekong basin in South Asia is managed by the United Nations-sponsored Mekong Committee, which coordinates planning and gives a forum for discussion among the four basin nations. Unfortunately, the effectiveness of the agency has been reduced by war and limited funding. As another example, managers in the Zambezi basin are coordinated by the Zambezi River Authority, but face a need for more cooperation among the eight basin countries. Although the concept of integrated river basin planning is becoming more acceptable, the amount of cooperation among nations or regions is still quite low in many situations.

Allowing markets to allocate water use allows water to be diverted to its most efficient uses; that is, the most productive uses of water will be able to pay the most for it. Less productive uses of water will be utilized only if water prices are low enough to make these uses profitable, which will reduce waste. Water markets are also able to adjust demand as prices change, thus responding more rapidly to changing supply conditions.

The development of dam operations and hydropower systems is a complex issue, with many advantages and disadvantages. Effects on global warming are mixed. Dams can provide benefits independent of climate change, but also have negative effects. The negative effects are likely to become more pronounced under climate change scenarios.

Dam systems also have mixed effects on other systems. Dam construction can provide hydropower, which is advantageous as a renewable energy source, although flooding of vegetated areas is likely to emit CO_2 to the atmosphere.

Dams also provide storage, which increases water use options and provides safety against drought. On the other hand, dam construction has negative consequences for other aspects of water systems. High nutrient sediment is impeded from flowing to lower river reaches and deltas, where it contributes greatly to agriculture. Reservoirs also lead to greater evaporation, reducing the amount of water available for other uses.

When climate change scenarios are taken into account, benefits provided by dams are likely to be reduced. Lower flows will reduce the effectiveness of dams, both in terms of storage capacity and in terms of hydropower effectiveness. Such reduced flow was demonstrated in the Uruguay, Zambezi, Nile, and Mekong case studies. In the Mekong River case study, even scenarios of increased river flow could reduce hydropower effectiveness. The situation in the Nile basin illustrates a potentially extreme danger: river flow under climate change may be reduced so greatly that the main hydropower turbines are unable to be operated. Dam operations can also exacerbate other climate change impacts: dams capture sediments, which makes river deltas even more vulnerable to sea-level rise.

INFORMATIONAL IMPROVEMENTS

Information is an extremely valuable resource. In many situations, information is necessary to better predict the effects of climate change through global warming. Both physical models and computer models of river systems, ocean systems, health effects, demographic shifts, agriculture, and many others require information and use in practice in order to improve our ability to assess the effects of potential impacts of climate change. Information is also necessary to make qualitative assessments. Information is not free, but is usually far less expensive than the costs that arise from not preparing for change, and being able to prevent and/or adjust to change.

In many situations, information is useful for many other applications besides climate change impacts. Therefore, the development of improved data bases, field research, and improved models is very valuable regardless of the potential for climate change. Examples, some of which have been discussed above, include improved surveillance systems of the incidence and spread of diseases, comprehensive coastal data sets for evaluation of and improvements in technology and greater resolution of climatological and water resource data that will improve ability to make regional forecasts.

Policies that are justified by the risk of climate change

Policies in this category are only justified when significant climate changes over the next century are taken into account. Since, as discussed above, the regional direction and magnitude of climate change is uncertain and there may be a long time lag before noticeable climate change impacts are seen, large expenditures or drastic measures are generally not justified for such policies. One may think of this category as 'low regrets', in relation to the 'no regrets' policies of the second category. That is, if such measures are adopted and climate does not change, there will only be 'low regrets' because large expenditures or major policy changes were not made. Primary examples of such policies are marginal actions on predetermined policies and projects, and legal and contingency plans.

Examples of low regrets policies are listed in Table 5, and specific examples of marginal adaptation policy follow. The legal public trust doctrine for sea-level rise, marginal additions to planning in Malaysia, marginal changes to population policy in Indonesia, and marginal changes to urban growth planning in China are discussed.

MARGINAL ADDITIONS TO RECLAIMED LANDS IN MALAYSIA

The government of Malaysia is reclaiming low-lying lands for development. These reclaimed areas are vulnerable to sea-level rise. Midun and Lee (1995) advocate raising reclaimed lands higher to reduce their vulnerability to sea-level rise.

MARGINAL POPULATION MOVEMENT DECISIONS IN INDONESIA

The government of Indonesia has been involved in a long-term program that is trying to reduce population pressure on the most densely populated areas, specifically, Java, Bali, and Lombok. The program involves moving families to other regions, where they are given land, a house, farming equipment, and other assets. Consideration should be given in such a program to the susceptibility of the new regions to climate change. If the government is considering regions for new growth, consideration should be given to how the regions will fare economically and physically under climate change scenarios. This information can be used to help determine the best regions for population relocation. Little is lost if climate change does not occur. However, if climate change occurs, significant losses in agriculture, relocation and lives will have been avoided (Toth, 1991).

MARGINAL CHANGES TO URBAN GROWTH PLANNING IN CHINA

Studies by Wang *et al.* (1995) and Han *et al.* (1995a,b) estimated that a 1-meter rise in sea level would flood the entire cities of Shanghai and Tianjin. The estimated costs of raising dikes and flood walls enough to prevent this result are $281 to $703 million, not including drainage. Consideration should be given, under such circumstances, to attempting to

Table 5. *Policies that are justified by the risk of climate change*

Rolling easements	If and only if sea-level rise is observed, property owners are required to remove threatened coastal structures. (Titus, 1991)
Beach construction set-back	This provides a buffer against sea-level rise. It can impose costs on current generations through lost opportunities for development.*
Migration corridors	In planned land purchases for conservation areas, create migration corridors for species. New purchases of land not currently planned may not be justified in benefit–cost calculation.
Marginal increase in coastal defenses	If building coastal defenses, marginally increase the height of the structures to account for sea-level rise. Such additions are less expensive when made while construction is in progress than after the initial work is complete.
Marginal increase in dam size	In planned dam construction, consider marginally increasing dam size. This may be less expensive than adding dam capacity in the future, and it avoids future damages.
Maintain options to develop new dam sites	Keep options open to develop new dam sites should they be needed.
Plant more heat- and drought-resistant trees on southern range boundaries	If they are as valuable as species currently planted and can survive in the current climate, more heat- and drought-resistant trees are more likely to survive climate change. Less heat- and drought-resistant species may not survive until maturity, which may be critical given the long life cycle of trees.

Note:
* In Uruguay, the set-back areas are used for roads and parks (Claudio Volonté, personal communication).

direct future urban growth to areas not so directly susceptible. The marginal costs of such policy redirections are limited, and reduce the number of citizens exposed to risk of such a disaster. Choices such as this that redirect growth are likely to last a generation or more. The costs of policies of this type are limited, and provide the benefit of exposing fewer people to the risk of such a disaster. The costs of such policies are in contrast with policies that may be necessary, such as forced resettlement and/or expensive protection measures, if climate change is severe.

CONCLUSION

This chapter has discussed the need for adaptive preparation to climate change. Anticipatory adaptive policies are particularly important in circumstances where impacts are uncertain, irreversible, or threaten human health and safety, and where policies will face rising costs and decreasing benefits if delayed, policies are unlikely to be politically feasible later, or policy life will extend far into the future. Anticipatory policies should be examined by these criteria and others, including net benefits, flexibility, and equity. Some policies may result in mitigation and adaptation. Others may be justified independent of climate change. Still others may only be appropriate under the risk of climate change.

Most of the policies that would help minimize climate change impacts are justified for reasons other than climate change. For example, lower population growth, trade liberalization, and economically based water allocation can be justified based on current needs and climate. The potential for climate change provides a reason to reexamine these and other policies. Policy makers should examine whether policies that make sense under any scenario are already in place. If not, climate change presents another reason for adopting them. If they are in place or under consideration, policy makers should ask whether the current or planned policies are strong enough given the possibility of climate change. Climate change may appropriately cause reconsideration of these policies.

REFERENCES

Ausubel, J. H. 1991. A Second Look at the Impacts of Climate Change. *American Scientist* 79:210–221.
Coastal Zone Management Subgroup, Response Strategies Working Group, Intergovernmental Panel on Climate Change. 1992. *Global Climate Change and the Rising Challenge of the Sea: Supporting Document for the IPCC Update Report 1992*. National Oceanographic and Atmospheric Administration, Washington, DC.
Council for Agricultural Science and Technology (CAST). 1992. *Preparing U.S. Agriculture for Global Climate Change*. Task Force Report No. 119. Ames, Iowa: CAST.
Dansgaard, W., *et al.* 1993. Evidence for General Instability of Past Climate from a 250-kyr Ice-Core Record. *Nature* 364:218–20.

Glantz, M. H., ed. 1988. *Societal Responses to Regional Climatic Change*. Boulder, Colo.: Westview Press.
Han, M., L. Wu, J. Hou, C. Liu, G. Zhao, and Z. Zhang. 1995a. Sea-level rise and the North China Coastal Plain: a preliminary assessment. *Journal of Coastal Research*, Special Issue No. 14:132–50.
Han, M., J. Hou, and L. Wu. 1995b. Potential impacts of sea-level rise on China's coastal environment and cities: a national assessment. *Journal of Coastal Research*, Special Issue No. 14:79–95.
Houghton, J. T., G. J. Jenkins, and J. J. Ephraums, eds. 1990. *Climate Change: The IPCC Scientific Assessment*. Intergovernmental Panel on Climate Change. Cambridge: Cambridge University Press.
Houghton, J. T., B. A. Callander, and S. K. Varney, eds. 1992. *Climate Change 1992: The Supplementary Report to the IPCC Scientific Assessment*. Cambridge: Cambridge University Press.
Midun, Z., and S. Lee. 1995. Implications of Accelerated Sea Level Rise: A National Assessment for Malaysia. *Journal of Coastal Research*. Special Issue No. 14:96–115.
National Academy of Sciences (NAS). Adaptation Panel. 1992. *Policy Implications of Greenhouse Warming*. Washington, DC: National Academy Press.
National Academy of Sciences (NAS). 1979. *Carbon Dioxide and Climate: A Scientific Assessment*. Washington, DC: National Academy Press.
Office of Technology Assessment (OTA). 1991. *Change by Degrees: Steps to Reduce Greenhouse Gases*. OTA-O-82. Washington, DC: U.S. Government Printing Office.
Office of Technology Assessment (OTA). 1993. *Preparing for an Uncertain Climate*. OTA-O-563. Washington, DC: U.S. Government Printing Office.
Peters, R. C., and T. E. Lovejoy. 1992. *Global Warming and Biological Diversity*. New Haven, Conn.: Yale University Press.
Rosenberg, N. J., and P. R. Crosson. 1991. *Processes for Identifying Regional Influences of and Responses to Increasing Atmospheric* CO_2 *and Climate Change – The MINK Project: An Overview*. Report prepared for the U.S. Department of Energy. DOE/RL/01830T-H5. Washington, DC: Resources for the Future.
Seidel, S., and D. Keyes. 1983. *Can We Delay a Greenhouse Warming?* Washington, DC: U.S. Environmental Protection Agency.
Smith, J. B. 1990. From Global to Regional Climate Change: Relative Knowns and Unknowns about Global Warming *Fisheries* 15(6):2–6.
Smith, J. B., and D. Tirpak, eds. 1989. *The Potential Effects of Global Climate Change on the United States*. EPA-230-05-89-050. Washington, DC: U.S. Environmental Protection Agency.
Smith, J., A. Silbiger, R. Benioff, J. Titus, D. Hinckley, and L. Kalkstein. 1991. *Adapting to Climate Change: What Governments Can Do*. Monograph. Washington, DC: U.S. Environmental Protection Agency, Office of Policy, Planning and Evaluation.
Tegart, W. J. McG., G. W. Sheldon, and D. C. Griffiths. 1990. *Climate Change: The IPCC Impacts Assessment*. Intergovernmental Panel on Climate Change. Canberra: Australian Government Publishing Service.
Titus, J. G. 1990. Strategies for Adapting to the Greenhouse Effect. *Journal of the American Planning Association* 56(3):311–23.
Titus, J. G. 1991. Greenhouse Effect and Coastal Wetland Policy: How Americans Could Abandon an Area the Size of Massachusetts at Minimum Cost. *Environmental Management* 15(1):39–58.
Toth, F. L. 1991. *Policy Responses to Climate Change in Southeast Asia: From Adaptation to Prevention*. Laxenburg, Austria: International Institute for Applied Systems Analysis. (Draft).
Wang, B., S. Chen, K. Zhang, and J. Shen. 1995. Potential impacts of sea-level rise on the Shanghai Area. *Journal of Coastal Research*, Special Issue No. 14:151–66.
Washington, W. 1992. Reliability of the Models: Their Match With Observations. In *Climate Change and Energy Policy*, eds. L. Rosen and R. Glasser, 63–74. New York: American Institute of Physics.
Wigley, T. M. L., and S. C. B. Raper. 1992. Implications for climate and sea level of revised IPCC emissions scenarios. *Nature* 357: 293–300.
World Wildlife Fund. 1992. *Can Nature Survive Global Warming?* World Wildlife Fund International Discussion Paper. Washington, DC: World Wildlife Fund.

Index